SYSTEMIC YOYOS

Some Impacts of the Second Dimension

SYSTEMIC YOYOS

Some Impacts of the Second Dimension

YI LIN

CRC Press
Taylor & Francis Group
Boca Raton London New York

CRC Press is an imprint of the
Taylor & Francis Group, an **informa** business

AN AUERBACH BOOK

Auerbach Publications
Taylor & Francis Group
6000 Broken Sound Parkway NW, Suite 300
Boca Raton, FL 33487-2742

© 2009 by Taylor & Francis Group, LLC
Auerbach is an imprint of Taylor & Francis Group, an Informa business

No claim to original U.S. Government works
Printed in the United States of America on acid-free paper
10 9 8 7 6 5 4 3 2 1

International Standard Book Number-13: 978-1-4200-8820-5 (Hardcover)

This book contains information obtained from authentic and highly regarded sources. Reasonable efforts have been made to publish reliable data and information, but the author and publisher cannot assume responsibility for the validity of all materials or the consequences of their use. The authors and publishers have attempted to trace the copyright holders of all material reproduced in this publication and apologize to copyright holders if permission to publish in this form has not been obtained. If any copyright material has not been acknowledged please write and let us know so we may rectify in any future reprint.

Except as permitted under U.S. Copyright Law, no part of this book may be reprinted, reproduced, transmitted, or utilized in any form by any electronic, mechanical, or other means, now known or hereafter invented, including photocopying, microfilming, and recording, or in any information storage or retrieval system, without written permission from the publishers.

For permission to photocopy or use material electronically from this work, please access www.copyright.com (http://www.copyright.com/) or contact the Copyright Clearance Center, Inc. (CCC), 222 Rosewood Drive, Danvers, MA 01923, 978-750-8400. CCC is a not-for-profit organization that provides licenses and registration for a variety of users. For organizations that have been granted a photocopy license by the CCC, a separate system of payment has been arranged.

Trademark Notice: Product or corporate names may be trademarks or registered trademarks, and are used only for identification and explanation without intent to infringe.

Library of Congress Cataloging-in-Publication Data

Lin, Yi, 1959-
  Systemic yoyos : some impacts of the second dimension / Yi Lin.
    p. cm. -- (Systems evaluation, prediction, and decision-making)
  Includes bibliographical references and index.
  ISBN 978-1-4200-8820-5 (acid-free paper)
  1. System theory--Mathematics. I. Title.

Q295.L5 2008
003--dc22   2008024292

Visit the Taylor & Francis Web site at
http://www.taylorandfrancis.com

and the Auerbach Web site at
http://www.auerbach-publications.com

Dedication

To my father: Even though you are gone from us, you live in my heart forever.

To my mother: Without your lifelong teaching, I would not have become who I am.

To my wife and children—Kimberly, Dillon, Alyssa, and Bailey—for your love, support, and understanding.

Contents

Preface xiii
Acknowledgments xix
About the Author xxi
Book Overview xxiii

1   Introduction: The Yoyo Structure 1
    1.1   Systems: A Historical Review 2
    1.2   Whole Evolution of Systems: Where Systemic Yoyo Is From 7
    1.3   Applications of the Systemic Yoyo Model 11
    1.4   Organization of This Book 14

PART 1: THE SYSTEMIC YOYO: THEORETICAL AND EMPIRICAL FOUNDATIONS

2   Blown-Ups, Eddy Motions, and Transitional Changes 19
    2.1   The Concept of Blown-Ups 19
    2.2   Mathematical Characteristics of Blown-Ups 20
    2.3   Mapping Properties of Blown-Ups and Quantitative Infinity 22
    2.4   Spinning Current: A Physical Characteristic of Blown-Ups 23
    2.5   Equal Quantitative Effects 28
    2.6   Various Properties of Blown-Ups 32

3   Conservation of Informational Infrastructure: Empirical Evidence 35
    3.1   Introduction 35
    3.2   Physical Essence of Dirac's Large Number Hypothesis 36
    3.3   The Mystery of the Solar System's Angular Momentum 40
    3.4   Measurement Analysis of Movements of the Earth's Atmosphere 42

viii ■ Contents

    3.5   The Law of Conservation of Informational Infrastructure 43
    3.6   Impacts of the Conservation Law of Informational Infrastructure 44
    3.7   Other Empirical Evidence for Yoyo Structures 46

PART 2: A REVISIT TO NEWTON'S LAWS, UNIVERSAL GRAVITATION, AND THE THREE-BODY PROBLEM

4 Newton's Laws of Motion 51
    4.1   The Second Stir and Newton's First Law of Motion 52
    4.2   Eddy Effects and Newton's Second Law 53
    4.3   Colliding Eddies and Newton's Acting and Reacting Forces 57
    4.4   Equal Quantitative Effects and Figurative Analysis 61
    4.5   Whole Evolutions of Converging and Diverging Fluid Motions 64

5 Kepler's Laws of Planetary Motion 67
    5.1   Newton's Cannonball 67
    5.2   Kepler's Law of Harmonics and Its Generalization by Newton 72
    5.3   Universal Gravitation 74

6 The Three-Body Problem 77
    6.1   The Problem and Some Current Results 77
    6.2   Three Visible Bodies and Existence of n-nary Star Systems 79
    6.3   Three Bodies with at Least One Invisible 83
    6.4   Open Questions 86

7 Stirring Energy and Its Conservation 91
    7.1   Rotation and Stirring Energy 91
    7.2   Conservation of Stirring Energy and Three-Level Energy Transformation 93
    7.3   Energy Transformation Process and Nonconservative Evolution of Stirring Energy 95
    7.4   Governance Law of Slaving Energy of Newtonian First Push 97
    7.5   Interactions and Einstein's Mass-Energy Formula 98
    7.6   Solenoidal Fields and Problem on Universal Gravitation 99
    7.7   Conservability of Stirring Energy and Physical Significance of Energy Transformation 103
    7.8   Discussion 106
    Appendix: Stirring Energy and Its Conservation 107
        A7.1   Evolution Engineering and Technology for Long-Term Disaster Reduction 107

	A7.2	The Background 107
	A7.3	Basic Principles of Flood Evolution and Development and Flood Disasters 109
		A7.3.1 Law of Conservation of Stirring Energy 110
		A7.3.2 Three-Ring Quasi-Stability Problem of Urban Flood Movements 110
	A7.4	Feasible Technology for Urban Long-Term Flood Prevention and Disaster Reduction 113
		A7.4.1 The Computational Method for Q_1 114
		A7.4.2 Design of Artificial Urban Lakes and Problem of River Course Drainage 115
	A7.5	Analysis of Urban Long-Term Flood Prevention and Disaster Reduction Facilities 116
		A7.5.1 Computation on an Artificial Lake at Fuhe Bridge along Northern Outer-Ring Road 116
		A7.5.2 Estimate for an Artificial Lake along the Qingshui River on the Upper Reaches of the Nan River 118
	A7.6	Discussion on Long-Term and Long-Effect Technology 119

8 Time and Its Dimensionality 121
    8.1 Problems to Be Addressed 121
    8.2 The Physics of Physical Quantities 125
    8.3 The Nonquantification of Events 127
        8.3.1 Problems on the Physics of Physical Quantities 128
        8.3.2 Nonquantification of Events 130
    8.4 What Time Is 131
        8.4.1 The Problem of Time 131
        8.4.2 Time in China 132
        8.4.3 Time in the West 134
        8.4.4 What Time Is 136
    8.5 Material and Quantitative Parametric Dimensions 138
    8.6 Some Final Words 140

PART 3: ECONOMIC AND FINANCIAL FORCES

9 The Economic Yoyo 145
    9.1 Whole Evolution Analysis of Demand and Supply 146
    9.2 The Yoyo Evolution of an Economic Cycle 149

10 The Happy Family 153
    10.1 Becker's Rotten Kid Theorem 155

10.2 Two Other Mysteries of the Family 165
10.3 Never-Perfect Value Systems and Parasites 170
10.4 Bergstrom's Rotten Kid Theorem and the Samaritan's Dilemma ... 179
10.5 Maximization of Family Income and Child Labor 183
10.6 Final Words 186

11 Child Labor and Its Efficiency 189
11.1 Child's Disutility of Work 190
  11.1.1 One-Sided Altruism Model 190
  11.1.2 Adult Child's Altruism toward Parents Model 196
11.2 Different Efficiencies and Potentially Different Outcomes 201
11.3 Marginal Bans on Child Labor 204
11.4 Conclusion 212

12 Economic Eddies and Existence of Different Industry Sizes 215
12.1 Economic Yoyos and Their Flows 217
12.2 Simple Model for Perfect Capital Markets 219
12.3 Simple Model for Imperfect Capital Markets 222

13 A Fresh Look at Interindustry Wage Differentials 225
13.1 Financially Resourceful Companies 227
13.2 Companies with Limited Resources 229
13.3 Look Back at Some of the Existing Literature 231
13.4 The Law of One Price 234

14 Dynamics between Long-Term and Short-Term Projects 237
14.1 The Yoyo Model Foundation for Empirical Discoveries 238
14.2 CEO's Choices of Projects 243
  14.2.1 Price Behavior of Different Investment Projects 244
  14.2.2 Dynamics of Projects 247
14.3 Conclusion 253

PART 4: STRUCTURE OF HUMAN THOUGHTS AND INFINITY PROBLEMS IN MODERN MATHEMATICS

15 A Quick Glance at the History of Mathematics 257
15.1 The Beginning 260
15.2 First Crisis in the Foundations of Mathematics 261
15.3 Second Crisis in the Foundations of Mathematics 265
15.4 Third Crisis in the Foundations of Mathematics 268

16 Hidden Contradictions in the Modern System of Mathematics 275
16.1 The Concepts of Actual and Potential Infinities 276
16.2 Are Actual Infinities the Same as Potential Infinities? 280

16.3　Do Infinite Sets Exist?..281
　　16.4　The Cauchy Theater Phenomena...283
　　16.5　The Return of the Berkeley Paradox...285
　　16.6　The Fourth Crisis in the Foundations of Mathematics?.................288

PART 5: ROLLING CURRENTS AND PREDICTION OF DISASTROUS WEATHER

17　V-3θ Graphs: A Structural Prediction Method....................................295
　　17.1　The Fundamentals..295
　　17.2　Roles of Rolling Currents and Ultra-Low Temperature
　　　　　in Weather Evolution..299
　　17.3　The Design of V-3θ Graphs..306

18　Case Studies Using V-3θ Graphs...313
　　18.1　Suddenly Appearing Severe Convective Weather............................313
　　　　　18.1.1　Background Information..314
　　　　　18.1.2　The Blown-Ups Principle...314
　　　　　18.1.3　Structural Analysis Method and the V-3θ Graphs..........315
　　　　　18.1.4　Structural Characteristics of Suddenly Appearing
　　　　　　　　　Convective Weather...316
　　　　　18.1.5　Suddenly Appearing Weather over Major Metropolitans....317
　　　　　18.1.6　Regional Suddenly Appearing Extraordinarily Heavy
　　　　　　　　　Rain Gushes...322
　　　　　18.1.7　Discussion..324
　　18.2　Small, Regional, Short-Lived Fog and Thunderstorms...................325
　　　　　18.2.1　The Background Information..326
　　　　　18.2.2　Case Studies on Fog..327
　　　　　18.2.3　Discussion..338
　　18.3　Windstorms and Sandstorms..339
　　　　　18.3.1　Background Information..340
　　　　　18.3.2　Practical Applications...342
　　　　　18.3.3　Discussion..353
　　18.4　Abnormally High Temperatures...354
　　　　　18.4.1　Background Information..355
　　　　　18.4.2　High-Temperature Weather under a Subtropical High
　　　　　　　　　of the West Pacific...356
　　　　　18.4.3　High Temperatures under Cold High Pressures..............364
　　　　　18.4.4　Some Final Words..367

References...369

Index..381

Preface

About 30 years ago, as a sophomore in college majoring in mathematics, I tried my first submission of a research paper for possible publication. However, to my surprise, in no time, the paper was rejected. After suffering from more than 50 failed attempts to get my various works published in the following years, I eventually realized for the first time in my life that even though I was one of the top percent of my class, all the formal training I received through various levels of formal schooling had not really prepared me to accomplish what I dreamed about: becoming a first-class scholar. This dream I have had since when I was in third grade. What I observed during the period of my over 50 failed submissions was that all the then-recently-published works, which I had a chance to read, belonged to the kinds of activities of patching holes and gaps existing in well-accepted theories. One piece of evidence supporting such an observation is the fact that to validate the significance of one's research, the researcher only has to relate his or her work to one of the big names of the scientific history. Some researchers even go as far as prophesying what these great minds of history would say and do if they were working on what these current researchers are doing. When it turns out to be difficult for them to relate to any of the well-known names of the past, researchers commonly publish their works in "legitimate" journals, hoping that the legitimacy of the journals helps to justify the quality of their works. One of the tragic consequences of such a trend is that new and revolutionary theories need more time to become accepted, even if the theories have gained great successes and victories in applications. Also, another side effect of such a trend is that more scholars spend the most valuable years of their otherwise productive careers playing scientific games instead of seeking scientific truths for the benefit and advancement of humankind.

Out of pure luck, I was required to take a course entitled "Natural Dialectics" in my first year of graduate school. The reason I say that it was pure luck is because if I were the person making the course selections, I would have definitely not taken that course, because the subject matter belonged to the big category called philosophy. Just as most young people at the time, and for no obvious reasons, I liked some subjects and disliked others. That course was taught by an ambitious young assistant professor from the philosophy department. To make his teaching easier, he taught the class all about his most recent research on how the great minds in history came up with their ingenious ideas and theories, and how they made their thoughts and theories known through hard work against all odds and all prejudices and through the abilities of their theories to bring forward brand new and astonishing predictions and understanding of various events. From this course, I learned to question authorities, locate incidents that really matter in people's lives but are unexplainable by using current theories and beliefs, and to make up my own theories for the purpose of explaining the unexplainable. Then, the rest of the work is to prove why the new theories work, how they work, and what new results they can bring forward.

Because I grew up in a time of change, the common belief was that the information about the scientific frontiers was the key for anyone to succeed scientifically. So, I searched through the research papers and books available to me so that I could grasp at least a little bit. Soon, one of George Klir's wonderful papers caught my attention. The exact title of the paper I have forgotten long ago. However, the spirit of the paper motivated and excited me to read more along the same and related lines. That was about one year before I came to the United States to pursue a PhD degree in mathematics. And more importantly to my future career, that was the very exciting time that one of my submissions was finally accepted for publication.

Influenced by the spirit and desire of my supervisor Shutang Wang in China for pursuing new knowledge, I soon had a chance to read a great many books and papers by some of my heroes of the modern time. So, about four months after arriving in the United States, with a college friend, Dr. Yonghao Ma, I started to write my first research papers about general systems theory, and my very first paper with Yiping Qiu on applications of our theory in materials science. Encouraged by our theoretical results and successes of applications, I have continued my research along several closely related lines for all these years. At this junction, I would like to mention that my PhD degree supervisor, Dr. Ben Fitzpatrick, had influenced me a great deal in terms of quality control of my own works.

In 1995, I organized most of my works into a monograph entitled *General Systems Theory: A Mathematical Approach*, which was later published in 1999 by Kluwer Academic and Plenum Publishers, and soon adopted as a graduate textbook by colleagues at the Tokyo Institute of Technology, Case Western Reserve University, and other universities around the globe. In the process of writing up possible applications of my theory along with the theoretical successes, I discovered a weakness of the theory. Because it had been developed mainly on the basis of the modern Zermelo-Fraenkel (ZFC) set theory, not much has been established for the purpose of data analysis. That is, when some information, in terms of data, is given, what new results and insights can be produced out of my own theory, even though some unsettled problems of the past can be resolved beautifully?

To address this problem, through reading many and various books and commentaries and through numerous discussions with colleagues from all over the world, I was finally led to the following four criteria regarding how a scientific theory can possess a glorious and long-lasting life:

1. The theory must be readable by as many people as possible.
2. The theory must coincide with people's intuition.
3. The theory must possess a certain kind of beauty, which can be easily felt.
4. The theory must be capable of producing meaningful results and insights that excite the population.

For example, calculus satisfies all these conditions with its beauty and intuition placed on the Cartesian coordinate system. And Euclidean geometry is also a long-living theory, because it is intuitive, possesses a logical and visual beauty, and can be employed in people's daily lives. However, in the 1800s, when mathematicians finally succeeded in rewriting the theory in the fashion of rigorous logical reasoning on the basis of axioms and formal logic, most parts of the original visual and intuitive beauty were lost. Educators caught up with the new development quickly declared in the 1970s, for the first time in history, that such a geometry could be deleted from an average child's precollege education.

To improve my own theory, I took a hard look at probability theory, statistics, and various data analysis methods. What amazed me is the following episode of life: when one listens to a weather forecast nowadays, for example, he or she will be given the information about a possible forthcoming weather condition attached with a probability. For example, tomorrow afternoon there will be a 65 percent chance of scattered showers. Now, no matter what happens during the said afternoon, the forecast will always be correct. What amazed me more was the appearance of the chaos theory some years ago. Following the crowd of those chaos experts were some religious leaders, because they finally had found allies in the scientific community, for the first time since the end of the 1800s when the war between Darwin's theory of evolution and the theory of creation in six days was concluded, with the victor being the theory of evolution. This alliance was because the work of chaos had led to the conclusion that God had created the universe and people, and he did not want people to find out the ultimate secret of his creation. (At this juncture, experts in chaos theory may well claim that their concepts of chaos have nothing to do with any such consequences derived from their work by other parties.)

At this puzzling moment in my own scientific quest, I had no choice but to go back to the historical motivation on which the systems movement started in the late 1950s and early 1960s. What I found this time was that there had been a wealth of well-documented systemic thoughts. However, what I did not see much were works on how these thoughts had brought changes, progress, and solutions to problems widely studied in the classical science. That is, criterion 4 of a long-lasting theory was not met.

With this discovery in mind, I, with colleagues, started off on an exciting and challenging new journey: based on well-established systemic thoughts, to derive new results and conclusions in as many scientific disciplines as possible by making good use of the plentiful information widely available in our unprecedented information age in human history. Along this exciting but difficult journey, my colleagues and I have had a great many opportunities to look into various age-old problems in a new light, such as:

- A (systemic) theoretical foundation of laws of conservation, which have been the cornerstones for making chemistry and physics exact sciences
- Practical implications of classical chaos
- Principles of mathematical modeling
- Limitations of human sensing organs and the knowability of the natural world
- Predictability of weather systems
- Artifacts of various numerical schemes
- Existence of a new law of conservation, which unifies all major laws of conservation in physics and supports the Big Bang theory
- Discontinuity and exponential curve fitting
- Astro-singularities and disastrous earthly consequences
- Unascertained information theories and their applications
- Theories of time and space, etc.

In this book, I will present to you some of my current works, created either alone or by joint effort with my colleagues in recent years along our exciting journey. More specifically, the systems movement and its successes have been built on the basic concepts of systems and related foundations. However, when the studies on general systems are compared to any of the long-lasting, successful theories from the past, what is lacking in systems research is a common playground on which we can form our intuition and see the beauty of whatever we establish in terms of (general) systems. For example, most of the successes of calculus can be accredited to the introduction of the Cartesian coordinate system, and those of Euclidean geometry to the imaginary but quite real life-like plane and three-dimensional space. Now, what is the intuitive playground for the study of (general) systems?

To this end, with Yong Wu, I introduced the systemic yoyo model in 2002. The goal of this book is to show how this yoyo model and its methodology can be employed to study many unsettled or extremely difficult problems in modern science and technology. What is most promising, as shown in this book, is that this model and its methodology can be applied equally well to such exact sciences as Newtonian physics, planetary motions, and mathematics, and such inexact areas of knowledge as economics, finance, corporate governance, and structure of human thoughts. Other than these theoretical promises, this book will also show that this model and its methodology can in practice produce tangible economic benefits to people and societies by providing specific engineering designs for long-term disaster reduction and by establishing useful methods for forecasting zero-probability disastrous weather systems.

I hope you will enjoy reading and referencing this book in your scientific exploration and academic pursuit. If you have any comments or suggestions, please let me hear from you by joining several thousands of other colleagues who have communicated with me or other members of my research groups. I can be reached at jeffrey.forrest@sru.edu or jeffrey.forrest@iigss.net.

Yi Lin

Acknowledgments

This book contains many research results previously published in various sources, and I am grateful to the copyright owners for permitting me to use the material. They include the International Association for Cybernetics (Namur, Belgium), Gordon and Breach Science Publishers (Yverdon, Switzerland, and New York), Hemisphere (New York), International Federation for Systems Research (Vienna, Austria), International Institute for General Systems Studies, Inc. (Grove City, Pennsylvania), Kluwer Academic and Plenum Publishers (Dordrecht, Netherlands, and New York), MCB University Press (Bradford, UK), Pergamon Journals, Ltd. (Oxford), Springer-Verlag (London), Taylor and Francis, Ltd. (London), World Scientific Press (Singapore and New Jersey), and Wroclaw Technical University Press (Wroclaw, Poland).

I express my sincere appreciation to many individuals who have helped to shape my life, career, and profession. Because there are so many of these wonderful people from all over the world, I will just mention a few. Even though Dr. Ben Fitzpatrick, my PhD degree supervisor, has left us, he will forever live in me and my works. His teaching and academic influence will continue to guide me for the rest of my professional life. My heartfelt thanks go to Shutang Wang, my MS degree supervisor. Because of him, I always feel obligated to push myself further and work harder to climb high up the mountain of knowledge and to swim far into the ocean of learning. To George Klir—from him I acquired my initial sense of academic inspiration and found the direction in my career. To Shoucheng OuYang and colleagues in our research group, named Blown-Up Studies, based on our joint works; Yong Wu and I came up with the systemic yoyo model, which eventually led to completion of this book. To Zhenqiu Ren—with him we established the law of conservation of informational infrastructure. To Gary Becker, a Nobel laureate in economics—his rotten kid theorem has brought me deeply into economics, finance, and corporate governance. To Wujia Zhu, Ningsheng Gong, and Guoping Du—with them I successfully explored problems in the foundations of mathematics.

About the Author

Dr. Jeffrey Yi Lin Forrest holds all his educational degrees in pure mathematics and had one year of postdoctoral experience in statistics at Carnegie Mellon University. Currently, he is a guest professor at several major universities in China, including the College of Economics and Management at NUAA, a professor of mathematics at Slippery Rock University, Pennsylvania, and the president of the International Institute for General Systems Studies, Inc., Pennsylvania. He serves on the editorial boards of 11 professional journals, including *Kybernetes: The International Journal of Systems, Cybernetics and Management Science*, *Journal of Systems Science and Complexity*, and *International Journal of General Systems*. Some of Dr. Lin's research was funded by the United Nations, the State of Pennsylvania, the National Science Foundation of China, and the German National Research Center for Information Architecture and Software Technology. By the end of 2007, he had published well over 200 research papers and 18 monographs and special topic volumes. Some of these monographs and volumes were published by such prestigious publishers as Springer, World Scientific, Kluwer Academic, and Academic Press. Over the years, Dr. Yi Lin's scientific achievements have been recognized by various professional organizations and academic publishers. In 2001, he was inducted into the honorary fellowship of the World Organization of Systems and Cybernetics. His research interests are wide ranging, covering areas like mathematical and general systems theory and applications, foundations of mathematics, data analysis, predictions, economics and finance, management science, and philosophy of science.

Book Overview

Since the early 1990s, systems research has been seen as the second dimension of science, as proposed by George Klir, complementing classical science, the first dimension, in a completely different direction. However, in this second dimension, many concepts and results can be imagined vividly and derived rigorously without any common ground different from those used in the first dimension to show them visually. This deficit surely poses a great difficulty for systems research and applications. To meet this challenge of the second dimension, this book systematically presents a new systemic model, named yoyo, which can be employed as a systemic method to analyze problems as well as an intuition for systemic thinking. Its role in systems research is analogous to that of the Cartesian coordinate system in modern science, because it provides a platform for conceptual manipulation in systems research and helps to establish classical models to resolve problems from the first dimension.

This book consists of five parts and 18 chapters. The first part lays the theoretical foundation for the yoyo model and its empirical justifications. The second part shows how to employ this model to address and resolve some open problems in the hard sciences, such as Newtonian mechanics, planetary motions, the three-body problem, etc. The third part presents applications of this model in economics and finance. With this model established, a difficult problem such as Becker's rotten kid theorem (a Nobel Prize–winning result) holding true can be successfully addressed. In Part 4, the structure of human thoughts and infinity problems in the system of modern mathematics are considered, while showing the appearance of the fourth crisis of mathematics. In Part 5, the concept of rolling currents is presented and employed to practically predict weather changes, especially the arrival of disastrous weather conditions. Each case study presented in this part, and successfully addressed in the book, represents a difficult, unsettled problem in meteorology in particular, and modern science in general.

Chapter 1

Introduction: The Yoyo Structure

In the past 80 some years, studies in systems science and systems thinking have brought forward brand new understandings and discoveries to some of the major unsettled problems in conventional science. To address and resolve practical problems, which have been extremely difficult in modern science, new theories and methodologies have been established (for more details, see Lin, 1999; Klir, 1985 and references therein).

Due to these studies of systems science or to conscious or unconscious employments of systems thinking, a forest of specialized and interdisciplinary explorations has appeared; it shows the overall trend of development in modern science and technology of synthesizing all areas of knowledge into a few major blocks, and the boundaries of conventional disciplines become blurred ("Mathematical Sciences," 1985). Along with these specialized and interdisciplinary explorations, some cross-disciplinary studies have appeared (see, for example, Wu, 1990; Mickens, 1990). Underlying these explorations and studies, we can see the united effort of studying similar problems in different scientific fields on the basis of wholeness and parts, and of understanding the world in which we live by employing the point of view of interconnectedness.

In 1924, von Bertalanffy wrote:

> Because the fundamental character of living things is its organization, the customary investigation of individual parts and processes cannot provide a complete explanation of the phenomenon of life. This investigation gives us no information about the coordination of parts and processes. Thus the chief task of biology must be to discover the laws of biological systems (at all levels of organization). We believe that the

attempts to find a foundation at this theoretical level point at fundamental changes in the world picture. This view, considered as a method of investigation, we call "organismic biology" and, as an attempt at an explanation, "the system theory of the organism."

From this statement and seemingly unsolvable (by using traditional science) problems in practice, such as the prediction of zero-probability disastrous weather conditions, we see that an important concept of systems was formally introduced. As tested in the past 80 some years, this concept has been widely accepted by the entire spectrum of science and technology (for details, see Blauberg et al., 1977; Klir, 2001).

In this chapter, we briefly outline the historical background of systems science (for a detailed treatment on the history of systems theory, see von Bertalanffy, 1968, 1972), and why systems science can be seen as the second dimension of the whole of science, as argued by George Klir (2001). Comparing to the importance of the Cartesian coordinate system in modern science (Kline, 1972), where this coordinate system is the intuition and playground on which important concepts and results of modern mathematics and science are established, after presenting the main results of the blown-up theory (Wu and Lin, 2002), we introduce the systemic yoyo model. Then, we outline how, in the rest of this book, this yoyo model is applied as an intuition or playground, on which we can generalize Newton's laws of motion, provide a brand new look at the study of universal gravitation, Kepler's laws of planetary motion, and shed new light on the study of the three-body problem, while introducing the important concept of stirring energy. After this yoyo model is successfully employed in the studies of economic and financial forces, we will see how it is applied to studies on the structure of human thoughts. This chapter ends with a summary on how the ideas and principles of this yoyo model have been employed to improve the prediction accuracy of some of zero-probability disastrous weather. That is, this book is organized in such as way that after the theoretical and empirical foundations of the systemic yoyo model are established, the reader will see that the systemic yoyo model can be applied to not only exact science, such as physics and mathematics, but also inexact areas of learning, such as human thoughts and social studies, leading to tangible economic benefits by providing weather forecasts with greatly improved accuracy. Historically, the basic idea of this yoyo model originated in the 1960s in Shoucheng OuYang's effort of forecasting extraordinary torrential rains (Lin, 1998).

1.1 Systems: A Historical Review

Before discussing the historical background of systems science, let us first talk about what systems methodology is. Although scholars in the field understand it in different ways, the fundamental points underlying these different understandings are roughly the same. To this end, let us first look at typical opinions of two important figures in modern science: H. Quastler and L. Zadeh.

Quastler (1965) wrote:

> Generally speaking, systems methodology is essentially the establishment of a structural foundation for four kinds of theories of organization: cybernetics, game theory, decision theory, and information theory. It employs the concepts of a black box and a white box (a white box is a known object, which is formed in a certain way, reflecting the efficiency of the system's given input and output), showing that research problems, appearing in the afore-mentioned theories on organizations, can be represented as white boxes, and their environments as black boxes. The objects of systems are classified into several categories: motivators, objects needed by the system to produce, sensors, and effectors. Sensors are the elements of the system that receive information, and effectors are the elements of the system that produce real reactions. Through a set of rules, policies, and regulations, sensors and effectors do what they are supposed to do.

By using these objects, Quastler proved that the following laws could describe the common structure of the four theories of organization:

1. Interactions are between systems and between systems and their environments.
2. A system's internal movements and the reception of information about its environment stimulate its efficiency.

Zadeh (1962) listed some important problems in systems science: systems characteristics, systems classifications, systems identification, signal representation, signal classification, systems analysis, systems synthesis, systems control and programming, systems optimization, learning and adaptation, systems liability, stability, and controllability. The characteristics of his opinion are that the main task of systems science is the study of general properties of systems without considering their physical specifics. Systems methodology in Zadeh's viewpoint is an independent scientific endeavor whose job it is to develop an abstract, a foundation with concepts and frames to study various behaviors of different kinds of systems. Therefore, systems science should be based on a theory of mathematical structures of systems with the purpose of studying the foundation of organizations and systems structures.

Even though the concept of systems has been a hot spot of discussion in almost all areas of modern science and technology, and was first introduced formally in the second decade of the 20th century in biology (von Bertalanffy, 1924), as all new concepts in science, the ideas and thinking logic of systems have a long history. For example, Chinese traditional medicine, treating each human body as a whole, can be traced back to the time of Yellow Emperor, about 4,800 years ago, and Aristotle's statement that "the whole is greater than the sum of its parts" has been a fundamental problem in systems science. That is, over the centuries, humankind has been studying and exploring nature by using the thinking logic of systems. Only in modern times have some new contents been added to the ancient systems thinking. The methodology of studying systems as wholes adequately agrees with the development trend of modern science, namely, to divide the object of consideration into parts as small as possible and study all of the individual parts, seek interactions and connections between phenomena, and observe and comprehend more and bigger pictures of nature.

In the history of science, although the word *system* was never emphasized, we can still find many explanatory terms concerning the concept of systems. For example, Nicholas of Cusa, a profound thinker of the 15th century, linking medieval mysticism with the first beginning of modern science, introduced the notion of *coincidentia oppositorum*, the opposition or, indeed, fight among the parts within a whole, which nevertheless forms a unity of higher level. Leibniz's hierarchy of monads looks quite like that of modern systems; his *mathesis universalis* presages an expanded mathematics that is not limited to quantitative or numerical expressions and is able to formulate much conceptual thought. Hegel and Marx emphasized the dialectic structure of thought and the universe it produces: the deep insight that no proposition can exhaust reality, but only approaches its coincidence of opposites by the dialectic progress of thesis, antithesis, and synthesis. Gustav Fechner, known as the author of the psychophysical law, elaborated, in the way of the natural philosophers of the 19th century, on supraindividual organizations of higher order than the usual objects of observation. For example, in this work he spoke at length on life communities and the entire Earth—thus romantically anticipating the ecosystems of modern parlance. Here, only a few names are listed. For a more comprehensive study, see von Bertalanffy (1972).

Even though Aristotelian teleology was eliminated in the development of modern science, problems contained in it, such as "the whole is greater than the sum of its parts," the order and goal directedness of living things, etc., are still among the problems of today's systems scientific research. For example, what is a "whole"? What does "the sum of its parts" mean? These problems have not been studied in all classical branches of science, because these branches have been established on Descartes' second principle, to divide each problem into parts as small as possible, and Galileo's method, to simplify a complicated process into basic portions and processes (Kuhn, 1962).

From this superficial discussion, it can be seen that the concept of systems we are studying today is not simply a product of yesterday. Instead, it is a reappearance of some ancient thought and a modern form of an ancient quest. This quest has been recognized in the human struggle for survival with nature, and has been studied at various points in time by using the languages of different historical moments.

Ackoff (1959) commented that during the past two decades, we witnessed the appearance of the key concept of systems in scientific research. However, with the appearance of the concept, what changes have occurred in modern science? Under the name of systems research, many branches of modern science have shown the trend of synthetic development; research methods and results in various disciplines have been intertwined to influence the overall research progress, so one feels the tendency of synthetic development in scientific activities. This synthetic development requires the introduction of new concepts and new thoughts in the entire spectrum of science. In a certain sense, all of this can be considered as the center of the concept of systems. One Soviet expert described the progress of modern science as follows (Hahn, 1967, p. 185): "Refining specific methods of systems research is a widespread tendency in the exploration of modern scientific knowledge, just as science in the nineteenth century with forming natural theoretical systems and progresses of science as its characteristics."

Von Bertalanffy (1972) described the scientific revolution in the 16th century as follows: "The Scientific Revolution of the sixteenth to seventeenth century replaced the descriptive-metaphysical conception of the universe epitomized in Aristotle's doctrine by the mathematical-positivistic or Galilean conception. That is, the vision of the world as a teleological cosmos was replaced by the description of events in causal, mathematical laws." Based on this description, can we describe the change in today's science and technology as follows? While Descartes' second principle and Galileo's method were being simultaneously and continuously used, systems methodology was introduced to deal with problems of order or organization.

Should we continue to use Descartes' second principle and Galileo's method? The answer is yes, for two reasons. First, they have been extremely effective in scientific research and administration, where all problems and phenomena could be decomposed into causal chains, which could be treated individually. That has been the foundation for all basic theoretical research and modern laboratory activities. In addition, they won victories for physics and led to several technological revolutions. Second, modern science and technology are not utopian projects as described by Popper (1945), reknitting every corner for a new world, but are based on the known knowledge base; they are progressing in all directions with more depth, more applicability, and a higher level of sophistication.

On the other hand, the world is not a pile of infinitely many isolated objects, where every problem or phenomenon can be simply described by a single causal relation. The fundamental characteristics of this world are its organizational structure and connections of interior and exterior relations of different matters. The study of either an isolated part or a single causality of problems can hardly explain completely or relatively globally our surrounding world. At this junction, the research progress of the three-body problem in mechanics is an adequate example. So, as human race advances, studying problems with multicausality or multirelationship will become more and more significant.

In the history of scientific development, the exploration of nature has always moved back and forth between specific matters or phenomena and generalities. Scientific theories need foundations rooted deep inside practice, while theories are used to explain natural phenomena so that human understanding of nature is greatly enhanced. In the following, let us turn our attention to the discussion of the technological background for systems theory to appear. That is, we will look at the need for systems science to arise in the development of technology and as a requirement for higher-level production.

There have been many advances in technology: energies produced by various devices, such as steam engines, motors, computers, and automatic controllers; self-controlled equipment, from domestic temperature controllers to self-directed missiles; and the information highway, which has resulted in increased communication of new scientific results. On the other hand, increased speed of communication furthers scientific development to a different level. Also, social changes have brought more pressing demands for new construction materials. From these examples, it can be seen that the development of technology forces humankind to consider not only a single machine or matter or phenomenon, but also systems of machines or systems of matter and phenomena. The design of steam engines, automobiles, cordless equipment, etc., can be handled by specially trained engineers, but when dealing with the design of missiles, aircrafts, or new construction materials, for example, a collective effort, combining many different aspects of knowledge, has to be in place, which includes a combination of various techniques, machines, electronic technology, chemical reactions, people, etc. Here, the relationship between people and machines becomes more obvious, and uncountable financial, economic, social, and political problems are intertwined to form a giant, complicated system, consisting of humans, machines, and many other components. The great political, technical, and personnel arrangement success of the American Apollo Project reveals the fact that history has reached such a point where science and technology have been so maturely developed that each rational combination of information or knowledge could result in unexpected consumable products.

A great many business problems require the location of optimal points of maximum economic effect and minimum cost in an extremely complicated network. Not only does this kind of problem appear in industry, agriculture, military affairs, and business, but politicians use similar (systems) methods to seek answers to problems like air and water pollution, transportation blockages, decline of inner cities, and crimes committed by teenage gangs.

In business, there is a tendency to design and manufacture fancier and more accurate products that bring in more profits. In fact, under different interpretations, all areas of learning have been faced with complexity, totality, and "systemality." This tendency denotes a sharp change in scientific thinking. By comparing Descartes' second principle and Galileo's method with systems methodology and thinking logic, and considering the development tendency, as described previously, appearing in the world of learning and production, it is not hard to see that because of systemic concepts, another new scientific and technological revolution will soon arrive. To this end, for example, each application of systems thinking points out the fact that the relevant classical theory needs to be modified somehow (see Klir, 1970; Berlinski, 1976; Lilienfeld, 1978). (Not all scientific workers have the same kind of optimistic outlook. Some scholars believe that this phenomenon is an omen that systems research itself is facing a crisis [Wood-Harper and Fitzgerald, 1982].) Klir (2001) looked at this phenomenon from a different angle. Because systems thinking focuses on those properties of systems and associated problems that emanate from the general notion of systemhood, while the divisions of classical science have been done largely on properties of thinghood, systems science and research would naturally transcend all the disciplines of the classical science and become a force making the existing disciplinary boundaries totally irrelevant and superficial. The cross-disciplinary feature of systems science and research implies that:

1. Researches done in systems science can be applied to virtually all disciplines of classical science.
2. Issues involving systemhood, studied in individual specialization of classical science, can be studied comprehensively and thoroughly.
3. A unifying influence on classical science, where there is a growing number of narrow disciplines, is created.

Therefore, the classical and systems sciences can be viewed as a genuine two-dimensional science. With the added advantage of the second dimension (the systems science in this book), we will show some important impacts of this second dimension on the first dimension (classical science).

1.2 Whole Evolution of Systems: Where Systemic Yoyo Is From

When we enjoy the magnificent benefits offered by modern science and technology, we still have to think about the limitations of and problems existing in the science and technology we inherited from the generations before us. From the primeval to modern civilizations, humankind has gone through a history of development for over several millions of years. However, a relatively well-recorded history goes back only as far as about 3,000 years. During this timeframe, the development of science mirrors that of human civilizations. And each pursuit after the ultimate truth is a process of getting rid of the stale, taking in the fresh, and making new discoveries and new creativities. Each time authority is repudiated, science is reborn. Each time the "truth" is questioned, scientific prosperity appears. That is, each time people praise authorities, they are in fact praising ignorance.

At the turn of 21st century, with his profound insights, independent creativity, and courage, Shoucheng OuYang proposed the blown-up theory of nonlinear evolution problems based on a reversed thinking logic, factual evidence, and over 30 years of reasoning and practice. It is found that in terms of formalism, nonlinear evolution models are singularity problems of mathematical blown-ups of uneven formal evolutions, and in terms of physical objectivity, nonlinear evolution models describe mutual reactions of uneven structures of materials, which is no longer a problem of formal quantities. Because uneven structures are eddy sources, leading to eddy motions instead of waves, the mystery of nonlinearity, which has been bothering humankind for over 300 years, is resolved at once both physically and mathematically. What is shown is that the essence of nonlinear evolutions is the destruction of the initial value automorphic structures and appearance of discontinuity. It provides a tool of theoretical analysis for studying objective transitional and reversal changes of materials and events.

In the ancient scientific history of the Western civilizations, there existed two opposite schools on the structure of materials. One school believed that materials were made up of uncountable and invisible particles, named atoms; the other school believed that all materials were continuous. A representative of the former school is the ancient Greek philosopher Democritus (about 460–370 BC), and of the latter, the ancient philosopher Aristotle (384–322 BC). Because abundant existence of solids made it easy for people to accept Aristotelian continuity, the theory of atoms was not treated with any validity until the early part of the 19th century, when J. Dalton (1766–1844) established relevant evidence. In principle, Leibniz (1663) and Newton's calculus originated in Aristotelian thoughts. Along with calculus, Newton constructed his laws of motion on the computational basis of calculus and accomplished the first successes in applications in celestial movements under unequal quantitative effects. With over 200 years of development, classical mechanics has gradually evolved into such a set of analysis methods based on continuity that even nearly a century after quantum mechanics and relativity theory were established, the thinking logic and methods developed on continuity are still in use today.

Due to differences in the environmental conditions and living circumstances, where the West was originated from castle-like environments and the East from big-river cultures, with agriculture and water conservation as the foundation of their national prosperities, the ancient Eastern civilizations were different from those of the West. So, naturally, Chinese people have been more observing about reversal and transitional changes of weather and rivers. Because fluid motions are irregular and difficult to compute exactly, the *Book of Changes* and *Lao Tzu* appeared in China. The most important characteristic of the *Book of Changes* is its way of knowing the world through materials' images and analyzing materials' changes through figurative structures with an emphasis placed on materials' irregularities, discontinuities, and transitional and reversal changes.

As pointed out in the blown-up theory, in terms of mathematical symbolism, due to escapes in uneven forms from continuity, the evolution of any nonlinear model is no longer a problem of simply expanding the given initial values. What is significant here is that through nonlinear evolutions, the concept of blown-ups can represent Lao Tzu's teaching that "all things are impregnated by two alternating tendencies, the tendency toward completion and the tendency toward initiation, which, acting together, complete each other (Liang, 1996)," and agrees with non-initial-value automorphic evolutions, what the *Book of Changes* describes: "At extremes, changes are inevitable. As soon as a change occurs, things will stabilize and stability implies long lasting." (Wilhelm and Baynes, 1967).

Because nonlinearity describes eddy motions, there must exist different eddy vectorities and consequent irregularities. That is, the phenomenon of orderlessness is inevitable. When looking at fluid motions from the angle of eddies, one can see that the corresponding quantitative irregularities, orderlessness, multiplicities, complexities, etc., are all about the multiplicity of rotating materials. Therefore, there exist underlying reasons for the appearance of quantitative irregularities, multiplicities, and complexities. Those underlying reasons are the unevenness of time and space of the evolutionary materials' structures.

With conclusions of the blown-up theory, one can see that all eddy motions, as described with nonlinearities, are irregular. And regularized mathematical methods become powerless in front of the challenge of solving discontinuously quantified deterministic problems of nonlinear evolution models.

One important concept studied in the blown-up theory is that of equal quantitative effects, which describes the conclusion of quantitative analysis under quasi-equal acting forces. Although this concept was initially introduced in the study of fluid motions, it represents the fundamental and universal characteristic of materials' movements. What is important about this concept is that it reveals the fact that nonlinearities are originated from the structures of materials instead of nonstructural quantities.

On the basis of the blown-up theory, the concepts of black holes, Big Bangs, and converging and diverging eddy motions are coined together in the model shown in Figure 1.1. This model was established in Wu and Lin (2002) for each object and every system imaginable. In particular, each system or object considered in a study is a multidimensional entity that spins about its invisible axis. If we fathom such a

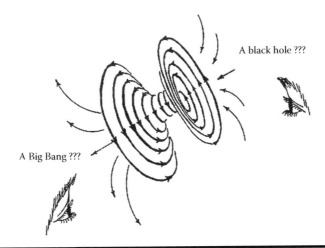

Figure 1.1 Eddy motion model of a general system.

spinning entity in our three-dimensional space, we will have a structure like that shown in Figure 1.1. The side of a black hole sucks in all things, such as materials, information, energy, etc. After funneling through the short narrow neck, all things are spit out in the form of a Big Bang. Some of the materials, spit out from the end of the Big Bang, never return to the other side, and some will. For the sake of convenience of communication, such a structure is called a (Chinese) yoyo due to its general shape. More specifically, what this model says is that each physical entity in the universe, be it a tangible or intangible object, a living being, an organization, a culture, a civilization, etc., can be seen as a kind of realization of a certain multidimensional spinning yoyo with an invisible spin field around it. It stays in a constant spinning motion, as depicted in Figure 1.1. If it does stop spinning, it will no longer exist as an identifiable system.

The theoretical justification for such a model is the blown-up theory (see Chapter 2 for more details). It can also be seen as a practical background for the law of conservation of informational infrastructures (see Chapter 3 for more details). More specifically, based on empirical data, the following law of conservation is proposed (Ren et al., 1998): for each given system, there must be a positive number a such that

$$AT \times BS \times CM \times DE = a \qquad (1.1)$$

where A, B, C, and D are some constants determined by the structure and attributes of the system of our concern, T is the time as measured in the system, S is the space occupied by the system, and M and E are the total mass and energy contained in the system.

Because M (mass) and E (energy) can exchange with each other and their total is conserved, if the system is a closed one, Equation (1.1) implies that when the time T evolves to a certain (large) value, the space S has to be very small. That is, in a limited space, the density of mass and energy becomes extremely high. So, an explosion (a Big Bang) is expected. Following the explosion, the space S starts to expand. That is, the time T starts to travel backward or shrink. This end gives rise to the well-known model for the universe as derived from Einstein's relativity theory. In our words, we have: each system goes through such cycles as ... → expanding → shrinking → expanding → shrinking → .... The geometry of this model from Einstein's relativity theory is given in Figure 1.1.

Practically, the yoyo model in Figure 1.1 is manifested in different areas of life. For example, each human being, as we now see it, is a three-dimensional realization of such a spinning yoyo structure of a higher dimension. To illustrate this end, let us consider two simple and easy-to-repeat experiences. For the first one, let us imagine we go to a sport event, say a swim meet. As soon as we enter the pool area, we immediately find ourselves falling into a boiling pot of screaming and jumping spectators, cheering for their favorite competing swimmers. Now, let us pick a person standing or walking on the pool deck for whatever reason, either for her beauty or for his strange look or body posture. Magically enough, before long, the person from quite a good distance will feel our stare and will be able to locate us in a very brief moment out of the reasonably sized and boiling audience. The reason for the existence of such a miracle and silent communication is because each side is a high-dimensional spinning yoyo. Even though we are separated by space and possibly by informational noise, the stare of one side on the other has directed that side's spin field of the yoyo structure into the spin field of the yoyo structure of the other side. That is the underlying mechanism for the silent communication to be established.

As our second example, let us look at the situation of human relationships. When individual A has a good impression about individual B, magically, individual B also has a similar and almost identical impression about A. When A does not like B and describes B as a dishonest person with various undesirable traits, it has been clinically proven in psychology that what A describes about B is exactly who A is himself (Hendrix, 2001). Once again, the underlying mechanism for such a quiet and unspoken evaluation of each other is because each human being stands for a spinning yoyo and its rotational field. Our feelings about another person are formed through the interactions of our invisible yoyo structures and their spin fields.

The presentation in this section is mainly based on Einstein (1983), Haken (1978), Jarmov (1981), Kline (1983), Liang (1996), Lin and OuYang (1996), Lin (1988, 1989, 1990), Lin et al. (1990), OuYang (1994), Prigogine (1967), Thom (1975), and Zhu (1985). For more details, please consult these references.

1.3 Applications of the Systemic Yoyo Model

Because spins are the fundamental evolutionary feature and characteristic of materials (Wu and Lin, 2002), as the first application of the yoyo model of general systems, a new figurative analysis method, composed of spin fields, is introduced. After establishing its theoretical and empirical foundations, this method is used to generalize Newton's laws of motion by addressing several unsettled problems in history. Through employment of the concept of equal quantitative effects, it is argued that this new method possesses some strength that pure quantitative methods do not have. On the basis of the characteristics of whole evolutions of converging and diverging fluid motions, the concept of time is revisited using this yoyo model. As further applications of the new figurative analysis method, we have a chance to walk through Kepler's laws of planetary motion and Newton's law of universal gravitation, and explain why planets travel along elliptical orbits, why no external forces are needed for systems to revolve about one another without colliding into each other as described by the law of universal gravitation, and why binary star systems, trinary star systems, and even n-nary star systems can exist, for any natural number $n \geq 2$. By checking the current state of research of the three-body problem, a brand new method is provided to analyze the movement of three stars, be they visible or invisible.

12 ■ Systemic Yoyos: Some Impacts of the Second Dimension

More specifically, after introducing a new method of figurative analysis, we have a chance to generalize all three laws of motion so that external forces are no longer required for these laws to work. As what is known, these laws are one of the reasons why physics is an exact science. As shown in this book, these generalized forms of the laws are equally applicable to social sciences and humanity areas as their classic forms in natural science.

In terms of energy, the concept of kinetic energy, which will be called irrotational kinetic energy, and the law of conservation of kinetic energy are established in modern science. For instance,

$$e = \frac{1}{2}mv^2 \text{ (Newton)} \tag{1.2}$$

and

$$E = mc^2 \text{ (Einstein)} \tag{1.3}$$

So, from the yoyo model, a natural question is: Because the square of speed constitutes kinetic energy, can the square of the angular speed of a rotation make up a different energy? By treating materials and objects as nonparticle spinning yoyos, the concept of stirring energy is introduced. Then, its possible conservations and some fundamental laws of evolution science are established. As a direct application of the concept of stirring energy, we look at the long-term technology of urban disaster reduction and prevention of floods, caused by suddenly appearing torrential rains. The results indicate that based on the concept of secondary circulations, by establishing urban artificial lakes, a high-capability flood prevention system can be materialized with very low costs, and this system also possesses the ability to improve the urban ecological environment. On the basis of Chengdu City's flood prevention facilities designed for major floods caused by torrential rains that are seen only once in 50 years, it is shown specifically how to improve the capability of these facilities to withstand such major floods that are recorded only once in 200 years, and how the capability of the flood prevention systems designed for once-in-200-years floods can be improved to materialize the goal of long-term flood prevention and disaster reduction. This application shows that evolution engineering can greatly relax the worries and reduce the potential troubles hidden in past flood prevention projects, such as high dams and high levees, and it can also help to recover the urban ecological balance and resolve such problems as shortage in water resources, etc., realizing tangible economic benefits with major ecological significance.

In terms of applications of the yoyo model in social science and humanity areas, it is shown that in a market of free competition, a concept as fundamental as demand and supply is about mutual reactions and mutual restrictions of different forces under equal quantitative effects. Hence, each economic entity can be naturally modeled and simulated as an economic yoyo or a flow of such yoyos. Due to the wide applicable scope and practical significance of the rotten kid theorem (Becker, 1974) and the excellent analysis of Bergstrom (1989) concerning the fact that this theorem is not generally true, after developing a yoyo model for various economic entities, a sufficient and necessary condition under which the rotten kid theorem holds true in general, in light of whole systems evolution, is established. After that, this result is employed to revisit the well-constructed examples by Bergstrom (1989): the lazy rotten kids, the nightlight controversy, the prodigal son, and the merit goods mentioned in (Becker, 1991). As a consequence, an astonishing corollary, which is named the theorem of never-perfect value systems, is derived. This theorem states that no matter how a value system is introduced and reinforced, the system will never be perfect. By connecting to Bergstrom's version of the rotten kid theorem (1989), we provide a much shorter proof explaining why transferable utility implies Becker's rotten kid theorem. By linking to Buchanan's Samaritan's dilemma, the clouds can be seen through with more clarity. Because "as long as the benevolent head continues to contribute his income to all, other members are also motivated to maximize the family income," as claimed by Becker's rotten kid theorem, the issue of child labor naturally comes into play. To this end, an example is constructed to show that the works of Baland and Robinson (2000) and Bommier and Dubois (2004) need to be fine-tuned. More specifically, child labor and children's human capital accumulation should not be mutually exclusive, as they had assumed.

By using the systemic yoyo model as the foundation to establish economic yoyos and their spin fields, one can create a qualitative means (called intuition) to foretell what could happen, how he should construct traditional models to verify the predictions, and why some of the economic observations must be universally true. In particular, if one looks at economic entities and financial organizations in the light of the systemic yoyo, he naturally models and simulates each economic entity, such as a person, family, firm, or economic sector (or industry), as an economic yoyo and its spin field, or a sea, or a flow of such yoyos. After establishing a yoyo model for various economic entities, a systemic method is developed to explore several important problems in economics and finance. For example, it is possible to look at the situation that instead of the kids, the parents are "rotten" and force their children to work. Then, efficiencies of the laissez faire triple (l_c^*, e_c^*, m_c^*) of child labor, formal schooling, and level of maturity are studied. What is important among the established results is that as long as a child's satisfaction is employed as a criterion for efficiency, the parents' desired levels of child labor, formal schooling, and maturity will always be inefficiently too high. Second, a profit maximization model is established to study interindustry wage patterns. And lastly, looking at different CEOs' choices of long- and short-term investment projects, one is able to analyze the price behavior of different investment assets, the dynamics among these assets, and how CEOs of firms gain control and influence over their monitoring boards of directors. These applications of the yoyo model show how the systemic yoyo methodology can be employed either as a tool of analysis to establish theoretical foundations for some important observations or as a road map for new discoveries and understandings in the research of economics and finance.

When looking at how people think, one can show the existence of the systemic yoyo structure in human thoughts. So, the human way of thinking is proven to have the same structure as that of the material world. After highlighting all the relevant ideas and concepts, which are behind each and every crisis in the foundations of mathematics, it becomes clear that some difficulties in human understanding of nature originate from confusing actual infinities with potential infinities, and vice versa. By pointing out the similarities and differences between these two kinds of infinities, one can then handily pick out some hidden contradictions existing in the system of modern mathematics. Then, theoretically, using the yoyo model, it is predicted that the fourth crisis in the foundations of mathematics has appeared. The value and originality of this application are that it shows the first time in history that human thought, the material world, and each economic entity share a common structure—the systemic yoyo. And it proves the arrival of the fourth crisis in mathematics by using systems modeling and listing several contradictions hidden deeply in the foundations of mathematics.

To show the reader that the yoyo model can be employed as not only the intuition and playground for theoretical studies in natural, social sciences and humanities areas, but also a useful tool for creating tangible economic benefits, we also look at how the thinking logic and the method of rotational materials can be employed to improve the forecasting accuracy of zero-probability disastrous weather such as suddenly appearing severe convective weather, small, regional, short-lived fog and thunderstorms, windstorms and sandstorms, and abnormally high temperatures. Most of these disastrous conditions have been extremely difficult for current meteorologists to predict.

1.4 Organization of This Book

This book contains five parts and 18 Chapters. In the first part, we will establish the rigorous theoretical foundation for the systemic yoyo model by looking at the concept of blown-ups, its mathematical and physical characteristics, and providing an explanation for quantitative infinity ∞. Alongside the theoretical foundation, empirical justifications, including the law of conservation of informational infrastructures, for such a model are also provided. In the second part, we look at applications of the yoyo model in classical mechanics, whole evolutions of converging and diverging fluid motions, Kepler's laws of planetary motion, the traditionally difficult three-body problem, and the existence of n-nary star systems. Then, we look at the concept of stirring energy, its conservation, and consequent evolution engineering and technology for long-term disaster reduction. To prepare for practical applications of the yoyo methodology, we also consider the questions of what time is and whether or not it occupies any material dimension.

In Part 3, we look at economic and financial forces from the perspective of nonlinearity and the yoyo model. After establishing the whole evolution analysis of demand and supply, we look at some mysteries of the family, child labor and its efficiency, existence of different industry sizes, interindustry wage differentials, the dynamics between long-term and short-term projects, and power struggles between boards of directors and CEOs. In Part 4, the structure of human thoughts and infinity problems in modern mathematics are studied. After a quick glance at the history of mathematics, the concepts of actual and potential infinities are introduced. On the bases of these concepts, some hidden inconsistencies in the modern system of mathematics are found. Then, it is shown that the fourth crisis in the foundations of mathematics has appeared. In Part 5, the concept of rolling currents and the phenomenon of ultra-low temperatures are introduced. A detailed explanation on how **V**-3θ graphs—a structural prediction method—are designed is provided. Then, as practical applications, we look at how these new concepts and methods can be materially employed to forecast suddenly appearing severe convective weather, such as torrential rains, small, regional, short-lived fog and thunderstorms, windstorms and sandstorms, and abnormally high temperatures. All of these applications are presented alongside real-life case studies.

THE SYSTEMIC YOYO: THEORETICAL AND EMPIRICAL FOUNDATIONS

Chapter 2

Blown-Ups, Eddy Motions, and Transitional Changes

Chapter 1 introduced the theoretical foundation of the systemic yoyo model. In this chapter, let us glance at the blown-up theory and related main results.

2.1 The Concept of Blown-Ups

When developments and changes naturally existing in the surrounding environment are seen as a whole, we have the concept of whole evolutions. And in whole evolutions, other than continuities, as well studied in modern mathematics, what seems to be more important and more common is discontinuity, with which transitional changes (or blown-ups) occur. These blown-ups reflect not only the singular transitional characteristics of the whole evolutions of nonlinear equations, but also the changes of old structures being replaced by new structures.

Reversal and transitional changes have always been the central issue and extremely difficult open problem of prediction science, because the well-developed method of linearity, which tries to extend the past rise and fall into the future, does not have the ability to predict forthcoming transitional changes and what will occur after those changes. In terms of nonlinear evolution models, blown-ups reflect destructions of old structures and establishments of new ones. Although these models reveal the limits and weaknesses of calculus-based theories, their analytic form can still be employed to describe to a certain degree the realisticity of discontinuous transitional changes of objective events and materials. So, the concept of blown-ups is not purely mathematical. By borrowing the form of calculus, the concept of blown-ups can be given as follows: for a given (mathematical) model, that truthfully describes the physical situation of our concern, if its solution $u = u(t; t_0, u_0)$, where t stands for time, satisfies

$$\lim_{t \to t_0} |u| = +\infty \qquad (2.1)$$

and at the same time, when $t \to t_0$, the underlying physical system also goes through a transitional change, then the solution $u = u(t; t_0, u_0)$ is called a blown-up and the relevant physical movement expresses a blown-up. For nonlinear models in independent variables of time and space, the concept of blown-ups is defined similarly, where blow-ups in the model and the underlying physical system can appear in time, space, or both.

Based on the evolutionary behaviors of the physical system before and after the transition, we can classify blown-ups into two categories: transitional and nontransitional. A blown-up is transitional if the development of the physical system after the special blow-up moment in time or space is completely opposite that before the blow-up. For example, if the evolution of the system grows drastically before the blow-up and the development of the system after the blow-up starts from nearly ground zero, then such a blown-up is transitional. Otherwise, it is called nontransitional.

2.2 Mathematical Characteristics of Blown-Ups

To help us understand the mathematical characteristics of blown-ups, let us look at the following constant coefficient equation:

$$\dot{u} = a_0 + a_1 u + \ldots + a_{n-1} u^{n-1} + u^n = F \qquad (2.2)$$

where u is the state variable and $a_0, a_1, \ldots, a_{n-1}$ are constants. Then, when $n = 2$, Equation (2.2) is a special Riccati equation that has been very well applied in the study of prediction science. A detailed analytical discussion (for the omitted technical details, please consult Wu and Lin [2002]) shows that under different conditions, the solution of Equation (2.2) can either be continuous and smooth or experience blown-ups or periodic transitional blown-ups. So, even for the simplest nonlinear evolution equations, the well-posedness of their evolutions in terms of differential mathematics is conditional. Here, the requirements for well-posedness are that the solution exists, and it is unique and stable (or continuous and differentiable). As a side note, please be aware that the condition of well-posedness is very important for the majority of the applied mathematics to work.

When $n = 3$, other than the situation that $F = 0$ has two real solutions, one of which is of multiplicity 2, Equation (2.2) experiences blown-ups. For the general case n, assume that Equation (2.2) can be written as

$$\dot{u} = F = (u - u_1)^{p_1} \dots (u - u_r)^{p_r} (u^2 + b_1 u + c_1)^{q_1} \dots (u^2 + b_m u + c_m)^{q_m} \quad (2.3)$$

where p_i and q_j ($i = 1, 2, \dots, r$ and $j = 1, 2, \dots, m$) are positive whole numbers, $n = \sum_{i=1}^{r} p_i + 2 \sum_{j=1}^{m} q_j$, and $\Delta = b_j^2 - 4c_j < 0$. Then, we have the following:

THEOREM 2.1

The condition under which the solution of an initial value problem of Equation (2.3) contains blown-ups is given by:

1. When u_i, $i = 1, 2, \dots, r$, does not exist, that is, $F = 0$ does not have any real solution
2. If $F = 0$ does have real solutions u_i, $i = 1, 2, \dots, r$, satisfying $u_1 \geq u_2 \geq \dots \geq u_r$,
   a. When n is an even number, if $u > u_1$, then u contains blow-ups.
   b. When n is an odd number, no matter whether $u > u_1$ or $u < u_r$, there always exist blown-ups.

Higher-order nonlinear evolution systems, in general, can be simplified into systems of nonlinear evolution equations in a form similar to that of Equation (2.2). Because the general case is difficult to analyze, let us look at the following second-order nonlinear evolution equation:

$$\ddot{u} = a_0 + a_1 u + a_2 u^2 + u^3 \quad (2.4)$$

where $\ddot{u} = \frac{d^2 u}{dt^2}$ and a_i ($i = 0, 1, 2$) are constants. Now, this equation can be reduced to the following system of first-order equations in two variables:

$$\begin{cases} \dot{u} = v \\ \dot{v} = a_0 + a_1 u + a_2 u^2 + u^3 \end{cases} \quad (2.5)$$

Multiplying the left-hand side of the second equation by v and the right-hand side by $\dot{u}$, and then integrating the resultant equation, provides the following:

$$v^2 = E = 2 \left(\frac{1}{4} u^4 + \frac{1}{3} a_2 u^3 + \frac{1}{2} a_1 u^2 + a_0 u + h_0 \right) \quad (2.6)$$

where h_0 is the integration constant. Now, the system in Equation (2.5) can be reduced to

$$\dot{u} = \pm \sqrt{E} \quad (2.7)$$

It can be shown that when $E = \frac{1}{4}(u^2 + b_1 u + c_1)^2$ and $b_1^2 - 4c_1 > 0$, Equation (2.7) has two solutions, one of which is continuous and bounded and the other containing a blown-up; other than this case, all other possible solutions of Equation (2.7) contain blown-ups. For the omitted technical details, consult Wu and Lin (2002).

As for evolution equations involving changes in space, they can directly and intuitively reflect the physical meanings of blown-ups. The one-dimensional advection equation is one of the simplest nonlinear evolution equations involving changes in space. Its Cauchy problem can be written as follows:

$$\begin{cases} u_t + uu_x = 0 \\ u|_{t=0} = u_0 \end{cases} \qquad (2.8)$$

where u stands for the speed of a flow, $u_t = \frac{\partial u}{\partial t}$ and $u_x = \frac{\partial u}{\partial x}$, with t representing time and x the one-dimensional spatial location. By using the method of separating variables without expansion, we can get

$$u_x = \frac{u_{0x}}{1 + u_{0x} t} \qquad (2.9)$$

Based on the definition of the degree of divergence, u_x is the first-dimensional degree of divergence. When $u_{0x} > 0$, that is, when the initial field is divergent, u_x declines continuously with time t until the diverging motion disappears. If $u_{0x} < 0$, that is, when the initial field is convergent, then the solution of $u_x \to \infty$ experiences a discontinuous singularity with time: $t \to t_b = -\frac{1}{u_{0x}}$. Evidently, when $t < t_b$, $u_x < 0$ evolves continuously. When $t > t_b$, $u_x > 0$. So, the convergent movement of the initial field ($u_{0x} < 0$) can be transformed into a divergent movement ($u_{0x} > 0$) through a blown-up. The characteristics of this kind of movement cannot be truly and faithfully described by linear analysis or statistical analysis. They reflect the fundamental characteristics of nonlinear evolutions.

2.3 Mapping Properties of Blown-Ups and Quantitative Infinity

One of the features of blown-ups is the quantitative ∞, which stands for indeterminacy mathematically. So, a natural question is how to comprehend this mathematical symbol ∞, which in applications causes instabilities and calculation spills that have stopped each and every working computer.

To address this question, let us look at the mapping relation of the Riemann ball, which is well studied in complex functions (Figure 2.1). This so-called Riemann ball, a curved or curvature space, illustrates the relationship between the infinity on the plane and the North Pole N of the ball. Such a mapping relation

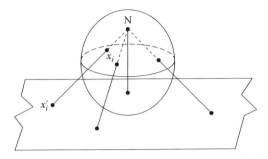

Figure 2.1 The Riemann ball relationship between planar infinity and three-dimensional North Pole.

connects $-\infty$ and $+\infty$ through a blown-up. Or in other words, when a dynamic point x_i travels through the North Pole N on the sphere, the corresponding image x_i' on the plane of the point x_i shows up as a reversal change from $-\infty$ to $+\infty$ through a blown-up. So, treating the planar points $\pm\infty$ as indeterminacy can only be a product of the thinking logic of a narrow or low-dimensional observ-control, because, speaking generally, these points stand implicitly for direction changes of one dynamic point on the sphere at the pole point N; in other words, the phenomenon of directionless, as shown by blown-ups of a lower-dimensional space, represents exactly a direction change of the movement in a higher-dimensional space. A similar discussion can be given to relate quantitative $\pm\infty$ (symbols of indeterminacy in one-dimensional space) to a dynamic movement on a circle (a curved space of a higher dimension).

Therefore, the concept of blown-ups can relatively and specifically represent implicit transformations of spatial dynamics. That is, through blown-ups, problems of indeterminacy of a narrow observ-control are transformed into a determinant situation of a more general observ-control system. This discussion shows that the traditional view of singularities as meaningless indeterminacies has revealed not only the obstacles of the thinking logic of the narrow observ-control, but also the careless omissions of spatial properties of dynamic implicit transformations.

2.4 Spinning Current: A Physical Characteristic of Blown-Ups

From the previous discussions, we can see that nonlinearity, speaking mathematically, stands (mostly) for singularities. In terms of physics, nonlinearity represents eddy motions. Such motions are a problem of structural evolutions, a natural consequence of uneven evolutions of materials. So, nonlinearity accidentally describes

24 ■ Systemic Yoyos: Some Impacts of the Second Dimension

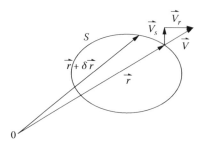

Figure 2.2   The definition of a closed circulation.

discontinuous singular evolutionary characteristics of eddy motions from the angle of a special, narrow observ-control system.

To support this conclusion, let us first look at the Bjerknes' circulation theorem (1898). At the end of the 19th century, V. Bjerkes discovered eddy effects due to changes in the density of media in movements of the atmosphere and ocean. He consequently established the well-known circulation theorem, which was later named after him. Let us look at this theorem briefly.

By *circulation* is meant a closed contour in a fluid. Mathematically, each circulation Γ is defined as the line integral about the contour of the component of the velocity vector locally tangent to the contour. In symbols, if $\vec{V}$ stands for the speed of a moving fluid, S an arbitrary closed curve, and $\delta\vec{r}$ the vector difference of two neighboring points of the curve S (Figure 2.2), then a circulation Γ is defined as follows:

$$\Gamma = \oint_S \vec{V}\, \delta\vec{r} \qquad (2.10)$$

Through some ingenious manipulations (for the omitted technical details, consult Wu and Lin, 2002), the following well-known Bjerknes' circulation theorem is obtained:

$$\frac{d\vec{V}}{dt} = \iint_\sigma \nabla\left(\frac{1}{\rho}\right) \times (-\nabla p) \cdot \delta\sigma - 2\Omega \frac{d\sigma}{dt} \qquad (2.11)$$

where σ is the projection area on the equator plane of the area enclosed by the closed curve S, p the atmospheric pressure, ρ the density of the atmosphere, and Ω the Earth's rotational angular speed.

The left-hand side of Equation (2.11) represents the acceleration of the moving fluid, which according to Newton's second law of motion is equivalent to the force acting on the fluid. On the right-hand side, the first term is called a solenoid in meteorology. It originated from the interaction of the p- and ρ-planes due to

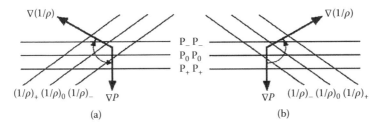

Figure 2.3 A diagram for solenoid circulations.

uneven density ρ, so that a twisting force is created. Consequently, materials' movements must be rotations, with the rotating direction determined by the equal p- and ρ-plane distributions (Figure 2.3). The second term in Equation (2.11) comes from the rotation of the Earth.

In general textbooks, Bjerknes' circulation theorem is applied to explain the formation of land-ocean breezes. In history, there existed people, such as B. Saltzman (1962), who have seen this circulation theorem as a major betrayal of classical mechanics. However, S. C. OuYang assigned this theorem a very high value, believing that in the effort of revealing the commonly existing and practically significant eddy effects of fluid motions, Bjerknes' circulation theorem has and will play an increasingly important role. This theoretical result has not only changed many opinions drawn on the classical theories of fluids, but also, more importantly, resolved the mystery of nonlinearity (Lin, 1998), which puzzled humankind for the past 300-plus years, and revealed the universal law of materials' evolutions. The significance of this theorem also includes:

1. In terms of dynamics, an uneven ρ implies eddy effects instead of the so-called effects of elastic pressure. The pressure gradient force $\left(-\frac{1}{\rho}\nabla p\right)$ should also be called the stirring gradient force of the density pressure. Therefore, $\left(-\frac{1}{\rho}\nabla p\right)$ is an eddy source. In the form of mathematics, it is nonlinear. So, nonlinearity implies eddy sources. Furthermore, uneven eddy motions are the most common form of materials' movements observed in the universe. Bjerknes' theorem has clearly shown this fact, even though he and scholars of the following generations did not recognize this implication until OuYang pointed it out.

2. In the universe, the common form of materials' movements is eddies, such as the eddy motions of the solar system, galaxies, nebula, etc., in the cosmic level; polar eddies, cyclones, anticyclones, etc., on Earth at the meso-level; and various rotating movements found in atomic structures, in terms of the microscopic level. That is why Kuchemann (1961) once claimed that "the tendon of moving fluids is eddies." What OuYang said is that the field of eddies is both a place with high concentration of fluids' kinetic energies and

a place with efficient transformations of kinetic energies into heat energies. In this sense, people are reminded of the fact that without eddy motions, there would not be any transformation of kinetic energies. Through concentration of kinetic energies and through transformations of these energies into heat energies, eddies actually consume kinetic energies. Such a rise and fall of energies determines the equilibrium of the heat-kinetic forces, internal to the eddies, and reflects the close relationship between the quantities of eddies and transformations of energies.

3. Well-known Chinese scholar Shi-jia Lu once pointed out that "the essence of fluids is eddies, since fluids cannot stand twisting forces and as soon as a twist exists, eddies appear" (unpublished notes). Because uneven densities create twisting forces, fields of spinning currents are naturally created. Such fields do not have uniformity in terms of types of currents. Clockwise and counterclockwise eddies always coexist, leading to destructions of the initial smooth, if any, fields of currents. When such evolutions are reflected in the continuous analytical methodology of any calculus-based narrow observ-control system, discontinuous singularities in solutions will naturally appear. Conversely, discontinuities of nonlinear differential equations offer us an analytic method for predicting discontinuous transitional changes of materials and the appearance of fields of eddy currents. However, to truly resolve this kind of problem of evolutions, we should give up the available analytic method, developed on the assumption of continuity, and search for new ideas and methods for a more effective approach for analyzing nonlinearity.

Next, let us look at the general dynamic system and how it is related to eddy motions. The following is Newton's second law of motion:

$$m\frac{d\vec{v}}{dt} = \vec{F} \qquad (2.12)$$

Based on Einstein's concept of uneven time and space of materials' evolutions, we can assume

$$\vec{F} = -\nabla S(t, x, y, z) \qquad (2.13)$$

where $S = S(t, x, y, z)$ stands for the time-space distribution of the external acting object. Let $\rho = \rho(t, x, y, z)$ be the density of the object being acted upon. Then, the kinematic equation (Equation (2.12)) for a unit mass of the object being acted upon can be written as

$$\frac{d\vec{u}}{dt} = -\frac{1}{\rho(t, x, y, z)} \nabla S(t, x, y, z) \qquad (2.14)$$

where $\vec{u}$ is used to replace the original $\vec{v}$ to represent the meaning that each movement of some materials is a consequence of mutual reactions of materials' structures. Evidently, if ρ is not a constant, then Equation (2.14) becomes

$$\frac{d(\nabla x \vec{u})}{dt} = -\nabla x \left[\frac{1}{\rho} \nabla S\right] \neq 0 \qquad (2.15)$$

which stands for an eddy motion. In other words, a nonlinear mutual reaction between materials' uneven structures and the unevenness of the external forcing object will definitely produce eddy motions.

On the other hand, because nonlinearity stands for eddy sources, it represents a problem about structural evolutions. This has essentially resolved the problem of how to understand nonlinearity, and has also ended the particle assumption of Newtonian mechanics and the methodological point of view of "melting shapes into numbers," which was formed during the time of Newton in natural sciences. What is more important is that the concept of uneven eddy evolutions reveals the fact that forces exist in the structures of evolving objects, and do not exist independently out of objects—what Aristotle and Newton believed, so that the movement of all things had to be pushed first by God. Based on such reasoning, the concept of second stir of materials' movements is introduced. At this junction, we need to point out that as early as over 2,500 years ago, Lao Tzu of China once said: "Tao is about physical materials. Even though nothing can be seen clearly, there exist figurative structures in the fuzziness" (English and Feng, *Tao Te Ching*, Chapter 21). It is no doubt scientific and epistemological progress that in our modern time, Einstein proposed that gravitations are originated from the unevenness of time and space, which had essentially ended the era of the Aristotelian concept of forces existing independently outside of materials. However, due to the fact that Einstein did not notice the unevenness of time and space of the object being acted upon, he did not successfully reveal the essence of nonlinear mutual reactions.

For example, in terms of the problem of the universe's evolution, Newton needed the hands of God. The first push of God has been vividly seen in the second law of motion. Einstein had realized the materialism of forces. However, based on his general relativity theory, it is concluded that the universe is originated from a Big Bang out of a singular point. Obviously, the singular point represents the transition of materials' evolution, which possesses more realisticity of materials than Newton's hands of God. Therefore, the scientific community has accepted such a Big Bang theory. However, the acceptance does not mean that such a Big Bang theory has revealed the true evolution of the universe. It is because the theory of Big Bang is established on the basis of the universe's background radiation (3K, where K is the absolute temperature index), which is assumed to be uniform in all directions. Next, should there be only one singular point? Because the so-called background radiation should also be originated from the unevenness of materials, can the universe's background radiation be calm? Even though a calm state could be reached temporarily, there should be a pre–singular point universe. Because unevenness is the fundamental property of spinning materials and the origin of all multiple levels of materials' eddies, the corresponding singular point explosions should also be multiple. So, the Big Bang theory, as a scientific theory, is still not plausible.

As a matter of fact, the concept of second stir can also be extended into a theory about the evolution of the universe. Because the concept of the second stir assigns forces with materials' structures, it naturally ends Newton's God. The duality of rotations must lead to differences in spinning directions. Such differences surely lead to singular points of singular zones. Through subeddies and sub-subeddies, breakages (or Big Bangs) are represented so that evolutionary transitions are accomplished. Evidently, according to the concept of second stir, the number of singular points would be greater than one, and the Big Bang explosions would also have their individual multiplicities. Therefore, the concept of second stir will be a thought left behind in the 20th century, concerning the physical essence of materials' evolutions, worthy of further and deeper thinking.

2.5 Equal Quantitative Effects

Another important contribution of the blown-up system is the introduction of the concept of equal quantitative effects. Even though this concept was initially proposed in the study of fluid motions, it has essentially represented the fundamental and universal characteristics of all materials' movements. What is more important is that this concept reveals the fact that nonlinearity is originated from the figurative structures of materials instead of nonstructural quantities of the materials.

The so-called equal quantitative effects mean the eddy effects with nonuniform vortical vectorities existing naturally in systems of equal quantitative movements, due to the unevenness of materials. Here, *equal quantitative movements* means the movements with quasi-equal acting and reacting objects, or under two or more quasi-equal mutual constraints. For example, the relative movements of two or more planets of approximately equal masses are considered equal quantitative movements. In the microcosmic world, an often seen equal quantitative movement is the mutual interference between the particles to be measured and the equipment used to make the measurement. Many phenomena in daily lives can also be considered equal quantitative effects, including such events as wars, politics, economies, chess games, races, plays, etc.

The Aristotelian and Newtonian framework of separate objects and forces is about unequal quantitative movements established on the assumption of particles. On the other hand, equal quantitative movements are mainly characterized by the kind of measurement uncertainty that when I observe an object, the object is constrained by me. When an object is observed by another object, the two objects cannot really be separated. At this junction, it can be seen that the Su-Shi principle of Xuemou Wu's panrelativity theory (1990), Bohr's (N. Bohr, 1885–1962) principle, the relativity principle about microcosmic motions, von Neumann's principle of program storage, etc., all fall into the uncertainty model of equal quantitative movements with separate objects and forces.

What is practically important and theoretically significant is that eddy motions are confirmed not only by daily observations of surrounding natural phenomena, but also by laboratory studies on structures of the universe ranging from small atomic structures to huge nebular ones. At the same time, eddy motions show up in mathematics as nonlinear evolutions. The corresponding linear models can only describe straight-line-like spraying currents and wave motions of the morphological changes of reciprocating currents. What is interesting here is that wave motions and spraying currents are local characteristics of eddy movements. This is well shown by the fact that linearities are special cases of nonlinearities. Please note that we do not mean that linearities are approximations of nonlinearities.

Because 99 percent of all materials in the universe are fluids, and because under certain conditions solids can be converted to fluids, OuYang (1998a, p. 33) pointed out that "when fluids are not truly known, the amount of human knowledge is nearly zero. And, the epistemology of the western civilization, developed in the past 300 plus years, is still under the constraints of solids." This statement no doubt points to the central weakness of the current theoretical studies, and also locates the reason why "mathematics met difficulties in the studies of fluids," as said by F. Engels (1939). It also well represents Lao Tzu's teaching: "If crooked, it will be straight. Only when it curves, it will be complete (English and Feng, 1972)."

In fact, humankind exists in a spinning universe with small eddies nested in bigger eddies or multiple eddies. Through subeddies and sub-subeddies, heat-kinetic energy transformations are completed. These nested eddies and energy transformations vividly represent the forever generation changes of blown-up evolutions of all things in the universe.

The birth-death exchanges and the nonuniformity of vortical vectorities of eddy evolutions naturally explain how and where quantitative irregularities, complexities, and multiplicities of materials' evolutions, when seen from the current narrow observ-control system, come from. Evidently, if the irregularity of eddies comes from the unevenness of materials' structures, and if the world is seen at the height of structural evolutions of materials, then the world is simple. And it is so simple that there are only two forms of motions: clockwise rotation and counterclockwise rotation. The vortical vectority of materials' structures has very intelligently resolved the Tao of Yin and Yang of the *Book of Changes* of the Eastern mystery (Wilhelm and Baynes, 1967), and has been very practically implemented in the common form of motion of all materials in the universe. That is where the concept of invisible organizations of the blown-up system comes from.

The concept of equal quantitative effects not only possesses a wide range of applications, but also represents an important omission of modern science, developed in the past 300-plus years. Evidently, not only are equal quantitative effects more general than the particle mechanic system with further reaching significance, but they also have directly pointed to some of the fundamental problems existing in modern science. For instance:

1. Equal quantitative effects can throw calculations of equations into computational uncertainty. Evidently, if $x \approx y$, then $x - y$ becomes a mathematical problem of computational uncertainty, involving large quantities with infinitesimal increments. Even though this end has been well known, in practical applications, people are still often unconsciously misguided into such uncertainties. For example, in meteorological science, one situation involves

$$2\vec{\Omega} \times \vec{V}_h \approx -\frac{1}{\rho}\nabla ph$$

where the left-hand side stands for the deviation force caused by the Earth's rotation and the right-hand side the stirring force of the atmospheric density pressure. However, scholars have tried for many decades to compute $\frac{dV_h}{dt}$ under the influence of such quasi-equal computational uncertainty. In fact, the concept of equal quantitative effects has computationally declared that equations are not eternal, or there does not exist any equation under equal quantities. That is why OuYang introduced the methodology of abstracting numbers (quantities) back into shapes (figurative structures). The purpose of doing this is to describe the formalization of eddy irregularities, which is different from regularized mathematical quantification of structures. That is why we should very well see that if 300 years ago, it was human wisdom to abstract numbers out of everything, then it would be human stupidity to continue to do so today.

2. Because the current variable mathematics is entirely about regularized computational schemes, there must be the problem of disagreement between the variable mathematics and irregularities of objective materials' evolutions. The corresponding quantified comparability can only be relative. And at the same time, there exists a problem with quantification where distinct properties cannot be distinguished because the relevant quantifications produce the same indistinguishable numbers. Because of this, it is both incomplete and inaccurate to employ quantitative comparability as the only standard for judging scientificality. For example, in terms of weather forecasting, it should be clear that difficulties we have been facing are consequences of the incapability of the existing theories in handling equal quantitative effects. It can also be said that the quantitative science, developed in the past 300-plus years, is incomplete and incapable of resolving problems about figurative structures.

3. The introduction of the concept of equal quantitative effects has not only made the epistemology of natural sciences go from solids to fluids, but also completed the unification of natural and social sciences. That is because many social phenomena, such as military conflicts, political struggles, economic competitions, chess games, races, plays, etc., can be analyzed on the basis of figurative structures of equal quantitative effects. However, what needs to be pointed out is that the current system of natural sciences is basically an extension of the research under unequal quantitative effects. At the same time, in some areas of research, there has been the problem of misusing unequal quantitative effects. These areas include, but are not limited to, Rossby's long waves in meteorology, the topographic leeward wave theory, the chaos theory, etc.

To intuitively see why equal quantitative effects are so difficult for modern science to handle by using the theories established in the past 300-plus years, let us first look at why all materials in the universe are in rotational movements. According to Einstein's uneven space and time, we assume that all materials have uneven structures. Out of these uneven structures, there naturally exist gradients. With gradients, there will appear forces. Combined with uneven arms of forces, the carrying materials will have to rotate in the form of moments of forces. That is exactly what the ancient Chinese (Lao Tzu) said: "Under the heaven, there is nothing more than the Tao of images." (English and Feng, 1972). This is in contrast to the Newtonian doctrine of particles: under the heaven, there is such a Tao that is not about images. The former stands for an evolution problem of rotational movements under stirring forces. Because structural unevenness is an innate character of materials, it is named second stir, considering that the term *first push* was used first in history (OuYang et al., 2000). What needs to be noted is that the terms *first push* and *second stir* do not mean that the first push is prior to the second stir.

Now, we can imagine that the natural world or the universe is composed entirely with eddy currents, where eddies exist in different sizes and scales and interact with each other. That is, the universe is a huge ocean of eddies, which change and evolve constantly. One of the most important characteristics of spinning fluids, including spinning solids, is the difference between the structural properties of inwardly and outwardly spinning pools and the discontinuity between these pools. Due to the stirs in the form of moments of forces, in the discontinuous zones, there exist subeddies and sub-subeddies (Figure 2.4, where subeddies are created naturally by the large eddies M and N). Their twist-ups (the subeddies) contain highly condensed amounts of materials and energies. In other words, the traditional frontal lines and surfaces (in meteorology) are not simply expansions of particles without any structure. Instead, they represent twist-up zones concentrated with irregularly structured materials and energies (this is where the so-called small-probability events appear and small-probability information is observed and collected, so such

32 ■ Systemic Yoyos: Some Impacts of the Second Dimension

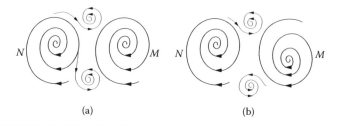

Figure 2.4 Appearance of subeddies.

information [event] should also be called irregular information [event]). In terms of basic energies, these twist-up zones cannot be formed by only the pushes of external forces and cannot be adequately described by using mathematical forms of separate objects and forces. Because evolution is about changes in materials' structures, it cannot be simply represented by different speeds of movements. Instead, it is mainly about transformations of rotations in the form of moments of forces ignited by irregularities. The enclosed areas in Figure 2.5 stand for the potential places for equal quantitative effects to appear, where the combined pushing or pulling is small in absolute terms. However, it is generally difficult to predict what will come out of the power struggles.

2.6 Various Properties of Blown-Ups

Summarizing what has been discussed above, when a mathematical model, which truthfully and adequately describes the physical situation of our concern, blows up at a specific time moment or a specific spatial location or both, and the underlying physical system also goes through a transitional change, then the solution of the model is called a blown-up solution and the relevant physical movement expresses a blown-up. Blown-up phenomena appear in life all the time, especially in studies involving evolutions.

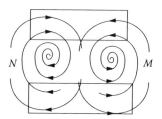

Figure 2.5 Structural representation of equal quantitative effects.

The mathematical characteristics of blown-ups can be briefly described in two cases.

Case 1: Blown-ups appear only with time. By analyzing nonlinear equations and more general nonlinear models, it is concluded that in most cases, nonlinearity implies blown-ups and the requirements for well-posedness of modern science (existence, uniqueness, and stability) do not hold true, meaning that most of the methods and thinking logic in modern science cannot be employed to resolve the relevant problems. Only under very special local conditions, there will not be any blown-up. That is when the available methods and theories in modern science can be applied.

Case 2: Blown-ups appear in both time and space. The results obtained here are very interesting. It is shown that if a system, when seen as a rotational entity, is initially divergent, then over time the whole evolution of the system is continuous and the divergent development of the system will eventually disappear smoothly and quietly. However, if the initial state of the system is convergent, then it is guaranteed that a moment of blown-up will appear at a definite time moment in the foreseeable future. After the system goes through a transitional change (blown-up), it will restart as a divergent system.

By introducing the concept of implicit transformations between a Euclidean space and a curvature space, it is shown that blown-ups in the Euclidean space are simply some transitional changes in the curvature space. And periodic blown-ups of the Euclidean space and rotational movements in the curvature space correspond very well, illustrating a method on how to resolve the problems of the quantitative ∞, numerical instability, and computational spills, by reconsidering the situations in a curvature space. By doing so, all these problems, which seem unsolvable in modern science, are avoided.

It is shown that the physical characteristics of blown-ups are spinning currents. By revisiting Bjerknes' circulation theorem, it is found that any force acting on a fluid must be a twisting force, creating spinning currents in the fluid. In addition to explaining land-ocean breezes, this theorem also implies that (mathematical) nonlinearity means eddy sources, and in turn eddy motions. This theorem actually provides a theoretical explanation for why the mostly seen form of motion in the universe is rotation, because uneven densities lead directly to eddy sources. And because of the duality of eddies in their rotational directions, the initial smooth fields, if any, have to be destroyed. One of the very significant outcomes of this analysis is that discontinuities and singularities existing in calculus-based models can actually be applied as a tool to predict forthcoming transitional changes. By looking at the general dynamic system, it is shown that Newton's second law of motion actually indicates that as long as the acting and reacting objects have uneven structures, their mutual reaction will be nonlinear and a rotational movement. So, the concept of second stir is introduced. On the basis of this new concept, new explanations and a generalization of the Big Bang theory are provided in terms of a new theory on the evolution of the universe.

With the theoretical background established for rotations to be the common form of movements in the universe, the concepts of equal quantitative effects and equal quantitative movements are introduced. Through using these concepts, it is analyzed that mathematical nonlinearity is created by uneven internal structures of the materials involved instead of the nonstructural quantities of the materials. It is found that equal quantitative effects lead to computational uncertainties and have the ability to unify natural and social sciences, because in both situations, figurative structural analyses can be employed. By looking at the geometry of equal quantitative effects, one can easily locate where small-probability events would appear and where small-probability information can be observed.

Here, we have the following important observation:

> **The principle of blown-ups:** Keep as much as possible the irregular information existing in available observations. Based on inevitable transitionality of evolutions and various possible transformations of fluids' rotations, figurative (structural) analysis methods should be employed to predict what is forthcoming. Any quantitative analysis, if considering applying it, should be kept to its reasonable level without being dominant over the figurative methods.
>
> **The principle of nonparticle materials' evolution:** Under the heaven, there is nothing but the Tao of images. Irregularities are the determinism of nature instead of informational uncertainties.

Let us conclude this section with a note on the references. Other than at specific locations where we have directly pointed to the source of information, the presentation of this section is mainly based on Glassey (1977), Guo (1995), Hess (1959), Holton (1979), Kuchemann (1965), Lin (2000), Lin and Fan (1997), Lin and OuYang (1998), Lin and Wu (1998), Lorenz (1993), OuYang (1994, 1995, 1998a, 1998b), OuYang et al. (1998, 2000), Saltzman (1962), X. M. Wu (1990), Y. Wu (1998), and Wu and Lin (2002). For more details and references, please consult these works.

Chapter 3

Conservation of Informational Infrastructure: Empirical Evidence

With the theoretical foundation of the systemic yoyo model discussed in the previous chapter, in what follows, we will look at some empirical evidence for the model. The presentation in this chapter is mainly from Ren et al. (1998).

3.1 Introduction

In the history of humankind, human intuitions were made "exact" by various laws of conservation, even though, at the times when they were proposed, there might not have been any theoretical or mathematical foundations for these laws.

Lin (1995) developed a theoretical foundation for some laws of conservation, such as the laws of conservation of matter-energy, fundamental particles, etc., on general systems theory (Lin and Fan, 1997). Addressing some problems related to Lin's work (1995) systematically showed that our understanding of its nature can be very much limited with our constant attempts to get closer to its limitations.

In this chapter, based on an intuitive understanding of the concept of general systems, one may feel that there should exist a law of conservation emphasizing the uniformity of spatial structures of various systems. Starting with this intuition, several areas of scientific research are searched, and similarities among these areas are found.

Based on these similarities, a law of conservation of informational infrastructure is proposed, and some possible impacts and consequences of this law are explored. Connecting to what has been presented earlier, it is expected that this presentation will help to provide the empirical evidence for the systemic yoyo model for each and every general system.

The following is the intuitive understanding of the concept of general systems, which is the heuristic foundation of the entire work presented in this chapter on the law of conservation of informational infrastructures.

From a practical point of view, a system is what is distinguished as a system (Klir, 1985). From a mathematical point of view, a system is defined as follows (Lin, 1987): S is a (general) system, provided that S is an ordered pair (M, R) of sets, where M is the set of objects of the system S and R is a set of some relations on the set M. The sets M and R are called the object set and the relation set of the system S, respectively. (For those readers who are sophisticated enough in mathematics, a relation r in R implies that there exists an ordinal number $n = n(r)$, a function of r, such that r is a subset of the Cartesian product of n copies of the set M.)

Now, a combination of these two understandings of general systems reveals the following: All things that can be imagined in human minds look the same as that of an ordered pair (M, R). Furthermore, relations in R can be about some information of the system, its spatial structure, its theoretical structure, etc. That is, there should exist a law of conservation that reflects the uniformity of all imaginable things with respect to:

1. The content of information
2. Spatial structures
3. Various forms of movements, etc.

In the following three sections, examples in several different scientific disciplines will be used to support this intuition of (general) systems.

3.2 Physical Essence of Dirac's Large Number Hypothesis

Dirac in 1975 proposed the well-known large number hypothesis. This hypothesis implies that the ratio of the static electrical force and the universal gravitation in a H_2 atom is given as follows:

$$\frac{e^2}{Gm_p m_e} = 2 \times 10^{39} \tag{3.1}$$

where e^2 is the static electric force in a H_2 atom, G the gravitational constant, m_p the mass of the proton, and m_e the mass of the electron. In terms of the atomic unit, the age of the universe is 2×10^{39}. Dirac considered these two nondimensional large numbers to be very close and believed that it could not be a coincidence and must stand for something fundamental.

Here, it can be seen that one important contribution, made by Dirac, is that he formally established a connection between the universe and the microscopic world. However, what is the most essential physical meaning of such a connection of the two worlds of different scales? To this end, Dirac concluded that the large number hypothesis implies that the gravitational constant G decreases as time advances and causes various matter to be created.

Now, let us explore the meaning of the large number hypothesis from a different angle. According to Allen (1976), in a H_2 atom, the static electrical force

$$e^2 = 2.307113 \times 10^{-19} \text{ (static electricity unit)}$$

the gravitational constant

$$G = 6.672 \times 10^{-8} \text{ dyne} \times cm^2/g$$

the mass of the proton

$$m_p = 1.6726485 \times 10^{-24} \, g$$

the mass of the electron

$$m_e = 9.109534 \times 10^{-28} \, g$$

the speed of light

$$c = 2.99792458 \times 10^{10} \, cm/s$$

and the radius of a classical electron

$$r_e = e^2/(m_e c^2) = 2.817938 \times 10^{-13} \, cm$$

Based on Pan (1980), let us take the Hubble constant:

$$H = 55 \, km \times s^{-1} \times Ma \, par \, sec^{-1} = 1.782428367 \times 10^{-18} \, s^{-1}$$

Now, from all these data values, one can compute:

1. In a H_2 atom, the ratio of the static electrical force and the universal gravitation is

$$\frac{e^2}{Gm_p m_e} = 2.269 \times 10^{39} \tag{3.2}$$

2. The ratio of the age of the universe (= $1/H$, approximately a Hubble age) and the time $\left(\frac{e^2/(m_e c^2)}{c}\right)$ needed for a light beam to travel through the distance equal to the radius of an electron is given as follows:

$$\frac{1/H}{e^2/(m_e c^2)/c} = 5.96866 \times 10^{40} \tag{3.3}$$

which can be rewritten as the ratio of the radius of the universe (= c/H, the Hubble distance) and the radius of an electron $\left(\frac{e^2}{m_e c^2}\right)$:

$$\frac{c/H}{e^2/(m_e c^2)} = 5.96866 \times 10^{40} \tag{3.4}$$

From Equations (3.2) and (3.3), or Equations (3.2) and (3.4), it can be seen that these two large numbers differ by only 10. Therefore, for numbers of such a magnitude, they can be seen as approximately equal. At this junction, one can tell that some undiscovered important physical essence might very well be implied by such a uniformity in the structural information of the microcosm and macrocosm. If these two quantities are seen as roughly equal, that is,

$$\frac{e^2}{Gm_p m_e} \approx \frac{1/H}{e^2/(m_e c^2)/c} \approx \frac{c/H}{e^2/(m_e c^2)} \tag{3.5}$$

cross-multiplying Equation (3.5) gives

$$Gm_p m_e \times \frac{1}{H} \approx e^2 \times \frac{e^2/(m_e c^2)}{c} \tag{3.6}$$

or

$$Gm_p m_e \times \frac{c}{H} \approx e^2 \times \frac{e^2}{m_e c^2} \tag{3.7}$$

Empirical Evidence ■ 39

The meaning of Equations (3.6) and (3.7) is that the product of the universal gravitation and the age of the universe (or the radius of the universe) approximately equals the product of the static electrical force (in a H_2 atom) and the time for a light beam to travel the distance of the radius of an electron (or the radius of an electron).

Because the measures for the universe are connected by the universal gravitation, and the measures for atoms in the microcosm are connected by electromagnetic force, the physical meaning of Equations (3.6) and (3.7) becomes clear: the product of physical quantities of the universal scale approximately equals the product of relevant physical quantities of the microcosm. That is, there is uniformity between the universe and the atomic world. Does this physical meaning of the large number hypothesis imply the following? Besides the unification of the four basic physical forces in the physical world—electromagnetic interaction, weak interaction, strong interaction, and gravitational interaction—there might be a grand unification in which the products of some relevant physical quantities in the universal scale and microscale approximately equal a fixed constant.

To this end, one has the following facts: According to Xian and Wang (1987), the radius of a H_2 atom (Bohr radius) is given by

$$r_y = 0.529 \times 10^{-8} \ cm$$

and the course velocity (reaction rate) of electromagnetic interactions is

$$V_{ce} = 10^{16} - 10^{19} \ s^{-1}$$

The forcing distance of strong interactions satisfies

$$r_q \leq 10^{-13} \ cm$$

and the course velocity of strong interactions is given by

$$V_{cq} = 10^{21} - 10^{23} \ s^{-1}$$

So, the following is obtained:

$$r_y \times V_{ce} = 0.529 \times 10^8 - 0.529 \times 10^{11} \ cm \times s^{-1} \qquad (3.8)$$

and

$$r_q \times V_{cq} \leq 10^8 - 10^{10} \ cm \times s^{-1} \qquad (3.9)$$

That is, we have obtained

$$r_y \times V_{ce} \approx r_q \times V_{cq} \qquad (3.10)$$

The meaning of Equation (3.10) is that the product of some physical quantities of electromagnetic interactions approximately equals that of relevant physical quantities of strong interactions. This end implies that there exists uniformity between electromagnetic interactions and strong interactions.

3.3 The Mystery of the Solar System's Angular Momentum

The mystery of our solar system's angular momentum can be stated as follows: in the solar system, the total mass of all planets accounts for 0.135 percent of the total mass of the solar system, while the total angular momentum of all planets accounts for 99.421 percent of the total angular momentum of the solar system. In other words, the mass of the sun amounts to 99.865 percent of the mass of the solar system, while the angular momentum of the sun only amounts to 0.579 percent of the total angular momentum of the solar system. Two natural questions are: How did this mystery of the solar system's angular momentum form? Are the masses and the relevant angular momentums related?

In this section, the masses and the relevant angular momentums will be treated in the same fashion as above: multiply them together, see Ren and Hu (1989) for more details. In Dai (1979), the masses and relevant angular momentums with respect to the center mass of the solar system of the sun and the planets in the solar system are given. So, the sum of all the products of the masses (m_p) and the angular momentums (J_p) of the planets can be obtained as follows:

$$\sum_p m_p J_p = 4.144 \times 10^{88} \; g^2 \times cm^2 \times s^{-1} \qquad (3.11)$$

and the product of the mass $m_\oplus$ and the angular momentum $J_\oplus$ with respect to the center of mass of the solar system of the sun can be obtained as follows:

$$m_\oplus \times J_\oplus = 4.143 \times 10^{88} \; g^2 \times cm^2 \times s^{-1} \qquad (3.12)$$

The numerical values in Equations (3.11) and (3.12) are very close and can be seen as the same. That is, we have seen the following law of conservation: the product of the mass and the angular momentum of the sun equals the sum of the products of the masses and the relevant angular momentums of all the planets. In this way, the mystery of the solar system's angular momentum can be resolved as follows: under the conditions of time and space of the solar system, there exists the conservation phenomenon of equal mass–angular momentum products between the sun and the planets. Hence, it is very possible that the stability of the solar system is realized through the law of conservation: equal mass–angular momentum products between the sun and the planets.

In terms of celestial mechanics, it can be shown (Ren and Hu, 1989) that in a system consisting of two celestial bodies (a two-body problem), the motion of the smaller-mass body can be obtained by solving the dynamic equation that describes how the smaller-mass body (m) circles around the larger-mass body (M):

$$J = mr^2 \frac{d\sigma}{dt} = m\sqrt{G(M+m)a(1-e^2)} \qquad (3.13)$$

where J stands for the orbital angular momentum, r the orbital vector, $\frac{d\sigma}{dt}$ the angular velocity, a the orbital radius, and e the orbital eccentric rate. By establishing a rectangular coordinate system with the origin located at the center of mass of the two-body system, one can derive the angular momentum J_1 of the center of larger-mass celestial body and the angular momentum J_2 of the smaller-mass celestial body as follows:

$$J_1 = \frac{Mm}{(M+m)^2} J \qquad (3.14)$$

and

$$J_2 = \frac{M^2}{(M+m)^2} J \qquad (3.15)$$

Because the center mass of the two-body system is always located on the line segment connecting the two celestial bodies, it now follows from Equations (3.14) and (3.15) that

$$MJ_1 = mJ_2 \qquad (3.16)$$

That is, the predicted law of conservation has been obtained: the product of the mass and the angular momentum with respect to the center of mass of the two-body system of the larger-mass celestial body (center body) is the same as that of the smaller-mass celestial body (circling body).

As an example, let us look at the two-body system of the Earth and the moon. Based on the numerical values of the masses and the angular momentums with respect to the center of mass of the Earth-moon system (Dai, 1979), we can calculate and obtain that

$$J_2 = 2.786 \times 10^{41} \ g \times cm^2 \times s^{-1}$$

$$J_1 = 3.4266 \times 10^{38} \ g \times cm^2 \times s^{-1}$$

$$MJ_1 = 2.0477 \times 10^{67} \ g^2 \times cm^2 \times s^{-1} = MJ_2 \qquad (3.17)$$

42 ■ *Systemic Yoyos: Some Impacts of the Second Dimension*

Equation (3.17) implies that in our Earth–moon system, the law of conservation of identical mass–angular momentum products holds true.

For a system of a large-mass celestial body and many small-mass circling celestial bodies, it can be shown (Xian and Wang, 1987) that the related multiplicative relationships of mass–angular momentums approximately hold true in the form of additive components. So, it can be seen that in a celestial body system of simple mechanics, there is a new law of conservation of mass–angular momentum products between the center celestial body and the circling celestial bodies.

3.4 Measurement Analysis of Movements of the Earth's Atmosphere

In meteorology, weather systems, representing movements of the Earth's atmosphere, are classified as large (lar), medium (mid), small (li), and micro (mic) scale systems. These systems of different classifications have their scales of spatial-level measurements (L) set at approximately 10^8, 10^7, 10^6, and 10^5 cm, their measures of vertical velocity (W) set at approximately 10^0, 10^1, 10^2, and 10^3 cm × s^{-1}, and the measures of their life spans (τ) set at approximately 10^6, 10^5, 10^4, and 10^3 s, respectively.

Similar to what has been done previously, let us now multiply relevant quantities for weather systems of the same classification and obtain the following:

1. For the products of spatial-level measures and relevant vertical velocities, we have:

$$L_{lar} \times W_{lar} \approx 10^8 \ cm^2 \times s^{-1}$$
$$\approx L_{mid} \times W_{mid}$$
$$\approx L_{li} \times W_{li}$$
$$\approx L_{mic} \times W_{mic} \quad (3.18)$$

2. For the products of life spans and relevant vertical velocities, we have:

$$\tau_{lar} \times W_{lar} \approx 10^6 \ cm^2$$
$$\approx \tau_{mid} \times W_{mid}$$
$$\approx \tau_{li} \times W_{li}$$
$$\approx \tau_{mic} \times W_{mic} \quad (3.19)$$

Equations (3.18) and (3.19) obviously mean that no matter which classification a weather system is in, the products of its spatial-level measure and vertical velocity are always approximately equal to a fixed constant. The same holds true for the product of the system's life span and vertical velocity. That is, in the Earth's atmosphere, there exists a law of conservation of products between different spatial measures and relevant vertical velocities, or between different life spans and relevant velocities.

3.5 The Law of Conservation of Informational Infrastructure

As is well known, laws of conservation are the most important laws of nature. For example, in classical physics, there are many established laws of conservation dealing with many important concepts, such as mass, energy, momentum, moment of momentum, electric charge, etc. The research of modern physics indicates that each moving object in an even and isotropic space and time, no matter whether the space is microscopic, mesoscopic, or macroscopic, a particle or a field, must follow the laws of conservation of energy, momentum, and moment of momentum. That is, a unification of space has been achieved, called a Minkowski space. In Einstein's relativity theory, the concepts of mass, time, and space have been closely connected, and the mass-energy relation realizes the unification of the concepts of mass and energy.

In the discussion above, it was shown that between the macrocosm and the microcosm, between the electromagnetic interactions of atomic scale and the strong interactions of Quark's scale, between the central celestial body and the circling celestial bodies of celestial systems, between and among the large, medium, small, and micro scales of the Earth's atmosphere, there exist laws of conservation of products of spatial physical quantities. Based on this fact, one can further conclude, after considering many other cases, that there might exist a more general law of conservation in terms of structure, in which the informational infrastructure, including time, space, mass, energy, etc., is approximately equal to a constant. In symbols, this conjecture can be written as follows:

$$AT \times BS \times CM \times DE = a \qquad (3.20)$$

or more generally,

$$AT^\alpha \times BS^\beta \times CM^\gamma \times DE^\varepsilon = a \qquad (3.21)$$

where α, β, γ, ε, and a are constants, and T, S, M, E and A, B, C, D are, respectively, time, space, mass, and energy, and their coefficients. These two formulas can be applied to various conservative systems of the universal, macroscopic, and microscopic levels. The constants α, β, γ, ε, and a are determined by the initial state and properties of the natural system of interest.

In Equation (3.20), when two (or one) terms of choice are fixed, the other two (or three) terms will vary inversely. For example, under the conditions of low speed and the macrocosm, all of the coefficients A, B, C, and D equal 1. In this case, when two terms are fixed, the other two terms will be inversely proportional. This end satisfies all principles and various laws of conservation in classical mechanics, including the laws of conservation of mass, momentum, energy, moment of momentum, etc. So, the varieties of mass and energy in this case are reflected mainly in changes in mass density and energy density. In classical mechanics, when time and mass are fixed, the effect of a force of a fixed magnitude becomes the effect of an awl when the cross section of the force is getting smaller. When the space and mass are kept unchanged, the same force of a fixed magnitude can have an impulsive effect, because the shorter the time the force acts, the greater will be the density of the energy release. When time and energy are kept the same, the size of the working space and the mass density are inversely proportional. When the mass is kept fixed, shrinking acting time and working space at the same time can cause the released energy density to reach a very high level.

Under the conditions of relativity theory, that is, under the conditions of high speeds and great amounts of masses, the coefficients in Equation (3.20) are no longer equal to 1, and Equation (3.21) becomes more appropriate, with the constants A, B, C, D, and a and the exponents α, β, γ, and ε satisfying relevant equations in relativity theory. When time and space are fixed, the mass and energy can be transformed back and forth according to the well-known mass-energy relation:

$$E = mc^2$$

When traveling at a speed close to that of light, the length of a pole will shrink when the pole is traveling in the direction of the pole, and the clock in motion will become slower. When the mass is sufficiently great, light and gravitation deflection can be caused. When a celestial system evolves to its old age, gravitation collapse will appear and a black hole will be formed. One can imagine, based on Equation (3.20), that when our Earth has evolved for, say, a billion or trillion years, the relativity effects would also appear. More specifically speaking, in such a great time measurement, the creep deformation of rocks could increase and solids and fluids would have almost no difference, so that solids could be treated as fluids. When a universe shrinks to a single point with the mass density infinitely high, a universe explosion of extremely high energy density could appear in a very short time period, and so a new universe is created!

3.6 Impacts of the Conservation Law of Informational Infrastructure

If the proposed law of conservation of informational infrastructure holds true (all the empirical data, as presented earlier, seems to suggest so), its theoretical and practical significance is obvious.

The hypothesis of the law of conservation of informational infrastructures contains the following facts:

1. Multiplications of relevant physical quantities in either the universal scale or the microscopic scale approximately equal a fixed constant.
2. Multiplications of either electromagnetic interactions or strong interactions approximately equal a fixed constant.

In the widest domain of human knowledge of our modern time, this law of conservation deeply reveals the structural unification of different layers of the universe so that it might provide a clue for the unification of the four basic forces in physics. This law of conservation may be new evidence for the Big Bang theory. It supports the model for the infinite universe with border, and the oscillation model for the evolution of the universe, where the universe evolves as follows:

$$\ldots \to \text{explosion} \to \text{shrinking} \to \text{explosion} \to \text{shrinking} \to \ldots$$

It also supports the hypothesis that there exist universes outside of our universe. The truthfulness of this proposed law of conservation is limited to the range of our universe, with its conservation constant being determined by the structural states of the initial moment of our universe.

All examples employed earlier show that to a certain degree, the proposed law of conservation holds true. That is, there indeed exists some kind of uniformity in terms of time, space, mass, and energy among different natural systems of various scales under either macroscopic or microscopic conditions or relativity conditions. Therefore, there might be a need to reconsider some classical theoretical systems so that our understanding about nature can be deepened. For example, under the time and space conditions of the Earth's atmosphere, the traditional view in atmospheric dynamics is that because the vertical velocity of each atmospheric huge-scale system is much smaller than its horizontal velocity, the vertical velocity is ignored. In fact (Ren and Nio, 1994), because the atmospheric density difference in the vertical direction is far greater than that in the horizontal direction, and because the gradient force of atmospheric pressure to move the atmospheric system 10 m vertically is equivalent to that of moving the system 200 km horizontally, the vertical velocity should not be ignored. The law of conservation of informational infrastructure, which holds true for all scales used in the Earth's atmosphere, might provide conditions for unified atmospheric dynamics applicable to all atmospheric systems of various scales. As a second example, in the situation of our Earth, where time and mass do not change, in terms of geological time measurements (sufficiently long time), can we imagine the force that causes the Earth's crust movements? Does it have to be as great as what is currently believed?

As for applications of science and technology, tremendous successes have been made in the macroscopic and microscopic levels, such as shrinking working spatial sectors, shortening the time length for energy release, and sacrificing partial masses (say, the usage of nuclear energy). However, the law of conservation of informational infrastructure might very well further the width and depth of applications of science and technology. For example, this law of conservation can provide a theory and thinking logic for us to study the movement evolution of the Earth's structure, the source of forces or structural information that leads to successful predictions of major earthquakes, and to find the mechanisms for the formation of torrential rains and the arrival of earthquakes (Ren, 1996).

Philosophically speaking, the law of conservation of informational infrastructure indicates that in the same way as mass, energy is also a characteristic of physical entities. Time and space can be seen as the forms of existence of physical entities, with motion as their basic characteristic. This law of conservation connects time, space, mass, and motion closely in an inseparable whole. So, time, space, mass, and energy can be seen as attributes of physical entities. With this understanding, the concept of mass is generalized, the wholeness of the world is further proved, and the thoughts of never diminishing mass and never diminishing universes are evidenced.

3.7 Other Empirical Evidence for Yoyo Structures

In this section, let us list a few other relevant experimental and clinical evidences for the systemic yoyo model. All the omitted details can be found in the relevant references.

Let us look at the first example, about how the systemic yoyo model is manifested in different areas of life. For example, each human being, as we see it, is a three-dimensional realization of such a spinning yoyo structure of a higher dimension. To illustrate this end, let us consider two simple and easy-to-repeat experiences. For the first one, let us imagine we go to a sport event, say a swim meet. As soon as we enter the pool area, we immediately find ourselves falling into a boiling pot of screaming and jumping spectators, cheering for their favorite swimmers competing in the pool. Now, let us pick a person standing or walking on the pool deck for whatever reason, either for her beauty or for his strange look or body posture. Magically enough, before long, our stare will be felt by the person from quite a good distance, and he or she will be able to locate us in a very brief moment out of the reasonably sized and boiling audience. The reason for the existence of such a miracle and silent communication is because each side is a high-dimensional spinning yoyo. Even though we are separated by space and possibly informational noise, the stare of one side at the other has directed that side's spin field of the yoyo structure into the spin field of the yoyo structure of the other side. That is the underlying mechanism for the silent communication to be established.

As the second example, let us look at the situation of human relationship. When individual A has a good impression about individual B, magically, individual B also has a similar and almost identical impression about A. When A does not like B and describes B as a dishonest person with various undesirable traits, it has been clinically proven in psychology that what A describes about B is exactly who A is himself (Hendrix, 2001). Once again, the underlying mechanism for such a quiet and unspoken evaluation of each other is because each human being stands for a spinning yoyo and its rotational field. Our feelings about another person are formed through the interactions of our invisible yoyo structures and their spin fields.

In theory, the foundation of the spinning yoyo model is the Bjerknes' circulation theorem; see Chapter 2 for more details. (For the completeness of our discussion here, let us highlight once again the relevant key points.) V. Bjerknes discovered eddy effects due to changes in the density of media in the movements of the atmosphere and ocean. Briefly, by *circulation* is meant a closed contour in a field. Mathematically, each circulation Γ is defined as the line integral about the contour of the component of the velocity vector locally tangent to the contour. If we let $\vec{V}$ stand for the speed of a moving fluid, S be an arbitrary closed curve, and $\vec{\delta r}$ be the vector difference of two neighboring points of the curve S, then the circulation Γ is written as follows:

$$\Gamma = \oint_S \vec{V} \vec{\delta r} \qquad (3.22)$$

By applying some genuine operations on this circulation Γ, Bjerknes obtained his circulation theorem:

$$\frac{d\vec{V}}{dt} = \iint_\sigma \nabla\left(\frac{1}{\rho}\right) \times (-\nabla p) \cdot \vec{\delta\sigma} - 2\Omega \frac{d\sigma}{dt} \qquad (3.23)$$

where the first term on the right-hand side originates from the intersection of the p- and ρ-planes due to uneven density ρ, so that a twisting force is created. Consequently, materials' movements must be rotations, with the rotation direction determined by the equal p- and ρ-plane distributions. The second terms comes from the rotation of the Earth.

Traditionally, the circulation theorem has been employed to explain the formation of land-ocean breezes in meteorology. OuYang (Lin, 1998), however, considered it in general terms and believed that it reveals the common existence of eddy effects in fluid motions—what Kuchemann (1965) once claimed: "The tendon of moving fluids is eddies."

Because uneven eddy motions are the most common form of materials' movements observable in the universe—for example, at the cosmic level, the eddy motions of the solar system, galaxies, nebulae, etc.; at the macrocosmic level, polar eddies, cyclones, anticyclones, etc., on the Earth; and at the microscopic level, various rotational movements found in all kinds of particle structures—it can be concluded that the circulation theorem explains the formation mechanism of all eddy flows in the universe.

A REVISIT TO NEWTON'S LAWS, UNIVERSAL GRAVITATION, AND THE THREE-BODY PROBLEM

Chapter 4

Newton's Laws of Motion

At the Sixth International Conference on Mathematical Modeling, held in St. Louis, Missouri, April 4–7, 1987, Yi Lin presented a paper entitled "Can the World be Studied from the Point of View of Systems?" Since that time, he and his colleagues worked on this question on both the theoretical front and the front of various applications. From the angle of theoretical development, they explored what made some theories successful and long lasting. From the angle of practical applications, they pondered over what theories have worked and what have failed. Some of their discoveries along these lines were published in various papers, books, or in the form of edited volumes.

Among many things they fathomed is how calculus has become such an important theory in the modern age. Because it is related to the work presented in this chapter, let us briefly describe what is found in this area. To address problems in mechanics and astronomy, objects are abstracted into points without size and volume. By identifying numbers as points in the number line, or on the coordinate plane, or in a multidimensional Euclidean space, the concept of objects or particles can be bridged with that of numbers. So, calculus, consisting of various computational schemes, can be employed to study and solve many problems in mechanics, astronomy, and beyond. With its successes achieved in the past 300-plus years, nowadays, to a great extent, the scientific value or quality of a piece of scientific research is judged by how much calculus-based mathematics is employed. For more detailed discussions along this line, please consult Wu and Lin (2002).

By going through various applications, they found that calculus and calculus-based theories have suffered from major failures in the areas of predictions. The most noteworthy of these include meteorology and finance. In other words, in terms of forecasting zero-probability disastrous weather systems and predicting market rises and falls, none of the currently available theories, developed on the basis of calculus, have worked with consistency and desired accuracy. To face this challenge, their research group worked on how to expand their pool of methods and tools to nontraditional areas. Later Shoucheng OuYang, a senior member of the research group, invented a method that he called informational infrastructural analysis. By employing this method, their practical forecasting of zero-probability disastrous weather systems has achieved much better results than the traditional method. The key for such a success is accomplished on the basis that in addition to the methods developed on numbers and calculus, figurative analysis developed on the concept of structures is practically employed. By comparison, they found that figurative analysis, using graphs of the underlying structures, sometimes can contain and can make use of more information more reliably and more accurately than any method based on nonstructural quantities only.

Based on the discovery (Wu and Lin, 2002, or Chapter 2) that spins are the fundamental evolutionary feature and characteristic of materials, in this chapter, we will study a brand new figurative analysis method, composed of spin fields, and how to apply it to the study of Newton's laws of motion. More specifically, after introducing the new figurative analysis method, we will have a chance to generalize all three laws of motion so that external forces are no longer required for these laws to work. As what is known, these laws are one of the reasons why physics is an exact science. It can be expected that these generalized forms of the laws will be equally applicable to social sciences and humanity areas as their classic forms in natural science. To this end, please consult the remaining chapters in this book. The presentations in this chapter and the following two chapters are based on Lin (2007).

4.1 The Second Stir and Newton's First Law of Motion

Newton's first law says that an object will continue in its state of motion unless compelled to change by a force impressed upon it. This property of objects, their natural resistance to changes in their state of motion, is called inertia.

Based on the theory of blown-ups, one has to address two questions not settled by Newton in his first law:

> **Question 4.1.** If a force truly impresses on the object, the force must be from outside of the object. Then, where can such a force be from?
>
> **Question 4.2.** This problem is about the so-called natural resistance of objects to changes in their state of motion. Specifically, how can such a resistance be considered natural?

It is because uneven densities of materials create twisting forces that fields of spinning currents are naturally formed. This end provides an answer and explanation to Question 4.1. Based on the yoyo model (Figure 1.1), the said external force comes from the spin field of the yoyo structure of another object, which is a level higher than the object of our concern. The forces from this new spin field push the object of concern away from its original spin field into a new spin field. If there were not such a forced traveling, the said object would continue its original movement in its original spin field. That is why Newton called its tendency to stay in its course of movement, resistant to changes in its state of motion, natural.

Based on this discussion and the yoyo model (Figure 1.1) developed for each and every object and system in the universe, Newton's first law of motion can be rewritten as follows:

> **First law on state of motion:** Each imaginable and existing entity in the universe is a spinning yoyo of a certain dimension. Located on the outskirt of the yoyo is a spin field. Without being affected by another yoyo structure, each particle in the said entity's yoyo structure continues its movement in its orbital state of motion.

For Newton's first law to hold true, one needs an external force. When people asked Newton where such an initial force could be from, he answered: "It was from God. He waved a bat and provided the very first blow to all things he created" (Kline, 1972). If such an initial blow is called the first push, then the yoyo model in Figure 1.1 and the stirring forces naturally existing in each yoyo created by uneven densities of materials' structures will be called the second stir. Now, let us move on to look at Newton's second law.

4.2 Eddy Effects and Newton's Second Law

Newton's second law predicts what will happen when a force does act on an object. The object's velocity will change and the object will accelerate. More precisely, the law says that its acceleration $\vec{a}$ will be directly proportional to the magnitude of the total (or net) force $\vec{F}_{net}$ and inversely proportional to the object's mass m. In symbols, the second law is written:

$$\vec{F}_{net} = m\vec{a} = m\frac{d\vec{v}}{dt} \quad (4.1)$$

Even though Equation (4.1) has been one of the most important equations in mechanics, when one ponders it long enough, he or she has to ask the following:

Question 4.3. What is a force?

Question 4.4. Where are forces from and how do forces act on other objects?

54 ■ Systemic Yoyos: Some Impacts of the Second Dimension

To answer these questions (the following analysis appeared in Chapter 2; to make our presentation plausible here, we will repeat some relevant parts), let us apply Einstein's concept of uneven time and space of materials' evolution (Einstein, 1997). So, we can assume

$$\vec{F} = -\nabla S(t, x, y, z) \tag{4.2}$$

where $S = S(t, x, y, z)$ stands for the time-space distribution of the external acting object (a yoyo structure). Let $\rho = \rho(t, x, y, z)$ be the density of the object being acted upon. Then, Equation (4.1) can be rewritten as follows for a unit mass of the object being acted upon:

$$\frac{d\vec{v}}{dt} = -\frac{1}{\rho(t, x, y, z)} \nabla S(t, x, y, z) \tag{4.3}$$

If $S(t, x, y, z)$ is not a constant, or if the structure of the acting object is not even, Equation (4.3) can be rewritten as

$$\frac{d(\nabla x \vec{v})}{dt} = -\nabla x \left[\frac{1}{\rho} \nabla S \right] \neq 0 \tag{4.4}$$

and it represents an eddy motion. That is, when the concept of uneven structures is employed, Newton's second law actually indicates that a force, acting on an object, is in fact the attraction or gravitation from the acting object. It is created within the acting object by the unevenness of its internal structure.

By combining this new understanding of Newton's second law with the yoyo model in Figure 1.1, we get the models on how an object m is acted upon by another object M (see Figures 4.1 and 4.2).

Figure 4.1(a) depicts the scenario that the object m is originally situated in a diverging eddy before the converging eddy M pulls it over into the spin field of M. Figure 4.1(b) describes how the object m in a diverging eddy can be either pushed (direction (a)) or pulled (direction (b)) by M. When m is pulled along direction (b), it will be captured by the diverging eddy M. In this case, if the spin field M is not powerful enough, the object m will not be pulled to the spin field M out of its original field. Figure 4.1(c) and (d) are the models for the scenarios on how the object m is pulled away from its original converging eddy by another eddy of the harmonic spinning pattern. Here, two spinning patterns are said to be harmonic if, other than the spinning directions, the spin fields look the same.

To be more specific, the object M, exerting a gravitational pull (Figure 4.1(a) and (c)) or a gravitational push (Figure 4.1(b) and (d)) on an object m, must be much greater than m to have an obvious effect on the state of motion of the object

Newton's Laws of Motion ■ 55

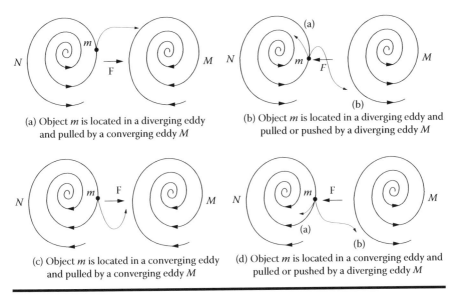

Figure 4.1 Acting and reacting models with yoyo structures of harmonic spinning patterns.

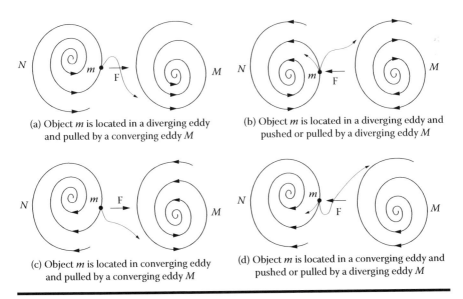

Figure 4.2 Acting and reacting models with yoyo structures of inharmonic spinning patterns.

m and to capture *m*. That is, the objects *M* and *m* are from different levels and have different scales. Otherwise, if they are of the same scale or level, instead of a dot, the object *m* will be the entire eddy *N* on the left-hand side in Figure 4.1(a)–(d). Then, *m* will not be pulled over to *M* in Figure 4.1(a). Instead, *m* will be a supplier of materials for the converging eddy *M*. In Figure 4.1(b), the objects *M* and *m* = *N* will be pushing each other away. In Figure 4.1(c), the converging eddies *m* = *N* and *M* might attract each other and become one, depending on whether their spinning directions and angles agree with each other. And in Figure 4.1(d), *M* serves as a supplier of the converging eddy *m* = *N*.

As for objects *m* and *M* with their yoyo structures spinning in inharmonic directions (or patterns), the acting and reacting models are given in Figure 4.2.

Now, by summarizing the discussion above, Newton's second law can be generalized as follows:

> **Second law on state of motion:** When a constantly spinning yoyo structure *M* does affect an object *m*, which is located in the spin field of another object *N*, the velocity of the object *m* will change and the object will accelerate. More specifically, the object *m* experiences an acceleration $\vec{a}$ toward the center of *M* such that the magnitude of $\vec{a}$ is given by
>
> $$a = \frac{v^2}{r} \quad (4.5)$$
>
> where *r* is the distance between the object *m* and the center of *M*, and *v* is the speed of any object in the spin field of *M* about distance *r* away from the center of *M*. The magnitude of the net pulling force $\vec{F}_{net}$ that *M* exerts on *m* is given by
>
> $$F_{net} = ma = m\frac{v^2}{r} \quad (4.6)$$

Let us now look at the justification for this generalization. Because it is assumed that the yoyo structure *M* spins at a constant speed around its center, one can imagine that each point *X* in the spin field of *M* travels at a constant speed. However, because the direction of the velocity of *X* is always changing, the velocity is not constant. So, there must be acceleration for *X*. Even though this acceleration does not change the speed of *X* in the spin field of *M*, it does change the direction of *X*'s velocity to keep it on its roughly circular path.

Let $\vec{V_1}$ be *X*'s velocity at time $t = t_1$ and $\vec{V_2}$ *X*'s velocity at time $t = t_2$. Then, $\Delta \vec{V} = \vec{V_2} - \vec{V_1}$ points roughly to the center of *M*, and so does the acceleration, because $\vec{a} = \frac{\Delta \vec{V}}{\Delta t}$. The magnitude of this acceleration is given by Equation (4.5), if the point X is a distance r away from the center of M. Now, by using Newton's second law, Equation (4.6) is established.

4.3 Colliding Eddies and Newton's Acting and Reacting Forces

Newton's third law is commonly known: to every action, there is an equal, but opposite, reaction. More precisely, if object A exerts a force on object B, then object B exerts a force back on object A, equal in strength but in the opposite direction. These two forces, $\vec{F}_{A\text{-}on\text{-}B}$ and $\vec{F}_{B\text{-}on\text{-}A}$, are called an action-reaction pair.

Similar to what has been done earlier, let us now analyze the situation in two different angles: (1) two eddy motions act and react to each other's spin field, and (2) one spinning yoyo is acted upon by an eddy flow of a higher level and scale. For the first situation, where two eddy motions act and react to each other's spin field, we have the diagrams in Figure 4.3.

By analyzing each spinning direction, it can be seen that in Figure 4.3(a), the acting and reacting pair is more of N exerting a force on M, and N serves as a supplier of materials, energies, etc., to the converging need of M. When both N and M are relatively stable, object M also exerts an opposite but reacting force on N. Otherwise, without such a reacting force, M would collapse and N become diluted. That is the moment both N and M would have to realign their balances and territories, or they both would simply dissolve and disappear.

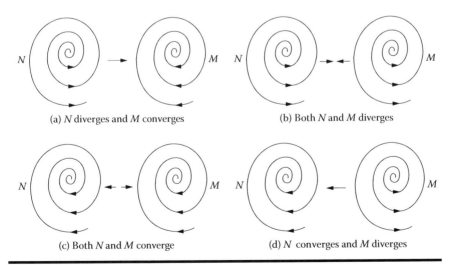

Figure 4.3 Same-scale acting and reacting spinning yoyos of the harmonic pattern.

58 ■ *Systemic Yoyos: Some Impacts of the Second Dimension*

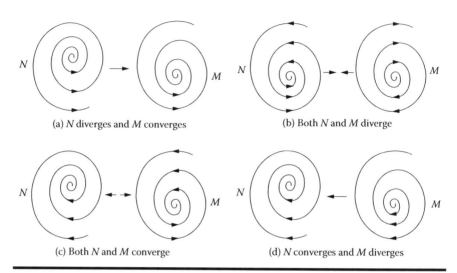

Figure 4.4 Same-scale acting and reacting spinning yoyos of inharmonic patterns.

When both objects *N* and *M* represent yoyo structures, as in Figure 4.3(b), they truly exert equal, but opposite, action and reaction forces on each other. Because their spin fields fight against each other, they will push each other farther apart. This is one situation Newton's third law is really talking about.

When objects *N* and *M* both converge harmonically, as shown in Figure 4.3(c), they exert an attraction or pull on each other. If they are relatively stable, their spin fields achieve temporary (roughly) balanced acting and reacting forces. However, sooner or later, these two spin fields will try to combine and have a tendency to become one spin field.

The situation in Figure 4.3(d) is the same as that in Figure 4.3(a), except that the roles of *N* and *M* are shifted.

For objects *N* and *M* with harmonic spin fields, we have the diagrams in Figure 4.4.

For the yoyo structures *N* and *M* in Figure 4.4(a), even though the situation is similar to that of Figure 4.3(a), where *N* diverges and *M* converges, due to the inharmonic spinning directions, *N* and *M* act and react to each other differently than in Figure 4.3(a). In our current case, *N* exerts a pushing force on *M* and does not easily serve as a supplier of materials, energies, etc., for *M*. When compared to Figure 4.3(b), the yoyo structures *N* and *M* in Figure 4.4(b) can work together more peacefully, because their spin fields do not fight against each other, even though they also push each other apart. For the situations in Figure 4.4(c), the spin fields *N* and *M* attract each other. But from their inharmonic spinning directions, it can be seen that they will not combine and will never become one. The situation in Figure 4.4(d) is exactly the same as that in Figure 4.4(a), with the roles of *N* and *M* switched.

Based on the analysis above, Newton's third law can be generalized for the case of two eddy motions acting and reacting to each other's spin fields as follows:

Third law on state of motion: When the spin fields of two yoyo structures N and M act and react on each other, their interaction falls in one of the six scenarios shown in Figure 4.3(a) to (c) and Figure 4.4(a) to (c), and the following are true:

1. For the cases in (a) of Figures 4.3 and 4.4, if both N and M are relatively stable temporarily, then their action and reaction are roughly equal, but in opposite directions, during the temporary stability. In terms of the whole evolution involved, the divergent spin field (N) exerts more action on the convergent field (M) than M's reaction peacefully in the case of Figure 4.3(a) and violently in the case of Figure 4.4(a).
2. For the cases (b) in Figures 4.3 and 4.4, there are permanent equal, but opposite, actions and reactions, with the interaction more violent in the case of Figure 4.3(b) than in the case of Figure 4.4(b).
3. For the cases in (c) of Figures 4.3 and 4.4, there is a permanent mutual attraction. However, for the former case, the violent attraction may pull the two spin fields together and have the tendency to become one spin field. For the latter case, the peaceful attraction is balanced by the opposite spinning directions, and the spin fields will coexist permanently.

That is, Newton's third law holds true temporarily for cases (a), permanently for cases (b), and partially for cases (c) in Figures 4.3 and 4.4.

Now, let us look at Newton's third law from the second angle: One spinning yoyo m is acted upon by an eddy flow M of a higher level and scale. If we assume m is a particle in a higher-level eddy flow N before it is acted upon by M, then we are looking at situations as depicted in Figures 4.1 and 4.2.

If the spinning yoyo m is located in a spin field N and affected by the converging spin field M (Figure 4.1(a)), due to the harmonic spinning directions of N and M, m experiences an action from M pulling it away from its current orbit, and a reaction from N to keep m on its current orbit. Because the influence of M is external, depending on the structure of the spin field of m, m might continue its rotation in the spin field N with its orbit altered, or it might be on its way to depart from N. If m does leave the spin field of N, it has two possibilities: it might fall into the spin field of M, or it might simply be shot out to an area unrelated to either N or M. When m takes the second possibility, it experiences an equal but opposite action from M and reaction from N.

In terms of the situation in Figure 4.1(b), due to the violent natures facing each other between the spin fields of N and M, the object m will be either kept in the spin field of N longer than it would be without the effect of M, or captured by the spin field of M, or it will fall in a subeddy created by both N and M (see Figure 4.5(a)).

60 ■ *Systemic Yoyos: Some Impacts of the Second Dimension*

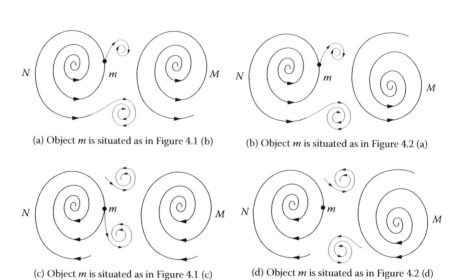

(a) Object *m* is situated as in Figure 4.1 (b)

(b) Object *m* is situated as in Figure 4.2 (a)

(c) Object *m* is situated as in Figure 4.1 (c)

(d) Object *m* is situated as in Figure 4.2 (d)

Figure 4.5 Object *m* might be thrown into a subeddy created by the spin fields of *N* and *M* jointly.

In terms of Figure 4.1(c), due to their conflicting spinning directions at the adjacent area between *N* and *M*, object *m* experiences both a push and a pull from the spin field of *M*. The act of pushing comes from the spinning direction of *M*, opposite to that in which *m* travels. The experience of pull is caused by the fact that *M* is a converging eddy motion. In this case, the acts of push and pull experienced by *m* are neither equal nor opposite to each other (see Figure 4.5(c)).

When *m* is located in a converging field *N* and acted upon by a diverging *M*, as shown in Figure 4.1(d), what *m* experiences is (1) more of a push toward the center of *N* without much of any reaction or (2) being thrown out of *N* and *M* along the spinning directions of *N* and *M* into their adjacent area.

The situation in Figure 4.2(a) is similar to that of Figure 4.1(b). The scenario in Figure 4.2(b) is similar to that of Figure 4.1(a), except that *m* can only be thrown out by both *N* and *M* along their spinning directions to areas not related to either *N* or *M*. This is when *m* experiences equal but opposite action and reaction from *N* and *M* individually. What the experience *m* goes through in Figure 4.2(c) is similar to that of Figure 4.1(d), except that both *N* and *M* apply a gravitational pull on *m*. Depending on the characteristics of *m*'s own spin field, it may stay in the field of *N*, or travel to the field of *M*, or be thrown out by both *N* and *M* along their common spinning direction in the adjacent area. When *m* is thrown out, *m* experiences an equal, but opposite, action and reaction from *N* and *M*, individually. Now, the situation in Figure 4.2(d) is similar to that of Figure 4.1(c), except that *M* helps to push *m* to converge to the center of *N* faster. See Figure 4.5 for more details.

Note: The subeddies created in Figure 4.5(a) are both converging, because the spin fields of *N* and *M* are suppliers for them and sources of forces for their spins. Subeddies in Figure 4.5(b) are only spinning currents. They serve as a middle stop before supplying the spin field of *M*. Subeddies in Figure 4.5(c) are diverging. Subeddies in Figure 4.5(d) are only spinning currents, similar to those in Figure 4.5(b).

Based on our analysis of the scenario that one object *m*, situated in a spin field *N*, is acted upon by an eddy flow *M* of a higher level and scale, we can generalize Newton's third law to the following form:

> **Fourth law on state of motion:** When the spin field *M* acts on an object *m*, rotating in the spin field *N*, the object *m* experiences equal, but opposite, action and reaction, if it is either thrown out of the spin field *N* and not accepted by that of *M* (Figure 4.1(a) and (d), Figure 4.2(b) and (c)) or trapped in a subeddy motion created jointly by the spin fields of *N* and *M* (Figure 4.1(b) and (c), Figure 4.2 (a) and (d)). In all other possibilities, the object *m* does not experience equal and opposite action and reaction from *N* and *M*.

4.4 Equal Quantitative Effects and Figurative Analysis

In the previous three sections, we have heavily relied on the analysis of shapes and dynamic graphs. To any scientific mind produced out of the current formal education system, the validity of such a method of reasoning will naturally seem questionable. To address this concern, let us start with the concept of equal quantitative effects. For a detailed and thorough study of this concept, please consult Wu and Lin (2002) and Lin (1998).

The concept of equal quantitative effects was initially introduced by OuYang (1994) in his study of fluid motions, and later, it was used to represent the fundamental and universal characteristics of all materials' movements (Lin, 1998). By *equal quantitative effects* is meant the eddy effects with nonuniform vortical vectorities existing naturally in systems of equal quantitative movements, due to the unevenness of materials. In this definition, *equal quantitative movements* means such movements where quasi-equal acting and reacting objects are involved, or two or more quasi-equal mutual constraints are concerned. For example, the relative movements of several planets of approximately equal masses are considered equal quantitative movements. In the current laboratories of physics, the interaction between the particles to be measured and the equipment used to do the measurement has often been seen as an equal quantitative movement. Many events in daily lives, such as wars, politics, economics, chess games, races, plays, etc., can also be seen as examples of equal quantitative effects.

More specifically, every time a measurement uncertainty exists, one faces the effect of equal quantities. For example, when we observe an object, our understanding of the object is really constrained by our background knowledge, our ability and limitation of human sense organs. When an object is observed by another object, the two objects cannot really be separated apart. So, such well-known theories as Bohr's principle, the relativity principle about microcosmic motions, von Neumann's principle of program storage, etc., fall into the category of equal quantitative effects.

What is significant about equal quantitative effects is that they can easily throw calculations of equations into computational uncertainty. For example, if two quantities x and y are roughly equal, then $x - y$ becomes a computational uncertainty involving large quantities with infinitesimal increments. Based on recent studies in chaos (Lorenz, 1993), it is known that for nonlinear equation systems, which always represent equal quantitative movements (Wu and Lin, 2002), minor changes in their initial values lead to dramatic changes in their solutions. Such extreme volatility existing in the solutions can be easily caused by changes of a digit many places after the decimal point. Such a digit place far away from the decimal point in general is no longer practically meaningful. That is, when equal quantitative effects are involved, we face either the situation where no equation can be reasonably established or the situation where the established equation cannot be solved with a valid and meaningful solution.

That is, the concept of equal quantitative effects has computationally declared that equations are not eternal, and that there does not exist any equation under equal quantitative effects. That is why OuYang introduced the methodological method of abstracting numbers (quantities) back into shapes (figurative structures). Of course, the idea of abstracting numbers back to shapes is mainly about how to describe and make use of the formality of eddy irregularities. These irregularities are very different from all the regularized mathematical quantifications of structures.

Because the currently available variable mathematics is entirely about regularized computational schemes, there must be the problem of disagreement between the variable mathematics and irregularities of objective materials' evolutions, and the problem that distinct physical properties are quantified and abstracted into indistinguishable numbers. Such incapability of modern mathematics has been shown time and time again in areas of practical forecastings and predictions. For example, because theoretical studies cannot yield any meaningful and effective method to foretell weather changes, especially about zero-probability disastrous weather systems, the technique of live report has been widely employed. (In fact, the study of chaos theory indicates that weather patterns are chaotic. A little butterfly fluttering its tiny wings in Australia can drastically change the weather patterns in North America (Gleick, 1987)). However, the area of financial market predictions has not been so lucky; the technique of live report cannot be applied as effectively. Due to equal quantitative effects, the movements of prices in the financial marketplace have been truly chaotic when viewed from the contemporary scientific point of view.

Newton's Laws of Motion ■ 63

That is, the introduction of the concept of equal quantitative effects has made the epistemology of natural sciences go from solids to fluids and completed the unification of natural and social sciences. More specifically, after we have generalized Newton's laws of motion, which have been the foundation to make physics an exact science in the previous three sections, these new laws can be readily employed to study social systems, such as military conflicts, political struggles, economic competitions, etc.

We have briefly discussed the concept of equal quantitative effects and inevitable failures of current variable mathematics under the influence of such effects. How about figurative analysis?

The usage of graphs in our daily lives goes back as far as our recorded history goes. For example, the written language of China consists of a huge array of graphic figures. In terms of figurative analysis, one early work is the book *Yi Ching* (known in English as *Book of Changes*). For now, no one knows exactly when this book was written and who wrote it. All that is known is that the book has been around since about 3,000 years ago. In that book, the concept of Yin and Yang was first introduced, and graphic figures are used to describe supposedly all matters and events in the world. When Leibniz (a founder of calculus) had a hand on that book, he introduced the binary number system and base p number system in modern mathematics (Kline, 1972). Later on, Bool furthered this work and laid down the foundation for modern computer technology to appear.

In our modern days, figures and figurative analysis are readily seen in many aspects of our lives. One good example is the number π. Because we cannot write out this number in the traditional fashion (in either the decimal form or the fraction form), we simply use a figure π to indicate it. The same idea is employed to write all irrational numbers. In the area of weather forecasting, figurative analysis is used each and every day in terms of weather maps. In terms of studies of financial markets, a big part of the technical analysis is developed on graphs. So, this part of technical analysis can be seen as an example of figurative analysis.

From the recognition of equal quantitative effects and the realization of the importance of figurative analysis, OuYang invented and materialized a practical way to abstract numbers back into shapes so that the forecasting of many disastrous zero-probability weather systems becomes possible. For detailed discussion about

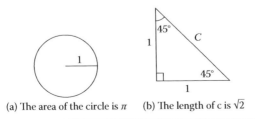

(a) The area of the circle is π (b) The length of c is $\sqrt{2}$

Figure 4.6 π and $\sqrt{2}$ figuratively and precisely.

this, consult Appendix D in Wu and Lin (2002) and Lin (1998). To simplify the matter, let us see how to abstract the numbers π and $\sqrt{2}$ back into shapes with their inherent structures kept. In Figure 4.6(a), the exact value π is represented using the area of a circle of radius 1. The precise value of $\sqrt{2}$ is given in Figure 4.6(b) by employing the special right triangle. By applying these simply graphs, the meaning, precise values, and their inherent structures of π and $\sqrt{2}$ are presented once and for all.

4.5 Whole Evolutions of Converging and Diverging Fluid Motions

When a figurative analysis like the ones in Figures 4.1 to 4.5 is used, one natural question is: What will happen to the whole evolution of each of the eddy motions if no external effect is influencing it? To address this question, let us study the whole evolutionary characteristics of eddy motions. To be specific, let us consider the following initial value problem for a rotational movement of fluid:

$$\begin{cases} (\Delta \psi)_t = J(\Delta \psi, \psi) \\ (\Delta \psi)|_{t=0} = \Delta \psi_0 \end{cases} \quad (4.7)$$

where $\psi(t, x, y)$ is a flow function, $u = -\psi_x$ and $v = -\psi_y$ the components of the flow's velocity; $J(\cdot, \cdot)$ is the Jacobi operator; and $\varsigma = \Delta \psi$ is the vorticity in the vertical direction. In Equation (4.7), we assume that the initial field ψ_0 is continuously differentiable and bounded on $-\infty < x, y < +\infty$. That is,

$$|\psi_0(x, y)| \leq R \quad (4.8)$$

for some fixed constant R and for any $x, y \in (-\infty, +\infty)$. Additionally, we assume that $\psi(t, x, y)$ is uniformly continuous on $t \in [0, T]$ and $x, y \in [0, L]$, $T < +\infty$, and take the boundary condition as follows:

$$\begin{cases} x = y = 0, & \Delta \psi(t, 0, y) = \Delta \psi(t, x, 0) = 1 \\ x = y = L, & \Delta \psi(t, L, y) = \Delta \psi(t, x, L) = \Delta \psi_L \end{cases} \quad (4.9)$$

By employing the nonexpansion method with separate variables, that is, assume

$$\begin{cases} \psi(t, x, y) = A(t)\Psi(x, y) \\ \psi_0(0, x, y) = A(0)\Psi(x, y) \end{cases} \quad (4.10)$$

we assume

$$\begin{cases} |\Psi_y| = |U(x,y)| \leq |c_1| \\ |\Psi_x| = |V(x,y)| \leq |c_2| \end{cases} \quad (4.11)$$

for any $x, y \in (-\infty, +\infty)$, where $u = AU$ and $v = AV$. Then, U and V are continuous and bounded. This end is different than the condition that u and v are continuous and bounded. Substituting Equation (4.10) into Equation (4.7) produces

$$\Delta\Psi \frac{dA}{dt} = A^2 \left[\Psi_y (\Delta\Psi)_x - \Psi_x (\Delta\Psi)_y \right] \quad (4.12)$$

Both terms in the bracket of Equation (4.12) are essentially the same. For the sake of convenience for our discussion, let us look at the first term. That is, we have

$$\Delta\Psi \frac{dA}{dt} = A^2 \Psi_y (\Delta\Psi)_x \quad (4.13)$$

By separating the variables, we can take

$$\frac{dA}{dt} = -\lambda A^2 \quad (4.14)$$

and

$$(\Delta\Psi)_x + \frac{\lambda}{\Psi_y} \Delta\Psi = 0 \quad (4.15)$$

From Equations (4.10) and (4.14), it follows that

$$A = \frac{A(0)}{1 + \lambda A(0) t} \quad (4.16)$$

Assume that $t_b = \dfrac{1}{A(0)\lambda}$. Then, if we multiply Equation (4.16) by $\Delta\Psi$, we have

$$\Delta\psi = A(t)\Delta\Psi(x,y) = \frac{A(0)\Delta\Psi}{1 + \lambda A(0) t} = \frac{\Delta\psi_0}{1 - t/t_0} \quad (4.17)$$

That is, Equation (4.17) is a problem with one blown-up.

If the initial state stands for a positive vorticity, that is, $\Delta \psi_0 > 0$, then when $t < t_b$, the movement of the fluid is a continuation of the initial positive vorticity. When $t = t_b$, the fluid movement experiences a blown-up. When $t > t_b$, because $1 - t/t_b < 0$, we have $\Delta \psi < 0$. That is, due to the reversal change of the blown-up, the initial positive vortical movement is transformed into one with negative vorticity. If the initial state represents a negative vorticity, that is, $\Delta \psi_0 < 0$, then similar to what was analyzed earlier, the fluid movement will also be transformed to a positive vortical movement through a blown-up at $t = t_b$.

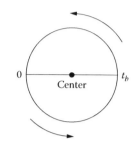

Figure 4.7 The circular time trajectory.

This conclusion reconfirms the yoyo model in Figure 1.1 and seconds the analysis given after Equation (1.1) in Chapter 1. More specifically, $t = t_b$ is the time moment when an object situated in a converging spin field (the black hole side) reaches the narrow neck tunnel. After going through to tunnel, the object is in a diverging spin field.

If the object starts in the diverging spin field (the Big Bang side) after the movement $t = t_b$ of a blown-up, the most diluted state, the object continues its journey into the converging spin field of the black hole side.

Because the concept of time used in our analysis in this section is still Newtonian absolute time, after the blown-up moment $t = t_b$, the time continues its journey to the conceptual infinity ∞. If we apply Einstein's concept of time and space, then our time here should simply be a measurement of the spinning motion. At the moment the Big Bang starts, the time clock begins to tick. When the evolution of any chosen materials reaches their most diluted state at $t = t_b$, the time clock starts to click backward. So, the trajectory of time can be seen as a closed circle (Figure 4.7). It begins at the moment of $t = 0$. As soon as it reaches the other side of the diameter at $t = t_b$, it begins to return to the beginning of time. After that, everything, maybe in a totally different appearance, will start all over again.

Chapter 5

Kepler's Laws of Planetary Motion

Based on the yoyo model and the figurative analysis method studied in the previous chapter, in this chapter, we will walk through Kepler's laws of planetary motion and Newton's law of universal gravitation; learn why celestial bodies travel along elliptical orbits and, without any external forces, do not collide into each other as described by the law of universal gravitation (instead, they revolve about one another), and learn the underlying mechanism for binary star systems to exist.

5.1 Newton's Cannonball

Based on observational data mostly inherited from Tycho Brahe, Johanne Kepler proposed three laws in the early 1600s regarding planetary motion:

Kepler's first law (the law of ellipses): The paths of the planets about the sun are elliptical in shape, with the center of the sun located at one focus.
Kepler's second law (the law of equal areas): An imaginary line drawn from the center of the sun to the center of a planet will sweep out equal areas in equal intervals of time.
Kepler's third law (the law of harmonics): The ratio of the squares of the periods of any two planets is equal to the ratio of the cubes of their average distances from the sun.

The second law implies that the speed at which a planet moves through space is constantly changing. The planet moves fastest when it is closest to the sun and slowest when it is farthest from the sun. Unlike the first two laws, which describe the motion characteristics of a single planet, the third law makes a comparison between the motion characteristics of different planets.

When Isaac Newton came along, he tried to develop an explanation on why all the planets revolve about the sun and why their orbits are elliptical. Here is a paraphrase based on a survey of various writings by Newton, and it can be found in many elementary physics books. Suppose a cannonball is fired horizontally from a very high mountain in a region devoid of air resistance. In the absence of gravitation, the cannonball would travel in a straight line. Yet, in the presence of gravitation, the cannonball would drop below this straight-line path and eventually fall to the ground. Now, suppose that the cannonball is fired horizontally again with a greater speed. In this case, the cannonball would still fall below its straight-line tangential path and eventually drop to the Earth. Only this time, the cannonball would travel farther before striking the ground.

Now, let us assume that there were such a speed at which the cannonball could be fired so that the trajectory of the falling cannonball matched the curvature of the Earth. In such a case, the cannonball would fall around the Earth instead of into it. The cannonball would fall toward the Earth without ever colliding into it, and subsequently become a satellite orbiting around the Earth in a circular motion. Then, at even greater launch speeds, the cannonball would orbit the Earth in an elliptical path. Based on where we are from, this explanation implicitly contains two problems:

> **Question 5.1.** Isaac Newton still needs a third party to provide an external force to throw a celestial body into space and start its initial movement.
>
> **Question 5.2.** Even with a greater initial speed at which the "cannonball" was thrown into space, according to Newton's analysis, the cannonball would only orbit the Earth along a bigger circular orbit, instead of an elliptical path.

To understand how the yoyo model (Figure 1.1) can be applied to resolve these two problems, let us first look at the so-called dishpan experiment. In the late 1950s, Dave Fultz (Fultz et al., 1959) constructed the following experiment: he partially filled a cylindrical vessel with water, placed it on a rotating turntable, and subjected it to heating near the periphery and cooling near the center. The bottom of the container is intended to simulate one hemisphere of the Earth's surface; the water, the air above this hemisphere; the rotation of the turntable, the Earth's rotation; and the heating and cooling, the excess external heating of the atmosphere in low latitudes and the excess cooling in high latitudes.

To observe the pattern of flows at the upper surface of the water, which was intended to simulate atmospheric motion at high elevations, Fultz sprinkled some aluminum powder. A special camera that effectively rotated with the turntable took time exposures so that a moving aluminum particle would appear as a streak, and sometimes each exposure ended with a flash, which could add an arrowhead to the forward and end of each streak. The turntable generally rotated counterclockwise, as does the Earth when viewed from above the North Pole.

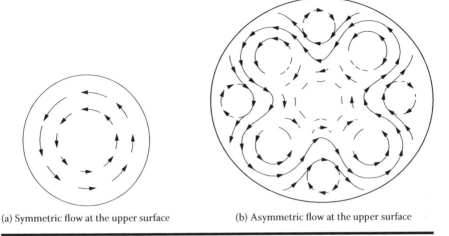

(a) Symmetric flow at the upper surface (b) Asymmetric flow at the upper surface

Figure 5.1 Patterns observed in Fultz's dishpan experiment.

Even though everything in the experiment was arranged with perfect symmetry about the axis of rotation, such as no impurities added in the water, and the bottom of the container was flat, Fultz and his colleagues observed more than they bargained for. First, both expected flow patterns appeared, as shown in Figure 5.1, and the choice depended on the speed of the turntable's rotation and the intensity of the heating. Briefly, with fixed heating, a transition from circular symmetry (Figure 5.1(a)) would take place as the rotation increased past a critical value. With the sufficiently rapid but fixed rate of rotation, a similar transition would occur when the heating reached a critical strength, while another transition back to symmetry would occur when the heating reached a still higher critical strength.

During roughly the same time frame, Raymond Hide did a similar experiment in England (Hide, 1953). Instead of a dishpan, Hide used two concentric cylinders with a fluid placed in the ring-shaped region between the them. He discovered similar transitions between symmetric and asymmetric flow patterns. In this experiment, the asymmetric flow was often regular and consisted of a chain of apparently identical waves, which would travel around the ring-shaped region without changing their shapes. What is more remarkable is that Hide found that a chain of identical waves would appear, but as they traveled along, they would alter their shapes in unison in a regular periodic fashion, and after many rotations of the turntable, they would regain their original shape and then repeat the cycle (Figure 5.2).

What is important about these experiments is that structures such as jet streams, traveling vortices, and fronts appear to be basic features in rotating heated fluids, and are not peculiar to atmospheres only. With this understanding in place, let us now see how we can use the yoyo model in Figure 1.1 to resolve Question 5.1.

As soon as the materials that eventually formed the sun, planets, etc., in our solar system came out of the narrow neck from the Big Bang side, they continued

70 ■ *Systemic Yoyos: Some Impacts of the Second Dimension*

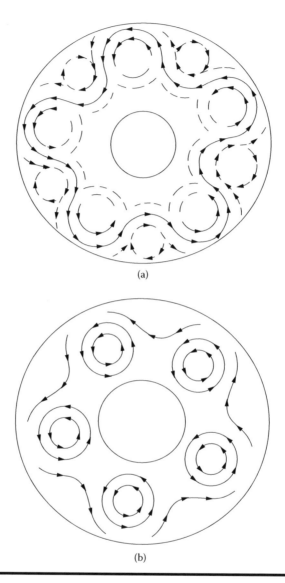

Figure 5.2 Streak patterns of the flow at the upper surface of Hide's dishpan experiment. (a) is followed by (b) after eight rotations, and (b) is followed by (a) after another eight rotations.

their spinning motion as when they were still in the narrow neck. Due to unevenness of distribution of these materials, or say, impurities, as indicated by the dishpan experiments, some local whirlpools were formed and led to the formations of planets. And the biggest chunk of materials in the middle is the sun, as we call it in our solar system. So, the yoyo model (Figure 1.1) implies:

Kepler's Laws of Planetary Motion ■ 71

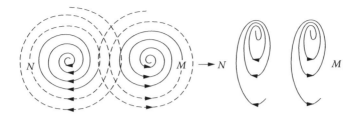

Figure 5.3 Imaginary action and reaction spin fields of our solar system *M* and a neighboring system *N*.

1. The sun also spins along with its spin field.
2. The reason why such planets as Jupiter, Saturn, Uranus, and Neptune are more gaseous than Mercury, Venus, Earth, and Mars is because they have traveled away from the sun, reaching the edge of the yoyo structure on the Big Bang side.
3. As Mercury, Venus, Earth, and Mars travel farther away from the sun, new planets will be created by materials continuously coming out of the narrow neck of the yoyo.
4. When a planet travels far away from the sun, it will no longer be a planet. It will dissolve and its components will either be taken by a system similar to our solar system or be sucked back into the yoyo structure of our solar system from the black hole side.

As for Question 5.2 and why the paths of the planets about the sun are elliptical in shape, with the center of the sun being located at one focus (Kepler's first law), it is because not too far from our solar system, there should be at least another similar celestial system, which coexists with our solar system side by side, as shown in Figures 4.3 and 4.4. As these systems act and react on each other, their would-be circular motions are pressured into elliptical motions. More specifically, let us look at the scenario as depicted in Figure 4.3(d), where we assume *M* is our solar system, with the sun located at the center of the spin field. With the spin fields of *M* and *N* interfering with each other (Figure 5.3), the would-be circular spin fields of *M* and *N* both become oval shaped, with their suns located on the upper foci of the elliptical orbits.

In fact, Figure 5.3 also partially explains Kepler's second law on why, if a particle travels along its elliptical orbit, its speed is different, depending on where it is. For example, let us take out one loop from the spin field *N* in Figure 5.3 and take a closer look in Figure 5.4.

Figure 5.4 The effect of the spin field of *M* on one loop of *N*.

72 ▪ Systemic Yoyos: Some Impacts of the Second Dimension

The downward motion along the loop of N on the right-hand side is helped by the spin field of M. At the same time, the upward traveling on the left-hand side of the loop of N has to overcome the encountering effect of M's spin field.

In terms of Kepler's third law, let us wait until the next section.

5.2 Kepler's Law of Harmonics and Its Generalization by Newton

When studying Kepler's laws of planetary motion, Newton tried to explain why the planets actually travel about the sun by employing the concept of gravitation. At the time, it was known that gravitation causes Earth-bound objects to accelerate toward the Earth at a rate of 9.8 m/s², and that the moon accelerates toward the Earth at a rate of 0.00272 m/s². To Newton, if the same force that causes the acceleration of an apple to the Earth also causes the acceleration of the moon toward the Earth, there then must be a plausible explanation for why the acceleration of the moon is so much smaller than that of an apple. The then-known facts told that

$$\frac{g_{moon}}{g_{apple}} = \frac{0.00272 \, m/s^2}{9.8 \, m/s^2} \approx \frac{1}{3600}$$

Intuition indicated to Newton that somehow the force of gravitation was diluted by distance. By comparing the distance from his apple to the center of the Earth to that from the moon to the center of the Earth, it was found that the moon in its orbit about the Earth is approximately 60 times farther away from the Earth's center than is the apple. This end implied that the force of gravitation between the Earth and any object is inversely proportional to the square of the distance, which separates that object from the Earth's center. Through further analysis, Newton established his law of universal gravitation:

$$F_{grav} = G \frac{m_1 m_2}{r^2} \tag{5.1}$$

where G is the universal gravitation constant, m_i is the mass of the ith celestial body, $i = 1, 2$, and r is the distance between the bodies.

With his law (8.1) in place, Newton provided an analytic proof for Kepler's third law as follows. Consider a planet of mass M_{planet} to orbit in nearly circular motion about the sun of mass M_{sun}. The net centripetal force acting upon this orbiting planet is given by

$$F_{net} = M_{planet} \times \frac{v^2}{r} \tag{5.2}$$

where v is the average speed of the orbiting planet and r the distance between the centers of mass of the planet and the sun. This net centripetal force is the result of the gravitational force that attracts the planet toward the sun. It can be represented by

$$F_{grav} = G \frac{M_{planet} \times M_{sun}}{r^2} \qquad (5.3)$$

Because $F_{net} = F_{grav}$, one has

$$M_{planet} \frac{v^2}{r} = G \frac{M_{planet} \times M_{sun}}{r^2} \qquad (5.4)$$

Because the speed of an orbit in nearly circular motion can be approximated as $v = \frac{2\pi r}{T}$, where T stands for the period of the planet orbiting about the sun for a complete circle, Equation (5.4) can be rewritten as

$$\frac{4\pi^2 r}{T^2} = G \frac{M_{sun}}{r^2}$$

That is,

$$\frac{T^2}{r^3} = \frac{4\pi^2}{GM_{sun}} \qquad (5.5)$$

Because the right-hand side of Equation (5.5) has nothing to do with the specific planet considered, the values of T and r on the left-hand side can be about any chosen planet. That is, the ratio T^2/r^3 equals the same value for all planets if the force that holds the planets in their orbits is the force of gravitation.

With his reasoning for Kepler's third law given, Newton considered the problem of why the sun is privileged to be stable when compared to the orbiting planets. By applying his third law of action and reaction, the sun should be moving about the planet, too. So, in the diagram in Figure 5.5, he named a point X as the center of mass of the two-body system. This point satisfies:

$$m_1 d_1 = m_2 d_2 \text{ and } d_1 + d_2 = r \qquad (5.6)$$

where r is the distance between the centers of the objects. With this concept in place, Newton generalized Kepler's third law as follows: Because to every action there is an equal and opposite reaction, in the planet-sun system, the planet does not orbit around a stationary sun. Instead, both the planet and sun orbit around the common center of mass for the planet-sun two-body system, satisfying

$$(m_1 + m_2) p^2 = (d_1 + d_2)^3 = r^3 \qquad (5.7)$$

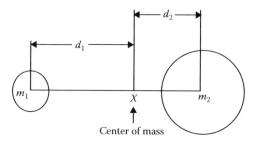

Figure 5.5 A two-body system.

where p is the planetary period with the sun as one mass and the planet as the other mass. Now, if one mass is very large compared to the other, then the sum of the two masses is always approximately equal to the larger mass. And if we take the ratios of Kepler's law for two different planets, the masses cancel out from the ratio and we are left with the original form of Kepler's third law:

$$\frac{p_1^2}{p_2^2} = \frac{r_1^3}{r_2^3}$$

On the other hand, if the two masses are equal to each other, then the center of mass of the two-body system lies equidistant from the two masses. And if the bodies are bound to each other by gravitation, each mass orbits the common center of the mass for the system lying between them. This situation occurs commonly with binary stars. In many binary star systems, the masses of the two stars are similar, and each star executes an elliptical orbit such that at any instant, the two stars are on opposite sides of the center of mass. That is, Newton's correction to Kepler's third law is very significant.

5.3 Universal Gravitation

Based on the discussions above, the following questions naturally appear:

Question 5.3. In Newton's law of universal gravitation (Equation (5.1)), why aren't the masses m_1 and m_2 pulled together to become one bigger mass? According to this law, when $r \to 0$, the gravitational pull F_{grav} between the masses m_1 and m_2 should be approximately ∞. So, no masses should be able to fight against such an infinitely large force of attraction.

Question 5.4. A similar question holds true for binary star systems. Why don't these stars simply become one star? Why do they still travel around their common center of mass in elliptical orbits?

Kepler's Laws of Planetary Motion ▪ 75

In the following, let us apply the figurative analysis introduced in the previous chapter to answer these questions.

Assume the masses m_1 and m_2 are the masses of the yoyo structures N and M in Figures 4.3 and 4.4. No matter which situation we are in, objects N and M act and react on each other with the gravitational pull F_{grav} given in Equation (5.1). However, at the same time, their spin fields keep them apart.

More specifically, for the scenario in Figure 4.3(a), when the eddy flows of N and M run into each other from the downside, they push each other away. In Figure 4.3(b), the flow directions of N and M are totally opposite of each other. In Figure 4.3(c), because the spin fields of N and M are both converging and their spinning directions are harmonic, the yoyo structures N and M have a tendency to combine and become one bigger eddy motion. Even though there is such a tendency, they still will not become one spin field (for details, see the discussion below). The scenario in Figure 4.3(d) is similar to that in Figure 4.3(a), except that the pushing between N and M occurs at the upper side of the adjacent area.

In Figure 4.4, due to the fact that all the spin fields of N and M do not rotate harmonically against each other, other than their gravitational pulls on each other, as given in Equation (5.1), there are uncompromising pushes between N and M. Unlike Figure 4.3(c), none of the scenarios in Figure 4.4 has a tendency to have the spin fields of N and M combined and become a unified spin field.

To address Question 5.4, we can see that each binary star system can only possibly come into existence out of the scenarios in Figure 4.3(b) and (c), and Figure 4.4(a) or (d). It is because in all these cases, the field of N spins against that of M. However, in Figure 4.3(c), because both N and M are converging fields, even though they might form a binary star system (yes, they will—see discussion below for details), to us, we cannot actually see them, because they are black holes. For the scenarios in Figure 4.4(a) and (d), because one of the spin fields is convergent, we do not see a binary star system either. That is, only scenario Figure 4.3(b) can lead to visible binary star systems.

Now we are ready to study why the binary stars N and M as shown in Figure 4.3(b) travel individually in their own nearly elliptical orbits. When we see the layout in Figure 4.3(b), the other sides of the yoyo structures of N and M look as they do in Figure 5.6. That is we have a pair of spin fields as shown in Figure 4.3(c).

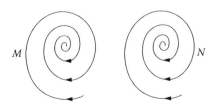

Figure 5.6 The other sides of the spin fields of N and M.

76 ■ *Systemic Yoyos: Some Impacts of the Second Dimension*

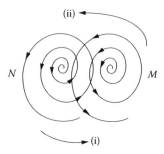

Figure 5.7 Spin fields of *N* and *M* repel against each other.

These two fields attract each other and have the tendency to combine into one spin field. So, the yoyo structures of *N* and *M* are pulled toward each other. However, when they are too close to each other, the diverging fields of *N* and *M*, as shown in Figure 4.3(b), start to repel against each other (Figure 5.7). The force of repellence comes from the opposite spinning directions of the fields of *N* and *M*. Under the influence of this force, *N* is pushed away along the (i) direction and *M* along the (ii) direction (Figure 5.7). When *N* and *M* travel away from each other to a certain distance, the attractions of the other sides of *N* and *M* (Figure 5.6) start to once again pull them together. Such an alternating effect of repellence and attraction keeps the stars *N* and *M* together and on their individual orbits.

Chapter 6

The Three-Body Problem

Continuing the discussions in the previous two chapters, in this chapter, we will see how the yoyo model and the figurative analysis method studied in Chapter 4 can be applied to explain why binary star systems, trinary star systems, and even n-nary star systems can exist, for any natural number $n \geq 2$. By checking the study of the three-body problem, we provide a brand new method to analyze the movement of three stars, be they visible or invisible. At the end, some open problems are cast for future research.

6.1 The Problem and Some Current Results

The well-known three-body problem is about how to compute the mutual gravitational interaction of three masses M_1, M_2, and M_3. This problem turns out to be surprisingly difficult to solve even in the so-called restricted three-body problem, corresponding to the simpler case of the three masses moving in a common plane satisfying that the mass M_3 is so small that it does not influence the circular motions of M_1 and M_2 about their centers of mass (Basdevant and Dalibrad, 2000).

To set up the problem analytically, let $M_1 = 1$ be the largest mass; $M_2 = \mu \ll 1$ be a mass in a circular orbit of semimajor axis a, half the distance across an ellipse along its long principal axes, about the center of mass of M_1 and M_2; and $M_3 = 0$ be a massless particle. Also pick such dimensions so that the gravitational constant $G = 1$. Then, the orbit period is given by

$$T \equiv 2\pi$$

and the mean motion is

$$n \equiv \frac{2\pi}{T} = 1$$

So, $n^2 a^3 = 1$. Because of the definitions of M_1 and M_2, the radii of their orbits are μ and $1 - \mu$, respectively. Now, enter a coordinate system that rotates with M_1 and M_2. In this system, M_1 has a fixed location at $(-\mu, 0)$, and M_2 at $(1 - \mu, 0)$. The equations of motion of M_3 are then given by

$$\ddot{x} = 2\dot{y} + x + (1-\mu)\frac{x+\mu}{r_1^3} - \mu\frac{x-1+\mu}{r_2^3} \qquad (6.1)$$

$$\ddot{y} = -2\dot{x} + y - (1-\mu)\frac{y}{r_1^3} - \mu\frac{y}{r_2^3} \qquad (6.2)$$

where $r_1 = \sqrt{(x+\mu)^2 + y^2}$ and $r_2 = \sqrt{(x-1-\mu)^2 + y^2}$. This is then converted to a Hamiltonian system with two degrees of freedom, with

$$q_1 = x, q_2 = y, p_1 = x - y, p_2 = y + x$$

$$H = \frac{1}{2}\left(p_1^2 + p_2^2\right) + p_1 q_2 - p_2 q_1 - \frac{1-\mu}{r_1} - \frac{\mu}{r_2} \qquad (6.3)$$

There is one integral of motion called the Jacobi integral, defined as follows:

$$C = -2H = x^2 + y^2 + \frac{2(1-\mu)}{r_1} + \frac{2\mu}{r_2} - \dot{x}^2 - \dot{y}^2 \qquad (6.4)$$

No other integral is known as of this writing. Making a mapping of $\dot{x}$ versus x for $C = 4.5$, there are four elliptical fixed points corresponding to periodic orbits around M_1 and M_2 in either direction. At $C = 4$, the trajectory is chaotic.

Primary resonance occurs when M_3 makes $J + 1$ orbits in the same time that M_2 makes J, or

$$n = \frac{J+1}{J}$$

where J is an integer. The period between conjunctions is then $2\pi J$. For $J \geq \mu^{-2/7}$, there is chaos.

6.2 Three Visible Bodies and Existence of *n*-nary Star Systems

In terms of practical use of the work discussed in Section 6.1, if there is such potential, we have to exam the unrealistic assumptions:

1. M_3 is massless; the mass of M_2 is much smaller than M_1.
2. M_2 travels along a circular orbit.

Even with such strong assumptions in place, we either run into the technical difficulty of not being able to get much or are trapped in chaos. In other words, it is necessary to introduce an alternative method to help out the current study of the three-body problem. In the following, let us see what results we can obtain with the figurative analysis method developed in Chapter 4.

First, let us consider three visible bodies, M_1, M_2, and M_3. This is the case of the traditional consideration. By *visible*, in terms of the yoyo model, we mean that the individual yoyo structures (Figure 1.1) of M_1, M_2, and M_3 have their Big Bang sides situated side by side against each other. With Figures 4.3 and 4.4 as our reference, we have four possibilities to consider (Figure 6.1), where the locations of

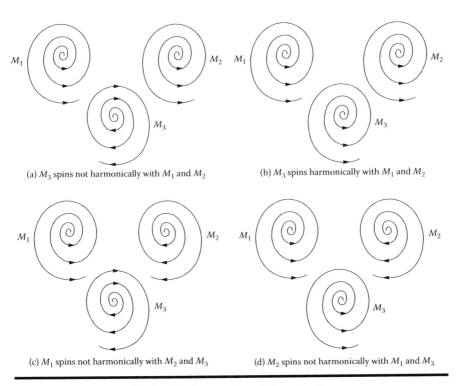

(a) M_3 spins not harmonically with M_1 and M_2

(b) M_3 spins harmonically with M_1 and M_2

(c) M_1 spins not harmonically with M_2 and M_3

(d) M_2 spins not harmonically with M_1 and M_3

Figure 6.1 Three visible bodies interacting with each other.

80 ■ *Systemic Yoyos: Some Impacts of the Second Dimension*

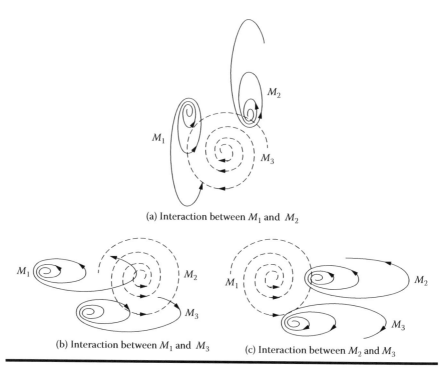

Figure 6.2 Pairwise interactions among M_1, M_2, and M_3.

M_1, M_2, and M_3 are relative. After analyzing and comparing the four possibilities in Figure 6.1, it can be seen that all the scenarios in Figure 6.1(a), (c), and (d) are essentially the same. So, without loss of generality, let us analyze cases (a) and (b) only.

For the scenario in Figure 6.1(a), let us first consider the shape of the spin fields of M_1, M_2, and M_3. When each of the spin fields exists alone without any interference from the others, the spins will be nearly circular. Interference between any two of them will force their circular spins to the shape of an ellipse (Figure 6.2). So, when the effect of the third spin field is added, the elliptical spin fields will be further twisted from their original circular shapes. For example, the ellipses of M_1 and M_2 in Figure 6.2(a) will be pressured further to the right, and the elliptical field of M_3 in Figure 6.2(b) and (c) will be further squeezed to the upside direction. So, a more accurate representation of the spin fields of M_1, M_2, and M_3, considering them interacting with each other at the same time, is given in Figure 6.3.

Now, let us turn our attention to the problem of how relatively M_1, M_2, and M_3 will travel with respect to each other. By noticing that M_1 and M_2 actually form a binary star system, our analysis becomes easier. Figure 6.4 shows how the masses M_1, M_2, and M_3 would move relatively to each other. Here, the original

The Three-Body Problem ■ 81

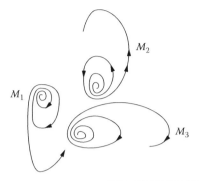

Figure 6.3 Interacting spin fields of M_1, M_2, and M_3.

orbit OM_1 of M_1 is pushed further to the left to RM_1, the original orbit OM_2 of M_2 is squeezed further to the right to RM_2, and the orbit RM_3 of M_3 is the combined outcome of the orbit M_1M_3 as expected by M_1 and the orbit M_2M_3 as expect by M_2.

If we look at the black hole side of the masses M_1, M_2, and M_3 (Figure 6.5), it can be seen that M_3 is only loosely attached to the binary star system (M_1, M_2), and a dynamic motion picture similar to that in Figure 6.4 holds true to the structure of Figure 6.5. That is, if M_3 stays with M_1 and M_2, M_3 will travel with M_2 (in a similar direction) more closely than with M_1. Chances are, if there appears a yoyo structure M_4 in the nearby region, M_3 might very well leave M_1 and M_2 and join M_4 to form a binary star system.

Now, let us analyze the three-body problem in Figure 6.1(b). In terms of the shapes of the spin fields of M_1, M_2, and M_3, due to the fact that M_3 in (b) spins in the opposite direction of that in Figure 6.1(a), the bulging right sides of M_1 and M_2 in Figure 6.3 will be located on the left sides, and the shape of the spin fields of M_3 will look the same as in Figure 6.3. Here, the figurative representation, similar to that of Figure 6.3, is omitted.

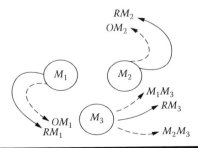

Figure 6.4 Altered traveling paths and directions of M_1, M_2, and M_3.

82 ■ *Systemic Yoyos: Some Impacts of the Second Dimension*

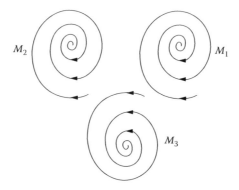

Figure 6.5 The other sides of M_1, M_2, and M_3.

In terms of relative movements between M_1, M_2, and M_3, we have a similar situation as in Figure 6.4, except that M_i will now travel along the dotted orbit OM_i, $i = 1, 2$, and the solid orbits RM_i, $i = 1, 2$, are the original ones when the interference of M_3 is absent.

Considering the pairwise interactions among the stars M_1, M_2, and M_3, we notice that we have three binary star systems: (M_1, M_2), (M_1, M_3), and (M_2, M_3). The other sides, the black hole sides, of M_1, M_2, and M_3 look like what is given in Figure 6.6. That is, we have a trinary star system, where each star M_i, $i = 1, 2, 3$, travels along its own nearly elliptical orbit with the common center of mass located at the center of all three orbits. When the visible sides (Figure 6.1(c)) get too close, their yoyo structures repel against each other. When they are really apart, their invisible sides pull their yoyo bodies together.

By generalizing this analysis, it can be seen that it is possible to have an *n*-nary star system, for any natural number $n \geq 2$, as long as the spin fields of the stars are in harmonic rotation with each other. Figuratively, for example, as long as all the

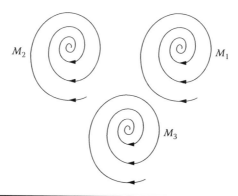

Figure 6.6 The black hole sides of the masses M_1, M_2, and M_3.

stars $M_1, M_2, \ldots, M_n$ spin in the same direction as the stars in Figure 6.1(b), we will have a stable *n*-nary star system.

6.3 Three Bodies with at Least One Invisible

Now, let us consider such three-body problems that at least one of the bodies is invisible. That is, we need to consider three kinds of three-body problems here:

1. All three bodies involved are invisible.
2. One of the three bodies involved is visible.
3. Two of the three bodies involved are visible.

For case 1, there are two different groups of three invisible body systems. A typical situation for each group is given in Figure 6.7. By comparing Figure 6.7(a) with Figure 6.6, and Figure 6.7(b) with Figure 6.5, we realize that the three-body

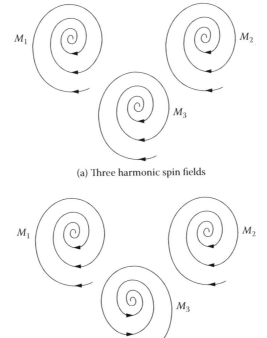

(a) Three harmonic spin fields

(b) M_3 spins not harmonically with the rest

Figure 6.7 Representatives of three-invisible-body problems.

84 ■ *Systemic Yoyos: Some Impacts of the Second Dimension*

problem of Figure 6.7(a) is the same as that of Figure 6.1(b), and that of Figure 6.7(b) is the same as that of Figure 6.1(a).

For case 2, where one of the three bodies involved is visible, we have three typical situations:

2.1. The two invisible fields attract each other and the visible field spins along.
2.2. The two invisible fields spin in opposite directions, and the visible field spins along with one invisible field and against the other.
2.3. The two invisible fields attract each other, and the visible field spins against both of the invisible fields (Figure 6.8).

Similar to all of the analyses we have done for three visible stars, for situation 2.1, M_2 and M_3 form an invisible binary star system, with the visible star M_1 pushed along with the movement of M_2. For situation 2.3, M_1 and M_2 form an invisible binary star system with the visible M_3 pushed away from both M_1 and M_2. For situation 2.2, due to inconsistent spinning directions, the invisible fields of M_2 and M_3

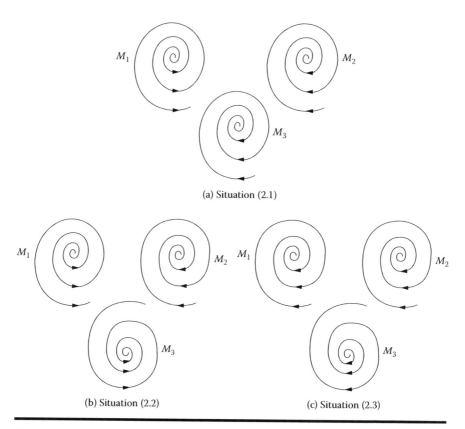

Figure 6.8 Typical trinary star systems with two invisible and one visible star.

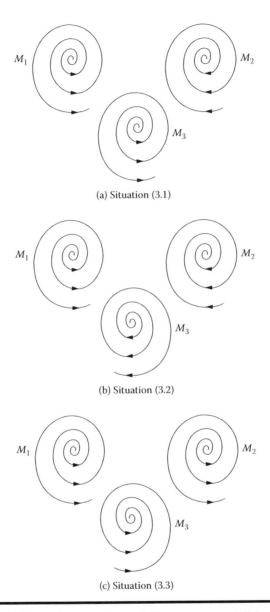

(a) Situation (3.1)

(b) Situation (3.2)

(c) Situation (3.3)

Figure 6.9 **Typical situations with two visible and one invisible star.**

will not have the tendency to combine, while M_1 and M_3 repel each other to their own individual and nearly elliptical orbits, with M_2 staying in its original place.

For case 3, where two of the three bodies involved are visible, we have the following three typical situations, as expected, because they are complementary to situations 2.1 to 2.3. A representative of each situation is given in Figure 6.9.

3.1. The two visible fields form a binary star system, and the invisible field spins along.
3.2. The two visible fields spin in opposite directions, and the invisible field spins along with one visible field and against the other visible field.
3.3. The two visible fields form a binary star system, and the invisible field spins against both of the visible fields.

A detailed analysis of situations 3.1 to 3.3 can be done, similar to what is given above, and will be omitted here entirely.

6.4 Open Questions

With the theoretical studies, empirical observations, and data as the foundation, the yoyo model (Figure 1.1) was introduced (Wu and Lin, 2002). So, our natural question here is:

> **Question 6.1.** For any specific system chosen, how can we materially measure the system's yoyo structure, its spinning direction, and relevant dimensions of this structure?

The importance of this question cannot be overemphasized. For example, by finding a way to measure the yoyo structure of disastrous weather systems, OuYang and his colleagues achieved better than usual results in their predictions of zero-probability events (Lin, 1998). My expectation here is, by being able to do the necessary measurements, we will be able to make better predictions about forthcoming events or the future course of true historical processes (Soros, 1998).

Practical answers to Question 6.1 might provide a means for us to determine the constants A, B, C, D, and a in Equation (2.1) for each chosen system. In that case, we will be able to study evolutionary characteristics of the system with added accuracy.

> **Question 6.2.** Other than the clear relationship between our spin fields, introduced in the yoyo model, and the concept of gravitational fields, what is the relationship between the yoyo spin fields and other known fields studied in physics?

Based on Einstein's concept of uneven time and space of materials' evolution (Einstein, 1997), the concept of gradients, and the fact that forces come from the structural unevenness of the materials of concern (Wu and Lin, 2002), we wrote down Equation (3.2). Here, a natural question is:

> **Question 6.3.** For a given external object M_1, acting on another object M_2, how can we actually write out or estimate the function $S = S(t, x, y, z)$ for the time-space distribution of M_1, and how about $\rho = \rho(t, x, y, z)$ for the density of M_2?

When objects M_1 and M_2 are arbitrarily chosen, only if the two functions S and ρ could be estimated would we have a chance to calculate the mutual reactions between M_1 and M_2, including the directions of their spin fields. Relating to the concept of yoyo spin fields, we may be dealing with a converging or diverging whirlpool. To be able to compute numerical details, using calculus-based theories, we might well need to know the speed of convergence or divergence of the whirlpool. That is;

> **Question 6.4.** For each given yoyo structure, how can we determine the speed at which it sucks in materials on one side and spits out materials on the other side?

A more definite answer to this question might help us determine the strength of gravitational pull or push a given yoyo structure exerts on another object located outside of the yoyo structure. More specifically, we have to ask:

> **Question 6.5.** For a given spin field, how can we construct a field function $F(S)$ that describes precisely the gravitational pull toward the center of the field at any specified location S in the spin field?

If such a field function can be constructed, we will have an opportunity to have Equations (4.5) and (4.6) greatly improved. The weakness of these equations is the variable v, which can be either difficult to measure or impossible to estimate in specific situations.

In terms of acting and reacting spin fields, as shown in Figures 4.3 and 4.4, we badly need to answer the following practically important questions:

> **Question 6.6.** For each given pair of acting and reacting spin fields, how can we quantitatively determine the strength of the acting and reacting forces and the resulting acting and reacting spinning effects on each spin field?
>
> **Questions 6.7.** When two spin fields N and M reach acting and reacting equilibrium on a small object m (Figure 4.5), how can we more precisely describe the motion of m in one of the possible subeddies?

Any of the subeddies described in Figure 4.5 has been seen as chaos when the method of equations is employed. So, any satisfactory answer to Question 6.7 will be groundbreaking with long-lasting scientific significance.

Because of the equal quantitative effects, we proved that figurative analysis is practically needed. However, the method of figurative analysis introduced in Chapter 4 needs to be greatly improved.

> **Question 6.8.** How can the method of figurative analysis studied in Chapter 4 be refined to a similar quality of plane geometry, where structures and quantities can be compared or computed precisely?

Evidently, if a yes answer to Question 6.8 can be established, we will be able to renew calculus to include the situations of discontinuity or nondifferentiability. As clearly shown in Lin (1998) and Wu and Lin (2002), these situations have made calculus and calculus-based theories invalid in front of some important practical applications.

Question 6.9. In terms of whole evolutions, how can we determine the time moment $t = t_b$ (Figure 4.7) when in a given spin field, a certain chunk of material has reached its most diluted edge of the field from the moment it came out of the narrow neck (Figure 1.1)?

Question 6.10. Again, in terms of whole evolutions, if a particle m resides in a yoyo structure M and the special time moments $t_m = t_{mb}$ and $t_M = t_{Mb}$ are the times in m and M, respectively, needed for a particle to reach the most diluted edge of its individual spin field, then are the times t_{mb} and t_{Mb} somehow related?

In terms of dishpan experiments, the solid borders might more or less contribute to the evolution of patterns of the fluid. If we use the results of these experiments to simulate our internal particle movements in a yoyo structure and its spin field, we have to ask:

Question 6.11. What will be the evolution of fluid patterns if the temperature along the periphery is cooling and the center heating?

Question 6.12. What will be the evolution characteristics of the fluid patterns if the temperatures are set as described in Question 6.11, but the periphery does not rotate? Or what if the periphery rotates faster than the center cylinder?

The situation described in Question 6.12 simulates more about the yoyo model (Figure 1.1) and the concept of spin fields.

Question 6.13. (continued from Question 6.12). What will be the evolutionary characteristics of the fluid patterns if, additionally, the simulated fluid spin field is converging or diverging to or from the center?

The yoyo model (Figure 1.1) and analysis (Chapter 5) indicate that as Mercury, Venus, Earth, and Mars travel farther away from the sun, new planets will be created by materials continuously coming out of the narrow neck of the yoyo structure of the solar system. Then, a natural question is:

Question 6.14. Have we seen any sign of new planets being born near the sun? Or does creating a new planet, like the natural evolutionary process, as described by Darwin in his theory of evolution, take a long time to materialize?

In Section 6.1, it is shown that by using calculus-based theories, when we study the three-body problem, not much can be done. And under practically meaningless restrictions, very little has been accomplished. At the same time, we run into the problem of chaos. As analyzed in Lin (1998), theoretical chaos does not mean the movement of the celestial bodies is chaotic. Instead, it represents the fact that the mathematical method used becomes invalid at those special cases or moments. By applying the figurative analysis introduced in Chapter 4, we have indeed achieved something new. But, what is still lacking is:

> **Question 6.15.** For both cases of the three-visible-body problem, Figure 6.1(a) and (b), how can we derive a more global view of their individual orbits for all the stars involved?
>
> **Question 6.16.** After proving the theoretical possibility to have an n-nary star system, for any natural number $n \geq 2$, the question is how to describe the individual orbit for each of the stars in the system?

To answer Questions 6.15 and 6.16 and several other problems listed here successfully, we need to first answer Question 6.8. This answer will lay down a more solid theoretical and analytical foundation for the newly introduced figurative analysis method to work with more satisfactory results.

Chapter 7

Stirring Energy and Its Conservation

In terms of squared speeds, the concept of kinetic energy, which will be called irrotational kinetic energy, and the law of conservation of kinetic energy are established in modern science. For instance,

$$e = \frac{1}{2} m v^2 \text{ (Newton)} \tag{7.1}$$

and

$$E = m c^2 \text{ (Einstein)} \tag{7.2}$$

So, a natural question is: Because the square of speed constitutes the kinetic energy, can the square of the angular speed of a rotation make up a different energy? (For more on models of rotation, see Lin, 2007.) In this chapter, on the basis of treating materials as nonparticles, we will study the concept of stirring energy (OuYang, [in press] a), its possible conservations, and some fundamental laws of evolution science (OuYang, [in press] b).

7.1 Rotation and Stirring Energy

Rotation is originated from the unevenness of materials' structures. And in the form of altering the materials' structures, rotation creates new materials and gives its expression in both the existence of relative structural stability and instabilizing evolutions. At the same time, we do see in nature materials, stable existence and instable evolution. More specifically, we see the existence of three-ringed stabilities. For instance, cosmic galaxy systems, star systems, and planetary systems can be stably observed, with the star systems being the second-level circulations concentrated with the main materials and energies of the universe. Correspondingly, the microscopic scales of materials consist of the molecular systems, atomic systems, and electronic systems, where what is interesting is that the atomic scale systems, as the microscopic second-level circulations, also possess high concentrations of energy, with the theory of nuclear energy proven. These observations seem to suggest that the quasi-stable existence of natural materials needs at least three levels of circulations, with the middle-level circulations holding huge amounts of energy. In particular, the star level of the cosmic system and the atomic level of the microscopic system represent high concentrations of nuclear energy.

If we expand this observation to the scale of human activities or the so-called meso-scale systems, we can observe the three-level circulation system of frigid zones, temperate zones, and torrid zones. Here, the temperate zones contain major amounts of energy. Each typhoon (or hurricane) is also a three-level circulation system, consisting of the typhoon eye, the region of torrential rains, and high-speed winds, and the outer-ring-shaped region of subtropical high pressures. Here, it is also the second-level circulation area that holds high concentrations of energy.

Based on the aforementioned observations, the inspiration of the stable mechanic device for three-ringed energy transformation, invented during the Bei Wei period of China (386–534 AC), and the observation that each natural river is always accompanied by lakes of various sizes to store floodwater, we can see that three-level circulations, shown in each energy transformation, have played the role of coordinating or restraining the energy transformation for the underlying stable existence, the destructive nature of non-three-level circulations to instable evolutions, and the dynamic equilibrium that without eddy motions there will be no kinetic energy transformation, and the amount of energy determines the internal heat of the eddy motions (OuYang, 1998).

However, in the past 300-some years, from Newton to Einstein, no one truly recognized the significance of rotations. Even after Einstein introduced the concept of curved space-time, he still employed the speed of zero-curvature space to represent materials' momentum (mv) and kinetic energy (mv^2 or mc^2) (Einstein, 1976). Even in angular momentum, angular speed is introduced similarly to the form of linear speed ($m\omega r$). Essentially, ωr is still the form of speed v. That is, the unit (m/s) of ωr is the same as that of linear speed v. That is, Newton, Einstein, and their followers did not provide the physical mechanism for the transformation and transfer of kinetic energies at the basic level of physics. In fact, the common and basic momentums and kinetic energies are the momentum ($m\omega$) and kinetic energy ($m\omega^2$) created by the angular speed of the rotational stir of the materials' structures existing in a curvature (non-Euclidean) space.

7.2 Conservation of Stirring Energy and Three-Level Energy Transformation

Conservation of energy is one of the three main achievements of modern science, which was established alongside the continuity of fluids (that is, treat fluids as low-resistance solids so that the movements of fluids can be studied as wave motions) and calculus. These three achievements constitute the fundamental theoretical system of modern science. In terms of this fundamental system established and developed in the past 300-some years, the achievement of Newton is the first push of his forces, and that of Einstein is the interaction between mass and energy. In essence, energies are also the works of forces. So, the problem of kinetic energies of the past 300-some years seems to have been solved, which is to express kinetic energy as the square of speed. What is missing is that no matter whether it is the conservation of kinetic energy or that of total energy, no existing laws tell us how energies change, in which forms energies transform and are transferred, and anything about the processes of energy transformation and transfer. Just as Newton's third law of motion, it does not provide us with any information on the form of motion and the process of interactions. Instead, it only spells out the impact or consequence of the interaction. However, no matter if we look at the life of a person, the history of humankind, or the history of the universe, we deal with a physical process and a problem on changes. Therefore, results on physical processes and laws on physical changes are important in terms of not only practical significance but also true laws of physics.

To help address this challenge, let us use formal logic. Because squared speeds represent kinetic energy, squared angular speeds should also stand for kinetic energy. Squared (linear) speed (v^2) and squared angular speed (ω^2) stand for different concepts, so they should play different roles. The unit for speed is m/s, while that for angular speed is 1/s, indicating unevenness in changes and distributions of the movement. In terms of quantities, angular speed measures the rotation of materials, while speed is the measurement of linear distance traveled by the object within the unit time. That is,

$$\vec{\omega} = \vec{i}\omega_x + \vec{j}\omega_y + \vec{k}\omega_z \tag{7.3}$$

Due to the unevenness of the special distribution of speed, the angular speed can be written as follows:

$$\vec{\omega} = \begin{vmatrix} \vec{i} & \vec{j} & \vec{k} \\ \dfrac{\partial}{\partial x} & \dfrac{\partial}{\partial y} & \dfrac{\partial}{\partial z} \\ u & v & w \end{vmatrix} \tag{7.4}$$

where the symbol → represents vector, and u, v, and w, respectively, the x-, y-, and z-components of the vector $\vec{V}$. For the horizontal two-dimensional plane, the angular speed in the vertical direction is

$$\omega_z = \frac{\partial v}{\partial x} - \frac{\partial u}{\partial y} \tag{7.5}$$

Introducing the flow function ψ gives

$$v = \frac{\partial \psi}{\partial x} \text{ and } u = -\frac{\partial \psi}{\partial y}$$

Similar to the traditional method of quantitative analysis, assume that ψ is twice continuously differentiable and can be represented as simple harmonic disturbances. Looking at the whole, let us introduce the combined disturbance of various scales:

$$\psi = \sum_n \psi_n$$

Then, we have

$$\nabla^2 \psi_n = -\mu_n^2 \psi_n \tag{7.6}$$

Considering the horizontal problem, take $V^2 = (\nabla \psi)^2$. From Equations (7.5) and (7.6) and the theorem of orthogonal plane divergences (OuYang et al., 2002), it follows that

$$\oiint_\sigma V_n^2 \, d\sigma = \oiint_\sigma \left(\sum_n (\nabla \psi_n)^2 \right) d\sigma = \oiint_\sigma \left(\sum_n \nabla \psi_n \cdot \nabla \psi_n \right) d\sigma$$

$$= \oiint_\sigma \left(\sum_n \nabla(\psi_n \nabla \psi_n) \right) d\sigma - \oiint_\sigma \left(\sum_n \psi_n \nabla^2 \psi_n \right) d\sigma$$

$$= \oiint_\sigma \left(\sum_n \mu_n^2 \psi_n^2 \right) d\sigma \tag{7.7}$$

and

$$\oiint_\sigma \omega_z^2 \, d\sigma = \oiint_\sigma \left(\sum_n (\nabla^2 \psi_n)^2 \right) d\sigma = \oiint_\sigma \left(\sum_n (\mu_n^2 \psi_n)^2 \right) d\sigma$$

$$= \oiint_\sigma \left(\sum_n \mu_n^2 \cdot (\mu_n^2 \psi_n^2) \right) d\sigma$$

Substituting Equation (7.7) into this last equation produces

$$\oiint_\sigma \omega_z^2 d\sigma = \oiint_\sigma \left(\sum_n \mu_n^2 V_n^2 \right) d\sigma \tag{7.8}$$

By comparing Equations (7.7) and (7.8), it can be seen that the closed line integral of the squared rotational angular speed contains the traditional closed line integral of the squared speed. That implies that even if we use the method of quantitative analysis, the concepts of linear speed and angular speed have different physical meanings. What is very interesting is that Equations (7.7) and (7.8) reveal a major negligence of the traditional conservation of kinetic energy, or that of total energy, which neither Newton nor Einstein expected. In particular, a conservation of stirring energy not only contains the conservation of speed kinetic energy, but also shows the way and procedure of the kinetic energy's transformation or transfer.

To analyze the physical meaning of stirring energies, a literature search shows that Fjörtoft (1953) mentioned the kinetic energy of nondivergent flows with formal similarity in terms of ζ^2 and v^2. Considering what he tried to describe, his work should be about the kinetic energy of rotational flows without clearly pointing out the difference between the physical indications of the vortical kinetic energy $\zeta^2/2$ and the speed kinetic energy $v^2/2$. Specifically, Fjörtoft employed the kinetic energy of nondivergent flows to design stable schemes for numerical computations to limit the instability of the energy. If we understand the kinetic energy of nondivergent flows as the stirring energy of rotations, then Fjörtoft had discovered the stirring energy in 1953.

7.3 Energy Transformation Process and Nonconservative Evolution of Stirring Energy

For a given three-ringed circulation, if we look at the second-level circulation, from the conservation of stirring energy in Equation (7.8), it follows that

$$\mu_1^2 v_1^2 + \mu_2^2 v_2^2 = c_1 = \text{const} \tag{7.9}$$

From Equation (7.7), it follows that

$$v_1^2 + v_2^2 = c_2 = \text{const} \tag{7.10}$$

Let Δv^2 stand for the changes in the speed kinetic energy from one time moment to the next neighboring moment; then the conservation of kinetic energy implies

$$\begin{cases} \mu_1^2 \Delta v_1^2 + \mu_2^2 \Delta v_2^2 = 0 \\ \Delta v_1^2 + \Delta v_2^2 = 0 \end{cases} \tag{7.11}$$

By eliminating Δv_1^2 and Δv_2^2, respectively, in Equation (7.11), we obtain

$$\left(\mu_2^2 - \mu_1^2\right)\Delta v_2^2 = 0 \text{ and } \left(\mu_1^2 - \mu_2^2\right)\Delta v_1^2 = 0 \qquad (7.12)$$

Because $\mu_1 \neq \mu_2$, Equation (7.12) implies that it must be $\Delta v_1^2 = \Delta v_2^2 = 0$. This end implies that with second-level circulations only, no energy transformation and transfer can be carried out, causing blockage or high concentration of energies, and consequently, instable evolution has to take place. From this explanation, it can be seen why each naturally existing river in the northern hemisphere has formed lakes on the right bank along its course to dredge locally blocked accumulation of energy.

If we introduce a third-level circulation, from Equations (7.7) and (7.8) we have

$$\begin{cases} v_1^2 + v_2^2 + v_3^2 = c_1 = \text{const} \\ \mu_1^2 v_1^2 + \mu_2^2 v_2^2 + \mu_3^2 v_3^2 = c_2 = \text{const} \end{cases} \qquad (7.13)$$

where c_1 and c_2 are constants. Let Δv^2 represent the changes in the speed kinetic energy from one time moment to the next neighboring moment. Then, to satisfy the law of conservation, we must have

$$\begin{cases} \mu_1^2 \Delta v_1^2 + \mu_2^2 \Delta v_2^2 + \mu_3^2 \Delta v_3^2 = 0 \\ \Delta v_1^2 + \Delta v_2^2 + \Delta v_3^2 = 0 \end{cases} \qquad (7.14)$$

By eliminating $\Delta\Delta i$, $\Delta\Delta i$, and $\Delta\Delta i$, respectively, in Equation (7.14), we obtain

$$\begin{cases} \left(\mu_1^2 - \mu_2^2\right)\Delta v_2^2 + \left(\mu_1^2 - \mu_3^2\right)\Delta v_3^2 = \mu_1^2 c_1 - c_2 = \text{const} \\ \left(\mu_2^2 - \mu_1^2\right)\Delta v_1^2 + \left(\mu_2^2 - \mu_3^2\right)\Delta v_3^2 = \mu_2^2 c_1 - c_2 = \text{const} \\ \left(\mu_1^2 - \mu_3^2\right)\Delta v_1^2 + \left(\mu_2^2 - \mu_3^2\right)\Delta v_3^2 = c_2 - \mu_3^2 c_1 = \text{const} \end{cases} \qquad (7.15)$$

Assume $\mu_1 > \mu_2 > \mu_3$ (the same results follow, if we let $\mu_1 < \mu_2 < \mu_3$). Then, the coefficients on the left-hand sides of the first and third equations in Equation (7.15) are positive, and those on the left-hand side of the second equation are negative. So, we have:

1. Both Equations (7.11) and (7.14) indicate that the conservation of the pure speed kinetic energy cannot limit the dissemination of energies. This is because outside the speed kinetic energy, there is still the spread of stirring energy.
2. When both Δv_1^2 and Δv_3^2 decrease, Δv_2^2 will increase. Conversely, when Δv_1^2 and Δv_3^2 increase, Δv_2^2 will decrease. That is, the three-ringed circulation,

through the second-level circulation, can complete its energy transformation and transfer with such a process, which is clearly shown.

3. If $\mu_2^2 - \mu_3^2 < 0$, when Δv_1^2 increases, so does the corresponding Δv_3^2. Conversely, when Δv_1^2 decreases, Δv_3^2 also decreases.

This analysis indicates that in the form of conservation of irrotational kinetic energy, the conservation of the total kinetic energy cannot be guaranted. Each transformation or transfer of stirring energy is carried out and completed through the second-level circulations, and between the first-level and third-level circulations there does not exist any energy transformation or transfer. That is, the previous discussion explains not only that the conservation of the irrotational kinetic energy is unable to restrict instabilities, but also that the existent unstable energies constitute the mechanism of transfer of realistic physical processes. This end reveals the physics of evolutionary processes.

7.4 Governance Law of Slaving Energy of Newtonian First Push

Continuing the discussion above, in this section, we will look at how the conservation of irrotational kinetic energy cannot by guaranteed by using equations of fluid motions. In particular, let us look at the following:

$$\dot{\vec{v}} + 2\vec{\Omega} \times \vec{v} = \vec{g} - \alpha \nabla P + \vec{F} \qquad (7.16)$$

where $\dot{\vec{v}} = \dfrac{d\vec{v}}{dt}$ is the acceleration vector, $\vec{\Omega}$ the vector for the earth's rotation, $\vec{g}$ the vector of gravitational acceleration, P the fluid pressure, $\alpha = 1/\rho$ the specific volume, ρ the density (the mass in the unit volume), and $\vec{F}$ the friction.

Due to the intuition of discussing kinetic energies, let us employ horizontal irrotational kinetic energy as our example with friction ignored. So, we have

$$\dot{\vec{v}}_h + 2\vec{\Omega} \times \vec{v}_h = -\alpha \nabla_h P \qquad (7.17)$$

where the subscript h represents the horizontal direction. Using $\vec{v}_h$ to dot-multiply Equation (7.17), called the scalar product operation of $\vec{v}_h$ and Equation (7.17), leads to

$$\frac{d}{dt}\left(\frac{v_h^2}{2}\right) = -\vec{v}_h \cdot (\alpha \nabla_h P) \qquad (7.18)$$

where according to the laws of vector operations, we have

$$\vec{v}_h \cdot (\vec{\Omega} \times \vec{v}_h) = \vec{\Omega} \cdot (\vec{v}_h \times \vec{v}_h) = 0$$

Evidently, the meaning of the conservation of irrotational kinetic energy is

$$\frac{d}{dt}\left(\frac{v_h^2}{2}\right) = 0 \text{ or } v_h^2 = C \text{ (a constant)} \tag{7.19}$$

To satisfy Equation (7.19) (the conservation of irrotational kinetic energy), we must require the right-hand side of Equation (7.18) to satisfy $\vec{v}_h \cdot (\alpha \nabla_h P) = 0$. For this end to hold, the absolute values of $\nabla_h P > 0$ and $\nabla_h P < 0$ must equal each other exactly. There is no doubt that only with the quantitative Equation (7.19), the form change or transformation process between $\nabla_h P > 0$ and $\nabla_h P < 0$ cannot be given. That is, the conservation of irrotational kinetic energy cannot be guaranteed.

The term $\vec{v}_h \cdot (\alpha \nabla_h P)$ on the right-hand side of Equation (7.18) is the slaving governance term of the kinetic energy of the Newtonian first push. To intuitively express its physical meaning, let us replace the special volume by mass, where erasing the subscript does not alter the physical meaning, so that we have

$$\frac{d}{dt}\left(\frac{v^2}{2}\right) = -\vec{v} \cdot (\alpha \nabla P) = -\frac{1}{m}\vec{v} \cdot \nabla P = \frac{1}{m}\vec{v} \cdot \vec{f} \tag{7.20}$$

Notice that the scalar product of vectors is a quantity, that is, $|\vec{v} \cdot \vec{f}| = |f v|$, where $v = \frac{ds}{dt}$ and s is the distance traveled. So, $\vec{v} \cdot \vec{f}$ stands for the changes in works done by the Newtonian pushing force, known as power. So, the physical meaning of Equation (7.20) is that changes in irrotational kinetic energy are exactly the power of the first pushing force. The corresponding equation (Equation (7.20)) is called the governance law of slaving energy of Newtonian first push (OuYang et al., 2002).

7.5 Interactions and Einstein's Mass-Energy Formula

If we take

$$\frac{de}{dt} = \vec{v} \cdot \vec{f}$$

from Equation (7.20), we can get

$$e = \frac{1}{2}mv^2 \tag{7.21}$$

which is Equation (7.1), the expression of irrotational kinetic energy of Newton's first push. If we introduce Newton's third law, which states that acting and reacting forces are equal, then the irrotational kinetic energy of interactions is given by

$$E = 2e = mv^2 \tag{7.22}$$

What is very interesting and strange is that if we use Einstein's speed c of light to replace v in Newton's classical mechanics, then Equation (7.22) immediately leads to Einstein's mass-energy formula (Equation (7.2)):

$$E = mc^2 \text{ (note the paradox: } c = \text{const}, a = \frac{dc}{dt} = 0, f = ma = 0) \quad (7.23)$$

Therefore, the mass-energy conservation, an outstanding achievement of modern physics, does not agree even with the basic knowledge of classical physics. What is made clear by Newton's second law is that the force is equivalent to acceleration; without acceleration there will not be any force. There is no doubt that Equation (7.22) agrees with forces being equivalent to accelerations, while Equation (7.23) becomes a kinetic energy without a force to produce works, because the acceleration is 0. So, a natural question arises: What kinetic energy is the kinetic energy that there is no force to do any work? Not only can those who do not have much education background not answer this question, but also those who are experts in physics cannot answer it. To this end, it is found that Einstein (1976) did not ever point out that Newton's second law was incorrect. Evidently, if Newton's second law is correct, then it has to be that the mass-energy formula is incorrect. On the contrary, if the mass-energy formula holds true, then Newton's second law, together with modern science, has to be thrown into the trash pen. Considering the acceptability of our reasoning and suggestions of colleagues, in our book *Entering the Era of Irregularity* (OuYang et al., 2002), we discussed the matter indirectly without completely breaking through the thin ice, so that the reader has his or her own space to imagine and understand what we try to say.

7.6 Solenoidal Fields and Problem on Universal Gravitation

It might be because V. Bjerknes' circulation theorem was not widely known in the scientific community that in the latter half of the 20th century the problem of nonlinearity became a hot topic and center of debate. As an intuition of physical problems, it can be seen that for the interaction between nonparticle objects, the probability for the interaction to act right on the centers of masses is extremely small. Evidently, each interaction without acting directly on the centers of masses will lead to rotational movements, and the corresponding nonlinear acting force will become a stirring force in the form of solenoids. Besides, even for the general form of mathematical nonlinearity, it can also cause rotational changes (OuYang et al., 2002). In this section, using mathematical vector operations, we can simply reveal the mathematical properties of nonlinearity and obtain the rotationality of solenoidal fields. For example, for the following vector operation formula,

$$\nabla(\vec{a} \cdot \vec{b}) = \vec{a} \times (\nabla \times \vec{b}) + \vec{b} \times (\nabla \times \vec{a}) + \vec{b} \cdot \nabla \vec{a} + \vec{a} \cdot \nabla \vec{b} \quad (7.24)$$

let us take $\vec{a} = \vec{b} = \vec{v}$. So, we have

$$\nabla(\vec{v} \cdot \vec{v}) = \nabla v^2 = 2\vec{v} \times (\nabla \times \vec{v}) + 2\vec{v} \cdot \nabla \vec{v}$$

This can be written as

$$\nabla \frac{v^2}{2} = \vec{v} \times (\nabla \times \vec{v}) + \vec{v} \cdot \nabla \vec{v} \qquad (7.25)$$

where, when $\frac{v^2}{2}$ on the left-hand side is multiplied by m, we obtain an irrotational kinetic energy. Taking its vorticity gives

$$\nabla \times \left(\nabla \frac{v^2}{2} \right) = 0$$

From Equation (7.25), we have

$$\nabla \times (\vec{v} \cdot \nabla \vec{v}) - \nabla \times \left[(\nabla \times \vec{v}) \times \vec{v} \right] = 0 \qquad (7.26)$$

For the convenience of our intuition, let us only take the vector equation for horizontal rotations. So, we have

$$\vec{v} = \vec{k} \times \nabla_h \psi \qquad (7.27)$$

where $\vec{k}$ represents the unit vector in the vertical direction and ψ is the flow function. So, we have

$$\nabla_h \times \vec{v} = \vec{k}\varsigma \qquad (7.28)$$

where $\varsigma = \frac{\partial v}{\partial x} - \frac{\partial u}{\partial y}$ stands for the vorticity along the vertical direction. Substituting Equations (7.27) and (7.28) into Equation (7.26) and taking what is in brackets of the second term of the horizontal vector produces

$$(\nabla_h \times \vec{v}) \times \vec{v} = \varsigma(\vec{k} \times \vec{k} \times \nabla_h \psi)$$

Notice the vector operation $\vec{a} \times \vec{b} \times \vec{c} = (\vec{a} \cdot \vec{c})\vec{b} - (\vec{a} \cdot \vec{b})\vec{c}$. So, the right-hand side of this last equation becomes

$$\varsigma(\vec{k} \times \vec{k} \times \nabla_h \psi) = -\varsigma \nabla \psi$$

From vector analysis, it follows that $\nabla \times (f\vec{a}) = \nabla f \times \vec{a} + f\nabla \times \vec{a}$. So, we have

$$\nabla_h \times \left[(\nabla_h \times \vec{v}) \times \vec{v} \right] = -\nabla_h \varsigma \times \nabla_h \psi \qquad (7.29)$$

Taking the horizontal vector of Equation (7.26) and substituting this vector into Equation (7.29) produces

$$\nabla_h \times (\vec{v} \cdot \nabla_h \vec{v}) = \nabla_h \psi \times \nabla_h \varsigma \tag{7.30}$$

Notice that the $\vec{v} \cdot \nabla_h \vec{v}$ term in the parentheses on the left-hand side of Equation (7.30) is a well-known nonlinear term (the same holds true even with the subscript removed), and the right-hand side is exactly a solenoidal, rotational field under a solenoidal stirring effect. So, our vector manipulation has very clearly declared that nonlinearity is a problem of physics about solenoidal fields with rotations; its one-dimensional mathematical form is the well-known $u \frac{\partial u}{\partial x}$ or $\frac{1}{2}\frac{\partial u^2}{\partial x}$. What needs to be clear is that quantities are only formal records and cannot really tell us the practical meaning of what is hidden behind the formality. For example, the essence of $\varsigma = \frac{\partial v}{\partial x} - \frac{\partial u}{\partial y}$ is really the angular speed ω. And, ω represents the uneven distribution of linear speeds, revealing the fact that rotational structurality is different than the irrotationality of linear speeds. In particular, quantities containing rotationality cannot be mingled with those that do not contain any rotationality. This end should be seen as common knowledge worthy of our special attention in quantitative analysis. For example, the linear expression $\frac{1}{2}\frac{\partial u^2}{\partial x}$ is a quantity with rotation, which can analytically lead to asymmetry and irreversible transformation from convergence to divergence. From two-dimensional nonlinearity, one can obtain the result that an internally spinning vortex can be transformed to an externally spinning vortex, and it is a reversible process (OuYang et al., 2002). Correspondingly, from Equation (7.26), it is not difficult to see that the three-dimensional nonlinearity stands for a solenoidal, rotational field, as shown by V. Bjerknes' circulation theorem and verified by the realistic fluids existing in the earth's atmosphere and flows in the ocean (OuYang et al., 2002).

Let us now, based on our discussion above, rewrite Newton's universal gravitation formula as follows:

$$\vec{F} = G \frac{\rho_1(x,y,z;t)\rho_2(x,y,z;t)}{R^3} \vec{R} \tag{7.31}$$

where $R = (r + r_d)$, r is the distance between the surfaces of P_1 and P_2, r_d is the sum of the radii of P_1 and P_2, and G is the gravitational constant. Taking the vorticity in Equation (7.31) leads to

$$\nabla \times \vec{F} = \frac{G}{R^3}\left[\rho_1\rho_2 \nabla \times \vec{R} + \left(\rho_1\nabla\rho_2 + \rho_2\nabla\rho_1\right) \times \vec{R}\right] \tag{7.32}$$

Evidently, in Equation (7.31), even as a mathematical quantity, $\rho_1(x,y,z;t)\rho_2(x,y,z;t)$ stands for a nonlinearity instead of the Newtonian variable of particle quantities. It is because of this that structures can alter mathematical properties. The corresponding rotationality of solenoidal stirs obtained by V. Bjerknes using the nonparticle method of circulative integrals is indeed a theorem that agrees with the reality. So, nonlinearity implicitly contains physical solenoidal rotational fields. Because of this result, we can tell how rich and colorful Einstein was in his imagination, even though he did not provide a flexible technique of operation. Because he could masterfully manipulate tensors, it is amazing that Einstein did not recognize the rotationality in nonlinearity from simple vector operations. Einstein studied curvature tensors following after Carton, and at the key moment of realizing vorticity tensors, he did not "push the door in the moonlit land." It might be due to his gentleman's elegance that Einstein only "knocked on the door," so that he could only stay outside of the door of universal gravitation.

At this junction, there is no harm for us to continue Einstein's associative thinking. That is, according to Equation (7.32), let us think in the following way. The so-called gravitation in universal gravitation is not a pulling force along a straight line; it seems to be the "carry-in" force of an inward spinning entity. Correspondingly, there should also be a "carry-out" force of the entity's outward spinning. The carry-in of the entity's inward spinning has been evidenced by the observation that when a celestial body falls, its path of falling looks like a curved page in an old book. The carry-in of high-speed flows is the "black hole," where rotations can indeed produce high-speed flows (Akimov and Shypov, 1997). As for the carry-out of outward spinnings, it should be a "white source," and the speed of carry-out, at least around the exit of the white source, can reach such a high level that we cannot imagine today. So, universal gravitation should be generally called a universal force. Whether it is a carry-in of an inward spinning or a carry-out of an outward spinning will be determined by the attributes of the entity. We need to be clear that this is only a conjecture based on solenoidal, rotational fields. However, Newton's mathematical formula for universal gravitation is indeed an eye-opener for our exploration of the unknown. The rotationality of solenoidal fields might still not be the ultimate answer. It is only an associative thought based on nonparticle structurality of materials. Maybe the gravitational anomaly has something to do with changes in the celestial body itself and the structure made up of the position of the body.

Also, from Equation (7.26), it can be seen that the condition of conservation of the kinetic energies is very strict, because it can be reached only under the condition that the inward and outward spins of the solenoidal field cancel out each other perfectly. What is different in this case, though, is that the kinetic energy is no longer irrotational. Instead, it is a stirring kinetic energy with rotations and almost zero probability for the exact cancellation to occur. Because of this, the physical reason why irrotational kinetic energies do not conserve in general becomes quite clear.

Besides, even though the rotationality of nonlinearity is derived out of some plain mathematical vector analysis, other forms of mathematics should not alter this mathematical property of nonlinearity. In particular, quantities, tensors, etc., are not exceptions, where quantities are zeroeth-order tensors and vectors first-order tensors. Therefore, the so-called mystery of nonlinearity should be resolved completely and end right here forever.

7.7 Conservability of Stirring Energy and Physical Significance of Energy Transformation

We will look at this topic from several different angles:

1. If the second-level circulation cannot materialize transformation and transfer of energies, energy blockage will have to be created, leading to instabilities in the accumulation of energy and triggering the third-level circulation to destroy the original second-level circulation to achieve the system's equilibrium and stability. To this end, both Yellow and Yangtze rivers in China can be looked at as two examples. For instance, the Yellow River is known for its floods. That is why Chinese history can be seen as one of struggling with floods. When it flows through the area of east longitude 100–110°, the Yellow River, located in the northern hemisphere, curves north and forms a great bend area on the left-hand side of the main river course. So, a current flowing in the opposite direction to the deviation force of the earth movement is created, where the deviation force inevitably pushes the current to flow to the right. That is why, along the Yellow River, it is very easy for floods to break through dykes and dams, causing great losses. For example, with a rate of flow of 6,000 m^3/s at the flood peak, Huayuan Kao dam in Zheng Zhou can be overrun. As for the corresponding Yangtze River, only looking to the east of Hubei, we can see such major lakes as Dongting Hu, Boyang Hu, and Tai Hu. Not only are these lakes located on the right-hand side of the Yangtze River, but also Xuanwu Hu of Nanjing was connected to the Yangtze River in ancient times. That is why it has been believed that even the tributaries of the Yangtze River can easily entertain the rate of flow of several tens of thousands of cubic meters of floods at their peaks. In particular, it is exactly because the huge lakes on the right-hand side of the Yangtze River constitute a second-level circulation to transfer energies that the capability of redirecting floods is greatly enhanced. In comparison, we can tell that from Huayuan Kao downward, there basically does not exist any large-scale lake along the Yellow River. That explains why it has been quite easy for the Yellow River to flood, causing great losses. It is because its main course does not coincide well with nature, where the river flows to and curves to the left against the effect of the earth's deviation force.

2. The transformation and transfer of circulative energies affirm the cause for materials' stable existence through the conservation of the quasi-three-ringed stirring energy. That might provide a theoretical explanation for Lao Tze's claim that out of three, all things are born; for such empirical beliefs about changes that nothing goes beyond three; etc. That is, these and similar beliefs are concepts not only known since ancient times, but also established on some well-understood principles of physics (even though these principles were not known in modern science).

3. Because energy transformations are completed through indirect (second-level) circulations, a natural need arises to question some of the methods and concepts used and studied in traditional theories, including direct damping (or cutoff), limiting instability, forcing dissipation, etc. Or it can be said that modern physics does not recognize the effect of indirect circulations. And not only can it not describe changes in physical processes, but also it has missed an important law on energy transformation—the conservation of stirring kinetic energy. This end indicates that many laws of physics are descriptions of events without touching on causes and processes of evolution.

4. Due to the capabilities of indirect circulations, between the first-level and third-level circulations there does not exist direct energy transfers. The inevitable consequence is that the first-level circulation moves, while the third-level circulation does not move. When the second-level circulation accommodates the retransfers of energy from the first-level circulation, it at the same time transfers to both the first- and third-level circulations, so that a transfer of energy is completed in the form of dispersion. This process should be that of realistic, physical dispersion, and the fundamental principle of dealing with water problems used by King Yu, the founder of the Xia Dynasty (ca. 21st to 16th century BC), and the father and son of the Li Bing family of ancient China.

5. Each three-level circulation transformation constitutes a quasi-closed system, completing transformations and transfers of energies. The fourth-level, fifth-level, and any other higher-level circulations at most receive or further transfer the remnant energy out of the three-level circulation without being able to constitute a system's quasi-stability. And the mechanism of physical adaptation process is that instable subcirculations readjust themselves to the primary three-level circulation. It is also because remnant energies exist out of the transformations of the three-level circulation that the cause for some planets to have satellites while others do not have, at the planetary level of the celestial circulations, might be the inevitable by-products of adjusting remnant energies out of the transformations of the three-level circulation. All subcirculations, which do not follow the three-ringed energy transformations, are not stable. And the instability of the fourth- or higher-level circulations would adjust toward the quasi-stability of the atmospheric three-ringed circulation. This end can be observed in the evolution of realistic atmospheric circulations and the experiment of spinning fluids (OuYang et al., 2002). As for the stability of the natural elements, made up of microscopic materials, and the instability of man-made elements, we have not seen much explanation provided by experimental or theoretical physicists. These phenomena might have something to do with the equilibrium problem of redistributing the separable energy transformations. Our idea of three-ringed energy transformation might provide a new way of thinking to attack this problem.

6. One can observe many three-ringed stabilities, such as (1) the stability in the cosmic scale, consisting of galaxy level (including the Milky Way), star level (including the solar system), and planets (including the earth); (2) the stability in atmospheric fluids, consisting of the frigid zones, temperate zones, and torrid zones; (3) the stability of typhoons' fields; and (4) the stability in the elementary elements of the microscopic world, consisting of molecular levels, atomic levels, and electronic levels. As for the bodily activities of animals and humans, optimal designs of mechanical devices, etc., they all reflect a quasi-three-ringed energy transformation and material transfer of the conservation of stirring energy. So, what is presented in this chapter does not seem to be accidental.
7. What needs to be pointed out is that stability is relative, and according to the point of view of evolution, there absolutely does not exist any stability. Any stability can be destroyed by the instability of the underlying systems' non-three-ringed circulations. However, because of the conservation of stirring energies, each instability must adjust itself toward a three-level circulation. And the readjusted three-level circulations might not be the same as the original system's three-ringed circulation. So, instabilities and conservation of stirring energies not only are the reason for the existence of materials' forms, structures, or attributes, but also reveal the evolutionary process from the start of the development to the end of a change. And in non-three-ringed circulations, changes in their direction of rotation can occur prior to their process of adjustment, which provides both the theory and methods for prediction practices of materials' evolutions.
8. What should be said is that three-ringed circulations constitute the systemic transformations of materials and energies, shown in the form of structural rotations. Their essence is no longer limited to certain phenomena of physics. Instead, it is a problem about laws of physics. It has only been because in the past 300-some years, our thinking has been confined by the concept of Euclidean spaces, that the rotational aspect of materials has been overlooked, so that we did not recognize that the general form of materials' kinetic energy is that of stirring energy, and did not realize that the concept of angular speed contains that of (linear) speed as a special case. If we extract and purify this as a principle of philosophy about nature, we have the following: the significance of quasi-three-ringed circulative energy transformation is no longer limited to the triangular stability of statics; it has already touched on the stability of materials' motions and the philosophical problem on evolutions where, at extremes, matter will evolve in the opposite direction. This is exactly one of the very important problems modern physics and natural philosophy have failed to address, and opens the door for a new physics on processes and evolution science (OuYang et al., 2001) in the name of conservation of stirring energies.

Because wholeness is a fundamental concept in systems science (Lin, 1999; Klir, 1985), it implies that the law of conservation of stirring energy is the existence law of systems stability, and that nonconservation at the same time constitutes the evolution law regarding systemic processes of changes. What is important is that the concept of stirring energy can contain traditional speed energy as a special case. So, it can be expected that the concept of stirring energy can help to transform modern science to evolution science through its nonconservations.

7.8 Discussion

The quasi-three-ringed stability principle of the conservation of stirring energies, which can be seen as a realization of the concept of multilevel systems (Lin, 1989, 1990), reveals the procedural aspects that all materials have places to come from and move to, and all events have a start and an end, and the transformation of energies. It not only contains the quantitative laws of modern science, but also reflects the laws of naturally existing events' processes, making it clear that events cannot be identified with quantities. This principle can be applied to product designs in engineering (there are such products in circulation already) or ecological environment planning, or employed to resolve problems regarding evolutionary predictions.

As for evolution predictions, it involves how instable energies and corresponding irregular events are understood. Irregular information comes from irregular events and has to be reflected in instable energies. So, the essence of limiting instable energies, as being done in theoretical studies, is to eliminate irregular events. That is, to make valid predictions, one has to keep and apply irregular information (OuYang and Chen, 2006). That is why the thinking logic and the methodological system of limiting energy flows and transformations in modern science constitute exactly a denial of evolution.

Corresponding to a scientific system, there should be relevant laws. In particular, in terms of the study of evolution problems, laws on processes have to be established. Because stirring energies reveal the transformation process of energies at the same time they describe the effects of instable energies, the significance of stirring energies is shown vividly in terms of the underlying physical mechanism. Besides, the concept of angular speeds enters curvature spaces in terms of the curvatures of the fields of flows. Due to this understanding, Einstein's treatment of the curvature problem by taking the speed of light as a constant has already violated the space-time curvature. In other words, in essence, the discussion in the form of speed kinetic energy (mc^2) has limited the problem of focus in Euclidean spaces. What is more important is that the assumption that the speed of light is constant has already made Einstein's mass-energy an energy of not doing any work, leading to the problem of whether or not the mass-energy formula and the general relativity theory hold true (OuYang et al., 2002).

What deserves our attention is that curvature spaces are not limited to the Riemann geometry of convex curvatures. There is also a problem with regard to concave curvature spaces. The corresponding modern dynamic equations are written in zero-curvature (Euclidean) spaces. So, even if the equations are correct, they still could not be introduced into realistic physical space. In particular, angular speed implicitly contains changes in (linear) speed. So, as long as the angular speed is a variable, it implicitly means a problem of changing acceleration. Modern science has admitted (Prigogine, 1980) that any problem involving changing accelerations belongs to a noninertial system. Therefore, the concept of stirring energy and its consequent evolution science must have gone beyond the range of modern science and helped to reveal many problems existing in the foundations of modern science.

Appendix: Stirring Energy and Its Conservation

A7.1 Evolution Engineering and Technology for Long-Term Disaster Reduction

In this appendix, we look at the long-term technology of the theory system of urban disaster reduction and prevention of floods, caused by suddenly appearing torrential rains. The results indicate that based on the concept of secondary circulations, by establishing urban artificial lakes, a high-capability flood prevention system can be materialized with very low costs, and this system also possesses the ability to improve the urban ecological environment. On the basis of Chengdu City's flood prevention facilities designed for such major floods caused by torrential rains that were only seen once in 50 years, it is shown specifically how to improve the capability of these facilities to withstand such major floods that were recorded only once in 200 years—and how the capability of the flood prevention systems designed for once-in-200-year floods can be improved to materialize the goal of long-term flood prevention and disaster reduction. This appendix shows that evolution engineering can greatly relax the worries and reduce the potential troubles hidden in past flood prevention projects, such as high dams and high levees, and it can also help to recover the urban ecological balance and resolve such problems as shortage in water resources, etc., realizing tangible economic benefits with major ecological significance. This presentation is based on Chen et al. (in press).

A7.2 The Background

After over 40 years of efforts by many first-class scholars of our modern time, including I. Prigogine, S. C. OuYang, etc., it is finally recognized that the quantitative analysis system, established since the time of Newton, is not an evolution science. OuYang, especially, established the detailed law of conservation of stirring energy and the problem of evolution technology and engineering design under this law of conservation. His work not only reveals the mutual constraints in materials' existence, but also provides an analysis method for evolution engineering (OuYang et al., 2001a, 2001b, 2002; Chen et al., in press). Even the quantitative method of estimation for durable good design and the traditional quantitative analysis method have also missed the evolutionary constraint of this conservation law of stirring energy. Considering that materials' evolutions are a fundamental problem of the natural world, even in terms of engineering design, there also exists such an evolution engineering problem of how to follow the principle of evolution. To this end, in the extended discussion on the law of conservation of stirring energy, a long-term-effect technology for treating water is proposed (Chen et al., in press).

In recent years, along with the rapid development of urban construction, man-made petrifaction effects have been expanding and enlarged, and the area of natural green ecological coverage has shrunk dramatically. In particular, lofty steel and concrete structures have strengthened the difference among local regions, causing less desirable weather conditions to suddenly appear, such as suddenly appearing torrential rains in urban areas, convective windstorms, abnormally high temperatures, etc. Combined with more ragged bottom cushions on the ground, these torrential and concentrated rains increased the demands on the drainage capability of the areas, creating greatly shortened time for creating, flowing, and cumulating rainwaters, so that it became more convenient for water flows to gather. As soon as a blockage in the drainage system appeared, the accumulated water caused floods. This phenomenon has been happening in recent years very often.

The available records show that in recent years both near-water and away-from-water cities have suffered from an increased number of flood-related disasters, while the level of underground water has dropped. And strong precipitation of several tens of millimeters has caused ground transportation systems to be paralyzed, due to disasters of water caused by clogged drainage systems. All these recently often-appearing problems have caught people's attention and are in urgent need of solutions.

Flood is a natural phenomenon. It must have its own laws of operation. And it should be a common knowledge that flows of automobiles are not the same as traffic accidents, and floods are not equal to flood disasters. So, there is a need to reconsider the recent flood disasters and inundations. In a 1980 (U.S.) report of flood prevention and disaster reduction, it was pointed out that in the past 60 years, over several tens of billions of dollars were invested in flood prevention and disaster reduction; however, the number of flood disasters have continued to increase, indicating that the flood prevention projects and disaster reduction facilities did not receive their material benefits. Only after the 1993 major floods (with a loss of US$12–18 billion) did people start to realize that they had to stay out of the flood zones (Li, 2006). After suffering through the 1995 floods in Holland, there appeared people who propose to give the right-of-way to floodwaters. As a matter of fact, in ancient China, there were many successful cases of flood prevention and water treatments that modern humans can learn from. For example, the key of the success of the Dayu treatment of water is to follow water's nature to dredge it. Li Yizhi once suggested storing floodwater to save its resource and dividing the flows to reduce the flood and redistribute the volume so that at the upper reaches there are accumulations, and at the lower reaches there are facilities to release the flood, so the excessive water is divided. That can be seen as a perfect method and system of flood prevention and water treatment. It was why in 256 BC there appeared the father and son team of Li Bing, the mayors of Qingshu, who adopted the water treatment method of "prevent floods, separate sand, and make use of water resources." These ancient Chinese were the pioneers of modern hydraulic engineering. The Dujianyan hydraulic project (also an irrigation system) is not only the monument of the time when it was constructed in 256 BC, but is still in operation at its full capacity today, after over 2,000 years. That great accomplishment motivated OuYang to study transformations of energy so that he found that the principle beneath the ancient hydraulic projects was exactly the conservation law of stirring energy, which he later introduced.

To a degree, Chinese history can be seen as one of treating waters. The ancient Chinese not only had a rich experience in dealing with water, but also had complete theoretical systems. In the West, for example, only after the United States suffered from 1927's major floods did Americans changed their belief that levees were invincible. That is, no levee can be sufficiently high to stop any flood ever recorded in history, or steady enough to stand any overflow or water erosion (Lu, 2000). It can be said that experts in China's community of waterpower completely have the ability to develop long-lasting effective technology to realize the situation of having floods without any disaster, with the floodwaters reserved for different purposes. However, the actual situation is different. Floodwaters are seen every year, and clogged drainage systems appear every season. Experts have been busy achieving near-term successes and rushing for immediate glories by making use indiscriminately of the idea of high dams and elevated lakes of the first push system. After over half a century of their efforts, it is gradually becoming clear that the high dams have created silted-up reservoirs and deteriorating ecological environments. Nearly 90 percent of the reservoirs with high dams and elevated lakes, constructed in the past half a century, have become ailing, leaving behind disasters after each torrential rain, which massively appear every year.

Because of what is described above, in this appendix, we see how the law of conservation of stirring energy can be employed to study the initial design and planning for the corresponding evolution engineering. That is, a long-lasting effective technology system is established for urban prevention of torrential rain floods and disaster reduction, while we will explain the evolution engineering of storing energy and reducing disasters of the kind of "following water's nature to dredge it." In the following, we will use the actual data of Chengdu City to specifically provide the detailed technical analysis.

A7.3 Basic Principles of Flood Evolution and Development and Flood Disasters

Floods are a problem facing natural sciences, but not a problem of the natural sciences. We should not be limited to the problem of life and economic losses seen in various local and regional flood disasters. What is more important is that we have to discover the fundamental laws governing the evolution and development of floods. To this end, on the basis of the concepts of evolutionary structures of materials, we will see how to establish a long-lasting and effective technology to prevent and control flood disasters by following along with nature.

A7.3.1 Law of Conservation of Stirring Energy

Each system constitutes a problem of whether or not it can be self-constrained, instead of a doctrine on particles, and is originated from the stable existence of materials. That is the foundation of natural sciences and also an essential unavoidable problem.

When the law of conservation of stirring energy is practically applied to evolution engineering, what is most important is that the stirring energy (ω^2) contains the traditional kinetic energy (v^2). What needs to be recognized is that the conservation of kinetic energy in the traditional first push system is an incomplete law of conservation of energy, and that limiting the propagation of energy is not the same as controlling the transportation of materials. This understanding reveals the reason why the law of conservation of kinetic energy cannot guarantee the conservation of energy. So, the law of conservation of stirring energy is an important development in modern science. For a more detailed discussion, please consult with Chen et al. (in press).

It can be said that in the past 300-plus years, no matter whether it is the conservation of kinetic energy or that of total energy, nothing has been said about how energies change and in which forms they are transformed or transferred, or the process of transformation and transferring. It is exactly like Newton's third law, which does not provide any information about the form of motion of the mutual reactions and the reaction process. Instead, it only tells about the results of the reactions and movements. It can also be said that neither Newton nor Einstein realized that traditional kinetic energy does not contain stirring energy (OuYang et al., 2001b). It is believed that even in the modern form of quantities, the introduction of the concept of stirring energy is a very important contribution and forms a major change in the fundamental concepts. That is, stirring energy contains kinetic energy as a special case and gives an expression to the transformation and transferring processes of kinetic energy. It is a law on results as well as the law and principle on evolutionary processes.

A7.3.2 Three-Ring Quasi-Stability Problem of Urban Flood Movements

As a form of fluids, floods and their movements have to follow the general laws of evolution of materials, and have their own specific rules regarding how materials and energies are transferred and transformed. For example, along the middle and lower reaches of Yangtze River, there have naturally formed three large lakes: Dongting Hu, Poyang Hu, and Tai Hu. They can be used to store and release. That is the reason why Yangtze River, as a natural river, does have floods; however, it does not bring disasters (Chen and Wang, 2001). On the other hand, in an effort to gain more land from the lakes, humans have neglected the deposits of silts, leading to the problem of floods. In particular, in terms of the plum rain weather, precipitation of 100–200 mm should not be considered abnormal. However, currently, such normal plum rains often cause flood disasters, which essentially points out that disasters might not originate in nature. So, the recent phenomenon that there are torrential rains every year, clogged drainage systems every year, floods every year, and flood disasters every year has to be a real problem.

So, let us once again state the structure of the flood water system that causes flood disasters:

1. Let Q_1 be the flood current strength at the gathering place of torrential rainwaters, which is related to the precipitation intensity of the torrential rain, the area of the water gathering place, the infiltration rate of the ground (or the degree of saturation of the bottom cushion), evaporation, and other factors. Based on the convention of hydrology, it is measured by the maximum runoff.
2. Let Q_2 be the capability for a lake, natural or artificial, to store flood. It stands for the ability of the secondary circulation of the stirring energy to completely transform the water flows and energy of a flood disaster. The stronger the storing capability of floodwaters, the greater the ability to prevent floods. The amount of floods in Yangtze River is more than 10 times greater than that of Yellow River. However, its number of flood disasters is lower than that of Yellow River. So, we have to admit that this lower number must have something to do with the storing capabilities of peak floods of Dongting Hu, Poyang Hu, and Tai Hu.
3. Let Q_3 be the draining capability of the flood course. This draining capability is mainly affected by the environmental conditions of the river course. When the river course is silted up, its draining capability will be decreased. The flood-releasing capability of a river course has to be related to the height and solidness of the embankment. Although increasing the height of the embankment can improve the flood-releasing capability, the higher embankment experiences much increased risk level of the dyke breaking up. Evidently, for a broken embankment, not only is its capability of releasing flood zero, but also its level of flood disaster is increased. That is, the expenses of flood prevention are increased while the loss of potential flood disasters becomes heavier. What is more important is that the ancient Chinese had recognized this fact over 4,000 years ago. However, today, people are still craving high embankments and high dams, so that the risk of suffering from flood disasters is greatly increased artificially. Also, the flood-releasing capability is controlled by an upper limit for high dykes. What an irony in such a time when people pursue after science and that they go against science!

Based on the analysis above and the law of conservation of stirring energy, let Q_1 be the primary circulation, Q_2 the secondary circulation, and Q_3 the third circulation. Take $E = Q_1 - Q_3$. If $E > 0$, then E is called the instable energy of the flood disaster. If $E \leq 0$, then there does not exist any flood disaster, simply called no flood disaster. Evidently, the symbols Q_1, Q_2, and Q_3 not only represent a structure of relationships among floodwater, storage of floodwater, and releasing of floodwater, but also describe the process of movement of floodwater and the related characteristics of transfers of energy. So,

$$Q_1 \leq Q_2 + Q_3 \qquad (7.33)$$

means the situation of having floods without any disaster, and

$$Q_1 \geq Q_2 + Q_3 \qquad (7.34)$$

for the situation where having any flood means a disaster.

From Equations (7.33) and (7.34), it can be seen that each flood system consists mainly of the three-ringed circulation of floodwater strength, storage capability, and releasing capability. Evidently, the basic condition for a flood disaster to occur is

$$Q_2 = 0 \text{ and } Q_1 > Q_3 \qquad (7.35)$$

As for the floods of suddenly appearing torrential rains, the variable Q_1 is mainly determined by the strength of the precipitation. Even if the meteorological forecast for a torrential rain is perfectly accurate, the flood disaster will still happen, due to $Q_2 = 0$. So, the problem of flood disasters is not entirely determined by the forecast of torrential rains and the relevant floodwaters. What is more important is how to recognize the rules that govern the movements of floodwaters, and the realization that floodwaters are not equal to flood disasters.

The current floodwater analysis for various flood-related designs, the variable Q_1, is determined by mostly using Pearson type III curves to predict future floodwater on the basis of the record waters seen once in 50 years, 100 years, or 200 years. For large reservoirs, the prediction is made using the record floods seen once in hundreds, thousands, or ten of thousands of years. Also in recent years, the hydrometeorological method of considering the maximum possible torrential rains is employed. Evidently, if we use higher embankments to improve the flood-releasing capability Q_3, then this plan is shown impractical by the historically affirmed fact that by doing so, the expenses of flood prevention are increased, while the losses of potential flood disasters are heavier.

According to the law of conservation of stirring energy and the Coriolis force (created by the deviation of the earth's rotation) acting on floodwaters, a secondary circulation Q_2 (in the northern hemisphere, it would be the right bank of the river current) is naturally formed, while the floodwaters do not directly rush to the third circulation Q_3. That is also why, although the amounts of flood peak flows in the Yellow River are not even close to those of some branches of the Yangtze River, the Yellow River has caused more flood disasters than the Yangtze River. That is, the main course of the Yellow River bends to the left due to the effect of the special terrains, while the Coriolis force makes flood currents go to the right, making it easy for floodwaters to break open the embankment along the river, causing flood disasters. That was why OuYang stated that any river that bends to the left in the northern hemisphere (the opposite holds true for the southern hemisphere) does not agree with the law of nature. That, in fact, has clearly pointed out the fundamental principle on how to treat and fix the Yellow River.

So, instead of considering the design of a high embankment based on the flood amount Q_3, it would be better to follow the natural geological structure to construct a secondary circulation (called artificial flood storage lakes). However, as of this writing, we still cannot see on the map any large-scale flood storage lake along the Yellow River downstream to Huagyuanko. Through the holding capability of the secondary circulation Q_2, potential disaster of floodwaters can be dissolved. This concept can not only materialize the goal of long-lasting and effective flood prevention and disaster reduction, but also transform potential flood disasters into usable water resources.

The law of conservation of stirring energy not only completes the conservation of energy in modern science, but also makes modern science walk into evolution science. What is more significant is that on the law of conservation of stirring energy, evolution engineering is established. So, this conservation law of stirring energy is also a practically effective principle and method useful to long-lasting and effective disaster prevention and reduction. That is, by using the holding capability of the secondary circulation (the indirect circulation), we can realize long-term and effective disaster prevention and reduction. And the ecological conditions of the environment can be consequently improved greatly.

A7.4 Feasible Technology for Urban Long-Term Flood Prevention and Disaster Reduction

To sufficiently understand how floodwaters move, we have to clearly know that floodwaters are not equal to flood disasters. Even if the forecasting for torrential rain, which causes Q_1, is not on target, through constructing or modifying Q_2, we can still reach the goal of reducing or eliminating the flood disaster or making sure that we have the flood without any aftermath disaster. That is,

$$Q_2 \geq Q_1 - Q_3 \qquad (7.36)$$

However, because specific situations will surely be constrained by particular environmental conditions, unsettled historical problems, or some traditional concepts and beliefs, we believe that there is still a need to make the aforementioned discussion and problems addressed known to as wide an audience as possible. Currently, many cities in China are pouring huge amounts of money into reconstructing their river courses around the cities. (For example, the budget for Chengdu City to reach its goal, named Year 2020, is ¥1.7 billion. All of this amount will be used to remake river courses throughout the city (Internal circulation, 2000)). So, it is necessary to place the technology of urban long-term and effective flood prevention and disaster reduction among the order of the day. To this end, in this appendix, we use specific examples to provide an analytic explanation by employing the law of conservation of stirring energy.

A7.4.1 The Computational Method for Q_1

There are many methods in hydrology on how to compute Q_1. By using the computational method of urban rainfall infiltration facilities (in America, the most applied is the Sjoberg-Martensson method, while the most used in Europe is the Geiger computational method, named after a German scholar), after having determined the reappearance time T of the next torrential rain, one can obtain the relationship curve (Figure 7.1) between the amount of runoffs and the life span of the rainfall by using the intensity formula of torrential rains, the area of the rainwater gathering place, and the corresponding average runoff coefficient. In Figure 7.1, the area enclosed by the coordinate axes and the curve equals the total amount of flood runoff of the rain. If we use Equation (7.36) to consider the holding capability of an artificial lake, then we have to study the following specific problem of design:

$$Q_1 = \int_0^{T_0} 3600 \frac{q_T}{1000}(C \times A + A_0) dt \qquad (7.37)$$

where Q_1 is the total amount of runoff (m³) with reappearance time T and the life span of the rainfall T_0 (in hours); t is how long in hours the rainfall has been

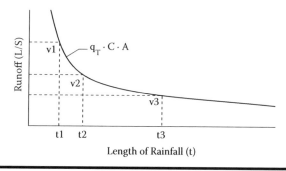

Figure 7.1 Relationship between life span of a rain and the amount of runoff.

going on; q_r is the intensity (1/s.ha [unit area second]) of the torrential rain, which has a reappearance time T and length t; A is the area of the rainwater gathering place (ha); A_0 is the area (ha) of the facility, which directly faces the rain and can be ignored; and C is the average runoff coefficient (for urban areas, 0.5–0.8; for countryside, 0.4–0.6).

So, we can simplify Equation (7.37) as follows:

$$Q_1 = Q_T \cdot t = 3600 \cdot C \cdot A \cdot t \frac{q_T}{1000} \qquad (7.38)$$

To simplify the computation, let us use Equation (7.38) to substitute for Equation (7.38). From Equation (7.38), corresponding to t1, t2, and t3 in Figure 7.1, we compute for v1, v2, and v3, respectively, and the areas enclosed by the axes and the curve. When comparing these values to those in Equation (7.37), there are some errors. After using the Sjoberg-Martensson analysis of the actual data, we obtain the makeup coefficient 1.25 (Wang and Li, 2001) to improve the outcomes of Equation (7.38). That is, we have

$$Q_1 = Q_T \cdot t = 1.25 \cdot \left(3600 \cdot C \cdot A \cdot t \frac{q_T}{1000} \right) \qquad (7.39)$$

which provides the total runoff Q_1 of the floodwater.

A7.4.2 Design of Artificial Urban Lakes and Problem of River Course Drainage

Other than flood prevention and disaster reduction, the construction of artificial lakes also needs to reach the goals of having no disaster even if there is a flood, storing flood and its energy, and maintaining the ecological balance. So, the design of an artificial lake should consider:

$$Q_2 \geq E \qquad (7.40)$$

where Q_2 stands for the holding capability of the planned artificial lake, and E the instable energy of the flood disaster.

As for the flood-releasing capability Q_3 of a river course, the traditional goal is to just release the floodwater. However, on the basis of the evolutionary law of conservation of stirring energy and the concept of balancing the ecological environment, the new goal for Q_3 is changed to the "responsibilities" the river, specifically if it is a city river, and its environment would be charged for. That is, under the condition of maintaining the ecological balance, it should be able to store water and energy as much as possible and perfect the overall water circulation system.

A7.5 Analysis of Urban Long-Term Flood Prevention and Disaster Reduction Facilities

In this section, we imagine that we would construct two artificial lakes, one at the location of Fuhe bridge along the northern outer-ring road, and the other at Longzhuayan of Qingshui River, an upper reach of Nan River, based on the actual situation of the flood-releasing river courses at around Chengdu City (Internal circulation, 2000). In the following, we will provide concrete analysis and detailed computations.

A7.5.1 Computation on an Artificial Lake at Fuhe Bridge along Northern Outer-Ring Road

Considering that the artificial lake will absorb the floodwater from the upper stream, the amount Q_3 of flood releasing of the river course at this location should be 0. So, we have

$$Q_1 = Q_2 = 1.25 \cdot \left(3600 \cdot C \cdot A \cdot t \frac{q_T}{1000} \right) \tag{7.41}$$

which is the computational formula for the design of the holding capability of an artificial lake.

Considering that the existing stone weir has a branch river that can hold and separate the floodwater from the upper stream, we start our computation of the amount of floodwater through Fuhe from the stone weir (the cross section of the intake sluice gate). We construct Table 7.1 for the statistics of maximum intensities of torrential rains for the rain gathering area of the entire section of Fuhe River, where the

Table 7.1 Statistics of Maximum Rain Intensities for the Entire Section of Fuhe River

Control Section	Rain Gathering Area (km²)	Flow Amounts of Various Frequencies (m³/s)		
		Once in 200 Years	Once in 100 Years	Once in 50 Years
The stone weir	0	0	0	0
Fuhe bridge	103.11	309	274	240
Dongfeng canal	108.9	326	290	254
Dongziko	135	405	359	315
Wangjianglou station	505	1,310	1,160	1,010
Huayang	744	1,900	1,700	1,490

computational formula (Internal circulation, 2000) (partially empirical and partially theoretical) for the maximum runoff Q_r (m³/s) of a flood peak is given below:

$$Q_r = 0.278 \times S \times F/\tau$$

where ψ is the runoff coefficient of the flood peak, defined by $\psi = 1 - \tau^n \times u/S$; S the intensity of the torrential rain, defined as the maximum rainfall amount in one hour (mm/h); F the area of the flow field (km²); n the formularization index of torrential rains; τ the time of converging flows in the flow field; and u the coefficient for producing a massive amount of water, that is, the average infiltration intensity (mm/h) in the flow field during the time of producing a massive amount of water. Also, the data for Dongfeng canal were collected at the entrance of the canal, those for Dongziko at Sa River entrance, those for Wangjianglou station at the emerging point of Nan River, and those for Huayang at Huayang water level station.

From Equation (7.41), it follows that we take $C = 0.57$. So, Table 7.2 implies that during the time period of the one-hour maximum precipitation, we have $t = 1$, $A = 103.11$, and $q_T = 114.5/3,600$. So, we can compute the total runoff amount for such a (sustained for one hour) torrential rain that is met once in 200 years for Fuhe bridge at the north outer-ring road of Chengdu as follows:

$$Q_1 = Q_2 = 1.25 \cdot \left(3600 \cdot C \cdot A \cdot t \frac{q_T}{1000}\right)$$

$$= 1.25 \times (0.57 \times 1145 \times 103.11 \times 10^6) = 8.4 \times 10^6 \text{ m}^3$$

For the holding capability of the artificial lake, we first construct Table 7.4 based on the historical records. Then, similar to what is done above, from Table 7.3, we can obtain the holding capability of our artificial lake to be 26.9×10^6 m³ for the torrential rain seen once in 200 years. When our artificial lake satisfies these calculated numbers for the runoff and holding capability, it will be able to play the role of adjusting and holding the floodwaters from the upper streams of Fuhe River.

Because the terrain of Chengdu slopes gently, the design for the prevention level of floodwaters of a general artificial lake is no more than 3 m deep. So, for our artificial

Table 7.2 Maximum Rainfalls (MM) within 24 Hours Actually Observed at the Urban Meteorological Station

Time Step	Once in 200 Years	Once in 100 Years	Once in 50 Years
1 hour	114.5	106	96
6 hours	223.2	201.6	180
24 hours	367.4	325.6	283.8

Table 7.3 Holding Capabilities for the Artificial Lake

Time Step	Once in 200 Years	Once in 100 Years	Once in 50 Years
1 hour	8.4	7.8	7.0
6 hours	16.4	14.8	13.3
24 hours	26.9	23.9	20.8

lake with a holding capability of 26.9×10^6 m^3, it will take about 8.97 km^2 of land, which produces a water surface area of the diameter of about 3.4 km. If our artificial lake project is connected with the water supply of the city and the prevention of a dropping groundwater level, the depth of our lake can reach somewhere between 15 and 20 m. In this case, at the flood season, our lake's flood prevention water level can reach 15 m, so that the flood prevention capability is improved five times, while the area of the lake surface is shrunk to 1.8 km^2. This final proposal, if actually implemented, can not only resolve the problem of flood disasters with long-lasting effect, but also help to supplement for the ground water system, the city water supply, maintaining an ecologically more balanced environment, among other benefits.

A7.5.2 Estimate for an Artificial Lake along the Qingshui River on the Upper Reaches of the Nan River

Table 7.4 lists the data for each part on the upper reaches of Nan River. We select Longzhuayan as the location for our artificial lake. Because the part of Nan River through the city has reached the goal of meeting the demand of the flood seen once in 200 years, it allows 40 percent of such flood to pass that part of the river. Based on what we did above, we can obtain a holding capability of 24.9×10^6 m^3, so that the area of the artificial lake will only be 1.66 km^2.

Under normal conditions, Chengdu City area would suffer from floods and related disasters, if the precipitation of the Chengdu plateau were more than 100 mm in 24 hours and the flow amount at Wangjianglou hydrologic station surpasses 400

Table 7.4 Statistics of Rain Gathering Areas of Parts of Qingshui River

Control Section on Qingshui River	Rain Gathering Area (km²)	Flow Amounts (m³/s) at Flood Peaks of Various Frequencies		
		Once in 200 Years	Once in 100 Years	Once in 50 Years
Bridge on outer-ring road	79.37	313	286	260
Longzhuayan	141	498	451	404

m³/s. If our proposed artificial lakes were constructed as planned above, the floodwaters at Fuhe River and Nan River would be adjusted and held up by the artificial lakes so that the flood flows could be reduced by as much as 600 m³/s, and the peak flood flows seen once in the 200 years at Wangjianglou hydrologic station and the Huayang water level station would be controlled under the level of that seen once in 50 years.

In other words, the facilities, designed for the once-in-50-year floods at Wangjianglou hydrologic station and Huayang water level station, would be able to withstand the once-in-200-year floods. And currently, facilities designed for once-in-200-year floods have the desired long-lasting effect. The terrain of Chengdu City slopes from the northwest direction to the southeast. At the same time, when the planned artificial lakes held up the desired amounts of floodwaters, they would reduce the pressure on Dongfeng canal and Sa River from excessive floodwaters. This function of the lakes would materialize the economic benefits of helping the hundreds of medium- to large-size business entities located in southeast Chengdu avoid disasters from floods.

A7.6 *Discussion on Long-Term and Long-Effect Technology*

In this appendix, we only use the flood prevention and disaster reduction of suddenly appearing torrential rains of Chengdu City as an example to illustrate our idea and method. Even though our results are only preliminary, they have shown some clear promise. From the point of view of evolution science, there will not be any absolute long-lasting effect. Our concept of technology of long-lasting effect discussed earlier is introduced corresponding to the concept of balancing strong forces of the first push system. That is why we have called our concept evolution engineering. The idea is to implement the engineering needs without depending on the balancing technology of strong forces. Instead, we suggest following nature and borrowing natural forces to release the forces. Or we would use the concept of energy transformations to maintain practically useful engineering facilities. So, speaking of the technology implied in our aforementioned long-lasting effective technology, it has a far-reaching benefit of being applied to many other areas of our daily lives besides flood prevention. For example, how to face twisting shear-stress forces in solids, resistances of fluids, and falling down and caving in, or collapses and withering of objects are not problems that quantitative analysis methods of the traditional inertial system can resolve. In terms of the estimations used in engineering designs, however, the computational methods on irregularities developed in the current quantitative analysis can, to a certain degree, be introduced to evolution engineering, while the specific procedures often become quite simple and practical. The reason for this is that on the basis of adopting new epistemological concepts, we will employ the actual information as much as possible.

The idea of following water's nature to dredge the water in essence is exactly the law of conservation of stirring energy and relevant technology applied to addressing the problem of floods. In ancient China over 4,000 years ago, the Chinese had already known that through a secondary circulation (an indirect circulation), energies or materials could be transported or delivered, and in the form of transportation, by holding and dispersing flood prevention and disaster reduction, could be effectively achieved. With several thousand years of practice, the long-lasting effect of this knowledge has been shown, and valuable experiences of unified arrangements and control of large rivers have been gained. In this appendix, we only use the present situation, that there are floods every year, which cause aftermath disasters every year, to illustrate the effectiveness of evolution engineering on the basis of the conservation law of stirring energy. In form, our work is platitudinous repletion of the ancient wisdom; in essence and reality, it involves a major change in our conceptual understandings. This is because we believe that the phenomenon that there are floods every year, which cause aftermath disasters every year, should be following up troubles left behind by modern science.

Along with the urbanization of the traditional towns and cities in China, drops in underground water levels have become a problem that has caught people's attention. In particular, due to the lower underground water levels, many cities along the coastal lines have suffered from invasions of seawater from underground. This problem should be seen as the number one most important issue facing the residents of these cities and towns, because the situation has become desperate and urgent. However, just by employing the secondary circulations (artificial urban lakes) of following water nature, this urgent problem can be easily resolved with a change in our basic concepts. More detailed discussion is omitted here.

We have to admit that the quantitative analysis system, established since the time of Newton, provides a set of effective methods for products and engineering designs. Because this system is developed on the basis that materials do not age and get damaged, it transports energies by releasing the strong forces through oscillations of resistance without noticing that nature transforms energy and transports materials in the form of rotational changes in materials' structures. So, to truly achieve the goal of flood prevention and disaster reduction, there is no need to employ violent struggles using strong forces to balance strong resistance. Instead, we should transport materials and transform energies through making use of the existing strong forces to revolve the existing strong forces. Hence, this end illustrates that the quantitative analysis system of the past 300-plus years has not touched on problems of evolution, and it has limited the development of science. Because of rotationalities, it is unavoidable for materials' evolutionalities to appear. So, it is not operationally possible to employ strong forces to resist evolutions. In particular, the method of using fortified embankments to stop floods will not work (at least in terms of large rivers). This fact also reveals that a slight difference in the basic concepts leads to enormously different consequences.

The essence of evolution engineering is to focus on or clearly address materials' evolution principles and to investigate, under the premise of evolutions, the theories and technologies of how to resolve disasters into benefits. In short, evolution engineering is about how to follow natural rotationalities to transform naturally existing destructabilities, instead of such an engineering system that resists the materials' destructabilities using means that are against the laws of nature.

Chapter 8
Time and Its Dimensionality

Due to the difficulties met by modern science in handling materials' changes and the practical need to resolve problems involving irregular events and information, in this chapter we will address the problem of time, including what time is, where time is located, whether or not time occupies a material dimension, etc. Then, combined with practical experiences, we will analyze the differences between events and quantities and those between material dimensions and parametric dimensions, and consider such a problem as the quantification of time. The analysis below indicates that modern physics is mainly a theory about quantitative physical quantities and parametric dimensions, which inevitably has to be a science that cannot really foretell what is forthcoming, combined with all the weaknesses of quantities. Our analysis shows that realistic materials, events, and information are not the same as quantities, and the transformation principle of materials and events does not follow the formal logical calculus of quantities. With practical evidence, it can be seen that the methods of structural transformation of events can shed light on problems regarding transformations of materials and events. This chapter is based on OuYang and Lin (in press a).

8.1 Problems to Be Addressed

Science can be seen as a process; its development is stimulated by calls of various problems. The so-called process has to involve the differences between the before and after (a sense of time) of events. If events possess only spatial differentialities, then we have a static, rich, and colorful world. If an event shows its difference along the order from before to after (time), then one sees the changes in the event. Speaking more comprehensively, changes in an event include variations in location, state, property or attribute, etc. Strictly speaking, changes in location, state, and attributes cannot be separated. Changes in location without involving the event's state, properties, etc., can only be human imagination or assumption to meet some specific needs. Because of the feeling of order, the problem about time has to appear. There is no doubt that if there did not exist any differentiality in each event from before to after, then the world would forever be a still picture; there would not be the past, the present, and the future. In this case, time would be a meaningless concept. So, to address what time is, we have to first consider the question: What is an event? Or, what is information? Or, how can events be described? That is, compared to time and the underlying materials, the concept of events should be more general and involves materials and their states, attributes, shown phenomena, etc. Out of the need of treatment using mathematical techniques, events are described by employed quantities in modern science so that such concepts as parametric dimensions, physical quantities, etc., are introduced. What needs to be pointed out is that quantities are only a property of events and are only limited to the postevent formality. An event can contain such properties as color, smell, taste, character (attribute), direction, state, quantitative formality, etc. Hence, physical quantities are only one quantified attribute of events. Now, a natural question is: Can quantified physical quantities be employed to substitute for the events from which the quantities were initially abstracted? It seems that in the past 300-plus years since the time of Newton, there did not appear any detailed explanation or argument relating to this question. Instead, it has become a convention and custom to employ quantified physical quantities in place of events. However, this problem reveals a fundamental weak link existing in the foundation of modern science.

There is no denial that physical quantification of events is necessary to develop the relevant mathematical tools, and applications of these tools indeed provide, to a certain degree, a description of the movements of the underlying materials and the related problem of oscillation. That is, quantities have made important and magnificent contributions to modern science. However, the scientificality of modern science is not limited to one aspect—quantities, but is more about the functionalities and effects of the events' attributes. Its signal of development should be about how to resolve the unsettled problems left behind from the past major scientific achievements and glories. In other words, the development of science comes from the calls of problems. Solving open problems and meeting new challenges are always the number one problem of science. Existing problems are also challenges to the magnificent achievements of the past. Or, as long as there exist unsettled problems, science has to be evolved further.

For our present time, the first problem, which constitutes a direct challenge to modern science, is about fluids' turbulences, which is relevant to materials' evolutions. It has been a long-lasting solid stronghold still not possible to conquer.

In particular, if we face the objective realities, we can say that all fundamental forms of materials' movements can be summarized as turbulences. This is because fluids are the main character among all materials, which constitutes almost 99 percent of the universe. If we attempt to understand materials' flows through movements, then the system of wave motions of modern science does not describe the basic movements. The more general and fundamental forms of movements are rotational motions. That is why we conclude that to understand turbulences, one has to resolve the problem of rotation. Even though modern science contains fluid mechanics as a branch, because the problem of rotation has not been resolved, it still cannot be called the science of fluids. Because fluids are afraid of twisting, as long as there is a twist, an eddy will appear. In fact, solids are afraid of twisting, too. Twist is a problem of stirring motions. Each stirring motion has to be rotational. Each rotation indicates variable acceleration, leading to a noninertial system, which is beyond the scope of modern science. Considering the wide-range existence of noninertial systems, the achievements and magnificence of modern science, in front of the problem of rotation, are only about the specific movements that keep the underlying materials unchanged. Even in terms of specific problems, modern science can only be seen as a school about wave motions. What is important is that modern science did not resolve rotations. And its incapability of solving the problem of rotation should not be used as an excuse to treat turbulences as problems of stochastics.

Next, materials' aging, damage, or breakage (called decrepitude or death) are a problem of common objective existence. The events that are described by using quantities are only a realization of regularized events, a human imagination, instead of a true regularization of the events. And the concept of regularization is exactly abstracted from the objective existence of irregular events. In other words, without irregular events, there would not be such a thing as regularized events. Therefore, when irregular events are seen as a problem, it is a call for further development of science as well as a challenge to modern science. Then, what are irregular events? Where are they from? What are their effects? No matter whether it is about theory or methodology, these are important problems we need to address. To this end, we will have to deal with such a science that is more intriguing than modern science—the evolution science. Evolution science specifically deals with the physics of processes and changes. What is different from modern science is that in evolution science, the future is not the same as the present, and there is a process bridging the present to the future. On the other hand, in the form of the present existence, modern science addresses an imaginary eternal existence, where the present invariance guaranteed by the initial value stability can lead to the never-ending time until eternity (or infinity). This end, Laplace, a follower of Newton, very well described by claiming that "knowing the initial values, I can tell you everything about the future" (Kline, 1972). And, Einstein's realization that "none of the classical mechanics, quantum mechanics, relativity theory, and the basic mathematical physics equations can provide the difference between the past, the present, and the future" (Einstein, 1976) admits that modern science does not touch on materials or evolution of events. It can be said that modern science merely lists the laws of resultant to quantitative analysis (OuYang et. al., 2002) without dealing with processes or paths. For instance, conservation laws of energy do not tell us what the conservation process of energy is; additionally, they do not touch on the mechanism of energy transformations. So-called entropy (it is in fact a condition to limit the increase in the quantity of heat [OuYang et. al., 2002]) is also a stable function that has nothing to do with a specific path. Like the truthfulness of the statement that all people are always walking to their eventual death, the concept of entropy becomes useless, because what people care about is how we all walk to the eventual death—the process of dying. As for why people always walk to their eventual death, modern science does not provide any answer. Statistics, a branch of modern science, does not have any essential difference from the system of dynamics, except that experts in this branch themselves believe it is different. Even though in form indeterminacy is used, its concrete procedure is still pursuing after the regularized determinacy by ignoring small-probability information using stable data series, changing "the future = the present" dynamics to "the future = the past" without providing any differences among the past, present, and future. That is why Einstein was an old diehard against chances. Irregular information is varying; that is why there does not exist randomness in evolution science (OuYang and Lin, 2006). In other words, the concept of randomness is a product of the system of quantitative analysis.

Third, lack of evolution processes is the same as "the past = the present = the future." So, prediction becomes meaningless, and it is unnecessary to have the specific science and professionals to concentrate on how to make forecasting. So, the concept of determinacy is introduced with respect to problems of changes. In other words, modern science, established by Newton, Einstein, and others, did not touch on predictions and is a theoretical system useful for the design and manufacture of durable goods. The reason why this system is still in use is because people have confused monitoring (what is done currently in the area of weather forecasting) with prediction. That is why predicting a transitional change becomes a difficult problem for the profession of forecasting. Therefore, foretelling what is forthcoming is a problem facing and challenging modern science, while reminding people about how modern science should be correctly treated. Because major natural disasters occur during transitional changes, lack of understanding of these changes essentially means that there is no prediction science currently available.

Combining and including rotational motions with what is known is the key for us to reach the proposed evolution science. In other words, the central problem to the three aforementioned challenges to modern science is to understand rotations and solve problems of rotations. The reason why we list physical quantities and time as specific and individual topics is because we want to reconsider, in light of rotations, the physical quantities and parametric dimensions of modern science, what time is, and whether time can be treated as a physical quantity.

8.2 The Physics of Physical Quantities

It can be said that because of the established concepts of physical quantities, modern science constitutes a system. Without physical quantities, there would be no modern science. The essence of the physical quantities is to quantify events. As for whether any of the quantifications of events truly represents the events does not, as of this writing, seem to have anybody who cares.

For example, mass (quantity) means an attribute of a matter or material, indicating whether the attribute is good, bad, superior, or inferior, providing a sense of order. However, modern physics messes up this sense of order of quality and keeps essentially only the quantity. Mass is one of the elementary quantities employed in modern physics with the direction of materials' attributes excluded. This end can more or less explain the purpose of the physics of physical quantities. Even though direction is an important property or factor of materials or events, because it is not a quantity, this factor has to be disregarded in the study of physical quantities. From the very start, when physical quantities were initially introduced in physics, they purposely eliminated the fundamental functions of events.

It can be claimed that without mass there would not be modern science. The concept of mass can be seen as the doorway to enter modern physics. Before Galileo, some scholars mentioned the concept of mass. However, limited by the Western stream of consciousness of the time, because the concept was not clearly identified with quantities, it was seen as irregular or not possible to absorb. In 1678, in his book *Mathematical Principles of Natural Philosophy*, Newton was the first person that clearly stipulated mass as the measurement of a material's density and volume. The reason we say "the first ... clearly" is because Newton intelligently converted the word *quality* into a specified concept of quantities, creating the start for the quantified physics of physical quantities, introduced for events, factors, information, etc. This end also paved the way for the later convention that quantities can be seen as parametric dimensions, so that time and any parameter of physical quantities can be identified with material dimensions. So, it is not too much of an exaggeration to say that without mass, there would not be modern physics. In modern-day textbooks of physics, mass is seen as a (quantitative) measure of the amount of matter contained in a body, indicating at least that modern physics is a quantitative theory.

Besides, elementary laws of physics, established on physical quantities, do not point out where the quantitative force f and acceleration a are from. Speaking more clearly, modern science does not tell people where force f and acceleration a are located. Evidently, acceleration a has to be dwelling on the moving material m. So, without materials, there would be no acceleration. As for force f, according to Mo Zi of ancient China, it should be originated in materials' structures. That is, f also exists in materials. The dualism of Western philosophy with separate objects and forces is essentially the separation of quantities and materials. The purpose for such a separation is that quantities, space, time, and attributes of materials can all be independent of the materials and can control or influence the materials. It can also be said that human feelings exist independently outside human beings and materials, and the message delivered here is that feelings dictate materials.

Modern science, established on physical quantities, has a very clear purpose. Quantities can be employed as such general concepts that they do not care about the properties of the underlying materials and time, and the differences between states and functionalities. Now, the question is: What can be the consequence and problem for quantities to represent all characteristics of events? To illustrate this end, without loss of generality, let us look at Newton's second law as our example. Evidently, this law is a mathematical formula established on physical quantities:

$$f = ma \qquad (8.1)$$

Even going along with the stream of consciousness of the time when this formula was initially established, we know that f stays "heavenly" constant and m has to be invariant after Newton made the material a particle, which is under the slaving effect of God. So, acceleration a has to be a constant. However, in reality, the acceleration changes generally. So, Newton's second law in the inertial system should be written as

$$f = m\bar{a} \qquad (8.2)$$

That is, $\bar{a}$ can only be the average of the realistic acceleration. This end indicates that modern science of the inertial system is such that it possesses no individuality. Because of the assumption of average acceleration, it exactly means problems of motion on unchanging materials. Otherwise, if m is a variable, the acceleration a has to be a variable, too. When a changes, it means variable acceleration, which not only violates Newton's initial intention, but also goes against what modern science has admitted (OuYang, in press): variable acceleration implies a problem of a noninertial system. So, from a deduction of formal logic, it follows that if the acceleration varies over time, then the mass should be a function of time, too. That is, we have

$$f = m(t)a(t) \qquad (8.3)$$

Here, the concept of time t is also a quantified physical quantity and has been expanded to the idea of parametric dimensions. If we do not treat t as a physical quantity, not only will Equation (8.3) not hold true, but also we will have to deal with the problem of whether the entire modern science system holds.

In short, the foundation of modern science sits on motions of unchanging materials and the inertial system. On the other hand, the foundation of evolution science will consist of:

1. Transformations of changes of materials
2. Noninertial systems

So, what is clear now is that the essence of nonlinearity, movements of variable acceleration, or the noninertial system of the mathematically not closed Equation (8.1) are all about the problem of materials' changes caused by rotations of materials.

Besides, because the force f comes from materials' structure, and in evolution science materials' structures also change, Equation (8.1) is both a nonlinear equation and a mathematically nonclosed equation. So, what Newton's second law tells us is that the event or information under the average acceleration, without any of its own characteristics, only describes the movement of unchanging materials, and that evolution science involves specific events or characteristic information, where materials change harmonically without following the inertial system. This end also makes it clear that the physics, representative of physical quantities, is indeed a science without dealing with changes due to the generality of quantities. What constitutes the true fundamental science is exactly the evolution science, where changes in materials are addressed.

Now, we see that physical quantities have made physics, in particular the theoretical physics, into a formal logical calculus of quantities. What we need to notice is that materials or events do not in general follow the manipulation of the formal logical calculus. However, without physical quantities and parametric dimensions, there would not be modern physics, leaving behind the locality of quantitative physics. In other words, modern science has not entered the physics of materials or events. If prediction science is also seen as part of physics, then the resultant theory will be a physics of event processes.

8.3 The Nonquantification of Events

Continuing the formal logic of modern science, if the average acceleration describes the generality of events, then variable acceleration will represent specifics. So, if the generality described by the average acceleration reflects the invariance of materials, then the specifics represented by the variable acceleration will stand for the changeability of materials. So, events that change correspond to specific (or peculiar) information. Not only are the variations of events with time the objective existence, but also they reveal the evolution of materials. Hence, investigations on specific (peculiar) information will inevitably lead to evolution science. To this end, we are obligated to reconsider irregular events from the angle of materials changes. That is how we concluded that irregular information is that about changes, and that evolution science does not contain stochastics and randomness (OuYang and Lin, 2006).

There is not doubt that these results pound greatly on the belief of indeterminacy of modern science. However, if we carefully ponder over that events are not equal to quantities, then these conclusions seem to be basic knowledge, because no matter how small an event's probability is, it is a basic fact about occurrence or existence. In fact, the averages widely employed in modern science, without any constraint, are such a minor nonreality that they can be ignored more convincingly than small-probability events.

What is more important is that it is exactly because irregular information is so vital regarding changes that we need to reconsider whether or not physical quantities can truly describe events and their information.

8.3.1 Problems on the Physics of Physical Quantities

Our previous discussion implies that the direct consequence of introducing physical quantities is to change the physics of materials' laws into quantitative physics that is only about quantities. This quantitative physics inevitably has to dwell on mathematical physics equations and related mathematical tools. As is well known, the deterministic problem of mathematical physics equations has to be quantitatively well posed. So, the first problem met by quantitative physics is numerical instability. And in Euclidean spaces, quantities have to face the problem of unboundedness. So, physical quantities run into the difficulty of being unable to handle the problem of not-well-posed quantities. Because modern mathematics did not provide any method to deal with not-well-posedness, such a lack of methods has to keep the physics of physical quantities at the stage of unchanging materials.

Next, because the physics of physical quantities represents eternal existence, many laws in modern physics have to be stated as quantitative conclusion laws. For example, Newton's third law, the conservation of momentum, the conservation of mass, the conservation of energy, and the later conservation of mass-energy, etc., are all written in the quantitative form. None of these laws explain where the force and energy are from or where they are stored, in which form movements and interactions of materials take place, through what process energies are conserved, except the phenomena being described in terms of quantitative conclusions.

By now, it looks like the statement "that without assumptions there will not be any science" comes out of the system of modern science itself. That is, in scientific explorations, assumptions and hypotheses are allowed. However, each hypothesis has to be feasible in terms of the problems of concern. For example, Einstein's assumption that in a vacuum, the speed of light is constant, no matter how one looks at it, baffles people. First, this hypothesis does not clarify whether the speed of light is that of photons or the propagation speed of light in the medium (currently, *speed of light* means the propagation speed of light). If the speed of light means the speed of photons, then the physics conclusion that photons do not have rest mass implies that $m = 0$. If the speed of light stands for the propagation speed in the medium, then in the vacuum, there does not exist any medium. So, we must also have $m = 0$. So, Einstein's mass-energy formula has to be

$$E = mc^2 = 0 \tag{8.4}$$

That is, energy cannot exist independently outside the materials. What is more important is that if the speed of light c is constant, we have

$$\frac{dc}{dt} = 0 \tag{8.5}$$

According to Newton's second law, acceleration is equivalent to the acting force and energy is the work of the force. So, even if $m \neq 0$, Einstein's mass-energy formula describes such an energy E that does not do any work. However, Einstein neither mentioned the existence of such an energy that does not do any work nor claimed that Newton's second law is incorrect. Instead, he emphasized specifically that without the invention of the inertial system and the concepts of mass and forces, there would not be physics. It seems that masters of physics should not have made such a low-level mistake. However, extended applications of physical quantities have already made the physics of physical quantities difficult to follow. That is why it becomes difficult for anyone to tell what is really going on, including such a great mind as Einstein.

There is no need to deny that quantities are postevent formal measurements. Laws of the physics of physical quantities not only stay at the level of formality, but also appear after the occurrence of events. This end has essentially reached the fact that the physics laws of physical quantities can only be laws about conclusions without the ability to uncover the causes, processes, and changes of the events. For example, the law of conservation of energy does not provide any information about in which form and in what procedure the conservation is achieved. The conservation law of mass-energy does not explain the way in which mass is torn. Newton's third law only states that the quantities of the acting and reacting forces are equal without spelling out clearly the fashion and process in which the forces interact with each other and where the acting force is from. The law of universal gravitation does not answer why the universality exists and what gravitation is. Even without questioning Einstein's energy that does not do any work, we still have to wonder how the speed of light can be constant when light travels in a curvature space: How can irrotational Riemann geometry exist in a curvature space? In fact, a simple vector analysis can show that nonlinearity is the rotationality of solenoidal fields (OuYang, in press). Then, how can linear manifolds be introduced into nonlinear tensors? What is important is that the consequent theory of gravitational collapse suffers from the problem of inconsistency. According to Newton's theory, when an object collapses to a point—a singular point—with infinite density, one can directly observe from the outside, and the singular point can only be related to materials, while from Einstein's general relativity theory, the singular point has something to do with the geometric structure of space-time, the corresponding time needed for the celestial body to collapse, equaling the quantitative ∞, and it is impossible to detect from the outside the existence of such a singular point. Evidently, even if we do not pursue after whose theory is correct, we still have to notice that singular point and infinity ∞ come from the quantitative concepts of Euclidean spaces. And none of these theories provide a description on how a celestial body collapses and any details on the process of collapse. They do not answer why universal gravitation can be universal, and if the gravitation here is an attraction. In particular, what is the time structure of space-time? In other words, can time be identified with space? Collapse is contracting movement of materials, and gravitational waves are propagations of energy without any materials' movement.

If, 300 years ago, it was intelligent for humankind to establish the physics of physical quantities on the basis of numbers, it will be the mental deficiency of humans if we believe, on the basis of modern physics, that the properties, states, and functionalities of materials or events can be identified with numbers. It is because the material world cannot have zero, and physics does not investigate nothing. However, mathematics has to have zero; otherwise, the quantitative manipulation of the entire mathematics would be paralyzed. The so-called quantitative singularities and infinity ∞ are problems of transition of events. Because quantities cannot deal with quantitative unboundedness, the physics of physical quantities inevitably has to experience the irresolvable problems of quantities.

There is no doubt that the physics of physical quantities has to be constrained by whatever deficiencies of quantities. In particular, quantities can be used to compute objectively nonexisting matters more accurately than those that actually exist. Scholars often treat averages as the generality. However, the probability for an average value to appear is smaller than the chance for small-probability events to occur. So, in general, averages do not objectively exist. Besides, events do not follow the logical calculus of quantities, which implies that the physics of physical quantities has to suffer from the problem of not completely agreeing with the reality, including the possibility of pushing physics into the investigation of nothing.

8.3.2 Nonquantification of Events

Over the years, we have studied such questions as: Where are quantities from? What are the cause and process of changes in quantities? Are quantities the only method for us to know the world? Why are quantities postevent formal measurements? Our affirmative answer to the last question no doubt indicates that from anything, quantities can be collected, that quantities cannot become the cause of any natural event, and that since the first day when the laws of the physics of physical quantities were established, they were laws without any explanation about the causes and processes. That is why the physics of physical quantities cannot foretell what is forthcoming, or quantities cannot predict. The formality of quantities surely represents the generality of numbers without being able to describe the specific changes and processes of events' properties, states, and functionalities. As soon as the book *Entering the Era of Irregularity* (OuYang et al., 2002a) was published, many readers and scholars highly appraised such conclusions as "quantities are post-event formal measurements."

What is different from the past is that it is now a well-known fact that quantifications can always experience irregular events that cannot be handled by using only quantities. So, it is natural to ask: What is an irregular event? To address this problem, one has to reconsider what information is.

If we understand information as symbols of events, then information will be names or labels of the events, where the events are objectively existent, while the information must exist only subjectively. The events represent a deterministic concept. Here, quantum events are a concept beyond the quantitative mechanics and naturally not the probabilistic wave of accidental events (OuYang and Lin, 2006, in press b). Quantities not only appear after events, but can also make the events disappear in the form of quantities. Even as a generality, averages are only approximations (realistic averages are also roving [OuYang and Lin, 2006]). Although events can be labeled using quantities, this does not mean that events follow the formal logic calculus of quantities. For example, integrals can be manipulated without paying much attention to the specific used paths. But that does not mean that the underlying events have nothing to do with the paths, along which they evolved. The counts of people can take the sequence of 1, 2, 3, ...; however, the corresponding numbers cannot be seen as the individuals being counted. The counts can be logically manipulated using the operations of addition, subtraction, multiplication, division, etc. But the specific people do not follow the rules of these operations. All this should be common knowledge. So, the physics of physical quantities must contain logical contradictions against some elementary properties of events. Events do possess the attributes of quantities. However, what is more essential are their properties, states, and functionalities, which cannot be well revealed by the formal logical manipulation of quantities. If quantitative averages can approximately describe generality, then not much, if any, quantities can do for specifics or peculiarities. Hence, quantities are not events, and the physics laws of physical quantities cannot substitute for the physical laws of events. Accordingly, we should establish a direct method of analysis for events. Characteristics of events are in structural properties. That is why the analysis method for events is called a structural method, leading to our structural predictions (OuYang, 1998; OuYang et al., 2002a). However, in this method, time cannot be identified with a parametric dimension of the quantities.

It seems that the understanding of the concept of time should be a basic sign of the maturity of science. Without the problem of time resolved, science essentially can only tell us what seems to be true. As an unsettled proposition in the history of science, we believe that the concept of time is a problem of physics, but more so of philosophy. To this end, we now turn to address what time is.

8.4 What Time Is
8.4.1 The Problem of Time

If we use a quantity to represent time, then we can see that time is the only physical quantity that appears everywhere, and nobody can escape the "punishment of time." At the same time, it is such a "thing" that is invisible and untouchable, leading naturally to the question: Where is time? Even so, the problem of time is quite special. If no one asks what time is, people do not really feel what it is. However, as soon as the question is posed, people suddenly discover that they do not know what to say and how to feel. In the past 300-plus years, almost all people feel that science and technology have accomplished extremely magnificent achievements, so that we can fly high to touch the moon and dive deep to tour the ocean floor. However, humankind has established such magnificent science and technology and yet still cannot separate the past, the present, and the future, and comprehend why events do not occur at the same time. The question of what time is seems to be easy, but we cannot handily find the answer. Worse, we do not even know where we should start to look for the answer.

8.4.2 Time in China

In the Chinese classic *The Book of Changes*, we find that the question of what time is has been answered unexpectedly. For example, there are at least two places in that book that mention the concept of time. One is hexagram 1 (quian)—the creative, and the other is hexagram 41 (sun)—decrease. In fact, hexagram 41 has already very clearly answered what time is. That is,

$$损益盈虚，与时偕行$$

which is translated by Wilhelm and Baynes (1967) as "decrease does not under all circumstances mean something bad, increase and decrease come in their own time." Translating this sentence to our modern language, we have that "time is reflected in the state of changes in materials' movements." The clearest implication is that time is not material and does not occupy any material dimension, or time cannot be identified with space. As for indirect mention of time, the first hexagram (quian) of *The Book of Changes* states that

$$天行健，君子自强不息$$

which is translated by Wilhelm and Baynes (1967) as "the movement of heaven is full of power, thus the superior man makes himself strong and unstirring." This sentence should have implicitly contained the concept of time. It is because the first three characters can be directly translated into "materials do not get destroyed," and the rest into "appear to be vividly alive in the changes of movement so that people can only try their best along with the changes in materials."

Next, although Lao Tzu, the great sage who has influenced people's thinking and intelligence for over 2,000 years, did not directly talk about time, if we attentively taste the first part of Chapter 25 of *Tao De Ching*, we can see that the concepts of time and space are essentially included. Here is English and Feng's (1972) translation: "Something mysteriously formed, born before heaven and Earth. In the silence and void, standing alone and unchanging, ever present and in motion. Perhaps it is the mother of ten thousand things." The Chinese original mentions "mixed with materials," which is translated into "something mysteriously … born" above. Whatever is mixed with materials, it must be a consequence of some stirs, which inevitably lead to rotations. That implies that the universe is not only curved, but also rotational. However, more than 2,000 years later, Einstein still employed irrotational Riemann geometry to study the universe. The corresponding "standing alone and unchanging, ever present and in motion" undoubtedly means that materials rotate endlessly, implicitly implying the passing of ages of time. That is why we have: time is from materials' rotational movements.

Besides, as of today, people are still using the term *universe* (*yu zhou*). This term was initially used by Shi Zheng in his book *Shi Zi*, written during the time period of warring states. We should be clear that the Chinese term *universe* has different meanings than the English words *cosmos* and *universe*. The Chinese *universe* (*yu zhou*) means "space" (*yu*) and "time" (*zhou*) together, while the English *cosmos* and *universe* only mean "space." The first character, *yu*, specifies materials occupying all imaginable fields in all directions; the second character, *zhou*, stands precisely for "coming from the past and heading into the future," representing the changes in materials' rotational motions and the sense of before and after existing in materials' changes, explaining why events cannot occur at the same time moment. However, *yu* comes before *zhou* and *zhou* follows after *yu*, indicating that space of the fields occupied by the materials is primary, while time does not occupy any material dimension. In the later time period during the warring states, there appeared another scholar, named Zhuang Zi. He introduced *yu shi ju hua*, where *shi* means "time," emphasizing more on materials' changes over time. He specified that no matter or event can stay unchanged forever, and all things change through the difference between the before and the after, so that the concept of time was introduced. Evidently, if materials did not possess any disparity in location and change in their movements, then time would become meaningless and it would no longer be necessary for us to treat it as a concept or proposition. Later in the Tang Dynasty, Li Bai, a great scholar and poet, more clearly pointed out that time does not occupy any material dimension, leading to the conclusion that time is a science in the culture.

In short, time in China can be summarized being reflected in the disparity existing between the before and the after experienced in the movements of materials or events. However, time itself is not a material.

What needs to be pointed out is that Mo Zi, a contemporary of Shi Zheng, also posed such a concept of *jiu yu*, where *jiu* stands for "time." What is different here is that later scholars in China mostly cited Shi Zi's *universe* so that *jiu yu* was washed away in the history. As for the words *shi jie*, which are still widely used in modern China, the word *shi* represents "time," while *jie* means "space." However, the currently accepted meaning of *shi jie*—"world"—practically contains space only.

8.4.3 Time in the West

The concept of time in the West can be traced back to Plato. He believed that the prototype of natural laws comes from absolutely static figures, which were both consciously and unconsciously inherited later by Newton and Einstein, causing modern science to evolve away from the investigation of materials' changes. Galileo's *Conservation between Two New Sciences* (1638) specified the physical time, in which he used "evenness and continuity" that foreshadowed the later development of the concept of time. Thirty years later, Isaac Barrow introduced quantitative time, treating the concept of time as a geometric straight line with only quantitative length. The version of time can be seen as the repeated summation of a series of time moments and also the flow of a time moment. It can also be represented as a circular ring. Newton continued Barrow's quantitative time and expanded it to absolute time and space. Just as what is stated at the start of his *Mathematical Principle of Natural Philosophy* (Newton, 2008), time is defined as follows: Absolute, realistic mathematical time; in terms of its properties and essence, it flows evenly forever without depending on any external matter. Evidently, in the understandings of time, the east and west are substantially different.

There is no doubt that the quantifications of time and space have something do to with the dualistic system of Western philosophical recognition. Not only can acting forces exist independently outside of materials, but also time and various physical quantities, introduced later, can float independently external to materials. What is worth our attention is why Newton started his *Mathematical Principle of Natural Philosophy* by first defining quantitative time. It seems that his aim was the consequent physics of physical quantities instead of natural philosophy. In other words, the introduction of the concepts of absolute time and space existing independently outside of materials is to meet the needs of the classical mechanical system. If time and space were not quantitative, Newton would then not be able to establish the mechanic system of particles. It should be clear that in the system of modern science, time is very important. It can be said that if time were not quantitative, modern science would be essentially ruined. So, it can be seen that time is an important proposition of science.

The quantifications of time and space not only confuse time with space in the history of science, but also lead to the concepts of parametric dimensions and physical quantities. Time becomes the fourth dimension, and physics the theory of physical quantities. Speaking more clearly, mathematics was originally a tool for the study of physics; with the introduction of parametric dimensions and physical quantities, mathematical principles dominated all physics laws, as evidenced by the later mathematical equations. In particular, the reason why Einstein identified time with space and treated three-dimensional evolution problems as four-dimensional existences no doubt is from time being seen as a material dimension in the form of a parametric dimension. So, the modern science of the inertial system constructs the classical mechanics on the basis of the quantitativeness of parametric dimensions and physical quantities, which is called the foundation of modern science. Evidently, the central idea is to pull out the physical properties and treat principles of quantities as physics or the laws of myriad things. Because of this, weaknesses and deficiencies of the quantitative principles will have to be brought into physics and become weaknesses and deficiencies of physics. Among all the greatest weaknesses is that quantities cannot handle transformations of eddy currents corresponding to the quantitative unboundedness. Therefore, we must provide information on the range of validity for modern science so that people can clearly see the boundary of those problems solvable by using modern science.

What should be known is that as soon as Newton's absolute time and space were publicized, he was severely criticized by his contemporary Leibniz, who pointed out that events are the essence. This opinion possesses not only the revolutionality of concepts, but also is ahead of Einstein in terms of their understandings of the problem of time. Even though Einstein also joined those who criticized Newton's absolute time and space, other than proposing curved spaces, he did not really stay out of the quantitativeness of time and did not provide any explanation on why the space is curved. Instead, he treated changes in materials as four-dimensional existences so that time is completely identified to space. On the basis of all these, Einstein (1997) believed that time is a character of human consciousness, becoming an illusion of man. That is, he had traveled farther away from the truth than Newton.

What modern science has not noticed is that even if we record time based on the Earth's rotation, time is also a variable of itself, which changes along with the changes in the Earth's rotation. Currently, the 24-hour day on Earth is partitioned into 86,400 portions, called seconds, which is only an approximation of the Earth's rotation. Over 600 million years ago, each day on Earth was shorter than 21 current hours. That is, the same time length of a day or a second changes through the ages.

To repudiate that time has an "arrow," in his late years Einstein (1997) had to admit that modern science, including his relativity theory, could not tell the difference between the past, the present, and the future. This end implies that he also recognized that time in modern science is only a parameter, and has nothing to do with time. This end explains why modern science is still standing outside the door of evolution science.

When the system of modern science extols its glories of the past 300-plus years, it seems that people have forgotten the unsettled problem of time, which makes the foundation of modern science shaky and leaves many problems unsolvable in modern science. For example, is time reversible? Why can different laws of physics use different scales of time? The law of radioactive decay uses the time measured by uranium 238 with half-life = 4.5 billion years. Does this time have the same meaning as those used in Newton's kinetics and the laws of gravitation? As of this writing, no answer has been obtained for this question. Even worse, people do not even know how to secure such an answer. These and many other fundamental problems directly lead to the problem of scientificality of modern science.

Even though in the 1980s, Prigogine maintained that time is irreversible and comes ahead of existence, he still did not correct the separation between physical quantities and materials (Prigogine, 1980).

That is, in the West, time is identified with quantities. Even though Leibniz pointed out that events are the essence, an opinion of a lofty realm, he did not have the power and influence to alter the Western stream of consciousness of the ages and consequent methods.

8.4.4 What Time Is

From the history of learning the concept of time, as summarized above, according to the knowledge behavior of modern humans, we can see at least the following attributes of time that are different from those of space:

1. Time comes from the changes existing in the rotational movements of materials. So, it reveals the differences among events' past, present, and future. Time directly shows the variance in properties, states, locations, and functionalities of events in the time order. No matter whether the relevant quantity changes, people can directly tell or distinguish materials' or events' past, present, and future based on the changes in properties, states, locations, and functionalities. And such essential problems as the properties, states, and functionalities of materials or events are exactly the areas modern science has not entered.

2. Time dwells in the rotational movements of materials so that it cannot exist independently outside the materials. Time is an attribute of materials so that it does not occupy any material dimension. In other words, as an independent concept, time is not a material and cannot independently exist or change without being associated with certain material. It is also a function of time itself. Because time comes from changes in materials' rotational movements, time also varies. That is, both time and materials' movements appear at the same time, which is the most important attribute of materials that modern science has missed. If the materials' attribute of time means that time is parasitic on materials, so that as long as there is material, the material will have to move, and as long as the material's movement exists, there will be time, then time cannot exist independently outside of materials, as described by Newton, Einstein, or Prigogine. Or we conclude that time cannot exist before existence.

3. At the same time when time possesses the attribute of direction, it also possesses the attribute of quantitativeness, where the direction points to only forward and cannot be reversed. Even when the spinning direction of materials is reversed, time still moves forward, together with the reversed materials' rotation, so that past events will never reappear again. Since after April 1955, even if we took a rocket flying at a speed faster than that of light, we still would not be able to travel through the time tunnel to have breakfast with

Einstein. Although this attribute of time was employed by Prigogine to introduce the third development period of science, Prigogine only imagined the torn time on top of the system of linearity—there is no negative time—and proposed his linear "arrowhead" using the dynamicality of time choice of space to substitute for Newton's staticality and Einstein's existence. However, Prigogine's time still exists independently outside of materials, leading to his time ahead of existence. Because modern physics, consisting of quantum mechanics, relativity theory, and others, does not possess any substantial difference from Newton's system of classical mechanics, these modern theories still maintain the eternal invariance of materials. But it seems that now is the time for science to rise to the true second or higher stages of development—the evolution science emphasizing changes in materials' movements. If we use what time is as the mark, then the directionality of time has made time into a vector, constituting a challenge to the entire system of science.

4. Time possesses the attribute of connecting events and shows the process differences between events' past, present, and future. That is, each material has its origin and destination, and its causal relationship and why events cannot occur at the same time. All these realizations and others form the basics of evolution science. Because materials' evolution does not get directly shown with quantities, materials' aging is exactly embedded in the materials' properties, states, and functionalities. Not only is modern science not deep enough to investigate materials' properties, states, and functionalities, but it also treats the relevant problems as quasi-problems, excluded from the Newtonian system (Bergson, 1963).

5. In terms of time itself, the concept does not possess any physical entity. Together with the multiplicity of materials, time's attributes of direction and quantitativeness also show their multiplicities. That is why Earthly time is different from Venus time; Norwegians are annoyed by their own "long life spans," so that, as of recently, they still keep the custom of "self-termination"; it might be due to the reason that they do not like to exist too long—bacteria reproduce rapidly. The corresponding "too long life span" and "too long existence" are not on the same timescale. So, time can neither be unified using the speed of the Earth's rotation nor follow the wish of humans to flow along the quantitative straight line or the geometric circular ring. Because of the differences in movement directions, changes in materials' structures appear, which can be recognized through the before-and-after differences in the structural properties, states, and functionalities of materials. So, the quantitative attribute of time can only be visible after instead of before materials' aging. Hence, quantitative time is merely the postchange results of materials or events.

6. Just because quantities are postevent formal measurements, Leibniz's belief that events are the essence is from a highly lofty realm of knowledge, which should be the pride of the Western scientific community. If scientific history

had gone along in the direction pointed out by Leibniz, evolution science would have appeared over 200 years ago, without having to fall into the current situation where modern science can only describe movements that have nothing to do with time or, speaking more oddly, the movements along the axis of time without time (Koyre, 1968).

Based on our discussion above, combined with our experiences in practice, we obtain the following conclusions:

- Time is originated from the rotations of materials' movements. That is why time shows the directionality and orderliness of events. However, time is not a material and cannot occupy any physical dimension. So, time cannot be identified with space. Or, according to the definition of physical dimensions, time cannot be treated as a coordinate dimension of materials or the parametric dimension of physical quantities.
- Time possesses the attributes of direction and quantity of rotation. However, these attributes are not an attribute of time itself. Instead, they are from the underlying materials or parasitic on the materials. Together with the changeability and multiplicity of the materials' movements, time also possesses the attributes of changeability and multiplicity.
- Each rotational direction presses the materials to age so that the materials' evolution plays out naturally.

By rotational directions, we mean un-uniformities in the directions of materials' rotations, leading to damages or breakups, caused by collisions, in materials' properties, states, or functionalities, and maybe damages or breakups in the materials' structures. That is why rotational directions press materials to age. The reason why we emphasize directions is that directions are not quantities but structures, and that in the study of evolution problems, directions are more important than quantities. Directionality appears ahead of the development of events and can be observed with less error than quantities. So, it can be a piece of useful information for foretelling what is forthcoming. This discovery has been used to make predictions of disastrous natural events. The outcome of practical predictions has been exceptional and way beyond our initial expectation. For details, consult OuYang (1998), OuYang et al. (2005), Zeng and Lin (2006), or later chapters in this book.

8.5 Material and Quantitative Parametric Dimensions

The best way to validate a theory is to see whether the conclusions of the theory are supported by practical applications. By a material dimension, we mean a component of a spatial decomposition of the field occupied by materials. Even though each material dimension can be labeled using quantities, its essence is not the same as quantities. On the other hand, for each quantitative parametric dimension, it is first quantitative, which does not stand for any independent, realistic entity. For example, the time variable, one of the earliest parametric dimensions employed in modern science, is not a concept of any realistic entity, but an attribute of some objective materials. It dwells on materials and is not a pure quantity, which makes it different from pure quantitative physical quantities. Besides, many physical quantities used in modern science are essentially not concepts of objective materials. Einstein identified time with space, which fundamentally confused the difference between material dimensions and parametric dimensions. It can be said that the dimensions of objective materials in Euclidean spaces are four-dimensional (in non-Euclidean spaces, they are more than four-dimensional), because the existence of any static entity needs support. So, each four-dimensional existence means that in the static state, there is a fourth supporting dimension. However, this supporting dimension is also an objective entity and cannot be identified with time of non-objective entities. So, our four-dimensional existence is different from Einstein's four-dimensionality. For example, in classical mechanics, force, acceleration, energy, etc., are attributes of materials. Because they exist on top of the existence of materials, they do not occupy any material dimension either. If we see physics as the knowledge of the principles of materials' transformations, then these principles of transformations cannot be identified with principles of physical quantities either. As for the physics of the principles on materials' transformations, it will have to produce such physicists that can foretell the future. In other words, in such a new era of learning, if one cannot predict what is forthcoming, he or she cannot be called a true physicist.

Now, we must mention the fact that in modern physics, the currently fashionable behavior is to identify events or information with physical quantities. That is, physical quantities are used to substitute for events. Of course, quantities have their own rules of logical manipulation. However, whether or not events or information follows these rules, modern science does not provide any affirmative or otherwise theoretical proof or experimental testing. Just like what is well known, quantities can only provide the generality without pointing to specifics. So, if events or information do not follow the rules of quantitative manipulations, which seems to be so, just as our practical experience suggests, then the physics of physical quantities cannot be identified with the physics of the principles of materials' transformations. For example, the chaos doctrine, which was once a hot topic of discussion in the latter half of the 20th century, was started off on the Lorenz chaos model, which was obtained from expansions of parametric dimensions. This model has nothing to do with the materials' dimensions of Salzmann's convection model (OuYang and Wu, 1998). In particular, a same physical quantity about the atmospheric stratification is separated into two parametric dimensions using both linear and nonlinear strata, respectively. This example shows how the physics of physical quantities has employed parametric dimensions to such a degree that it no longer has anything to do with the underlying physical problem, indicating confusion between physical quantities and parameters. Evidently, although the model established in this fashion experiences chaos, it is not the same as the original convection model standing for chaos. Besides, in 20th-century science, the problem of computational inaccuracy of quantities was even being seen as a scientific theory. So, now should be the time for us to consider the problem of how to make scientific investigations satisfying the scientificality. In the applications of parametric dimensions, it has been developed to such an abusive stage that physics is led to the study of nonmaterial fabrications, imagined spaces, and nonexistent time. So, one has to ask: From the start, has physics ever investigated existent matter?

That is why we used to name the physics of physical quantities a school of the physics of the principles of materials' transformations, because material dimensions cannot be mixed up with parametric dimensions. One of the reasons is that quantities cannot handle irregular events, which inevitably leads to the thought that quantitative averages cannot be used to substitute for the generality of events either. Also, in terms of events themselves, they should have their own direct method, which should be different from the quantitative formal logical analysis and should be able to deal with quantitatively unsolvable problems (OuYang et al., 2002a, 2005; OuYang, 1998).

8.6 Some Final Words

The development of science comes from the call of unsettled problems. The system of modern science also openly admits that it did not solve the problem of time. In particular, after parametric dimensions and physical quantities were introduced into modern science, quantities have been essentially employed to substitute for materials and events. In the last 200 to 300 years, no one seems to have seriously studied whether materials and events are truly identical to quantities, although the scientific community has already experienced the difficulty of being unable to deal with irregular events using quantities. Even so, still not enough attention is directed to attack this difficulty. The reason why we have placed so much emphasis on this problem is because of the tragic aftermaths of major natural disasters. To practically forecast these major disasters, we have to take risks to try out different methods and theories. In our practice of actually forecasting natural disasters, what we first discovered is that irregular information cannot be replaced by quantities and does not follow the manipulation rules of the quantitative logical calculus. Next, we discovered that quantitative averages are not the same as the generality of events, and they are only cognitive wishes of man. Exploring along these lines of thinking, we gradually found that the parametric dimensions of modern science suffer from various problems. For example, although the heat of the chaos doctrine is already cooled, all the corresponding quantitative instabilities, complexities, etc., as studied in the chaos theory, are originated from the fact that quantities cannot handle irregular events and information and directly or indirectly have something to do with parametric dimensions. In particular, one of the reasons why chaos becomes a theory is because of the applications of parametric dimensions (OuYang and Wu, 1998).

To meet the practical challenge of applications, over the past decades we have tested our method of nonquantitative, digitalized structure analysis (event analysis) (OuYang, 1998; OuYang et al., 2005; Zeng and Lin, 2006). Our study indicates that irregular information is the specific varied information caused by changes in the underlying events. The existence of irregular information directly shows that quantities and physical quantities cannot deal with rotational movements and transformations or changes in materials.

We should recognize that "events are not the same as quantities" is indeed pretty elementary common sense. However, people have been fighting to climb on top of "the giants' shoulders" without noticing that the "giants" are standing on a sand dune and can no longer support any forthcoming giants on their shoulders. That is why the giants have to find some more solid foundation!

ECONOMIC AND FINANCIAL FORCES 3

Chapter 9
The Economic Yoyo

In this chapter, we learn how the systemic yoyo model (Figure 1.1) and the new figurative analysis method (Chapter 4) can be applied not only to the study of the laws of motion, astronomy, and the three-body problem (Chapters 4 to 6), where Newton's laws have been historically considered one of the main reasons why physics is an exact science, but also equally to such inexact studies as economics, which is a part of social sciences. Even though situations studied in this branch of social sciences are fundamentally different from those considered in natural sciences, because humans are involved in each economic situation and their desires alter the evolution of the outcome, leading to unpredictable, chaotic consequences (Soros, 1998), what is presented in this part of the book shows that when each economic entity is seen as a rotating yoyo with a spin field around it, this fundamental difference seems to disappear and the seemingly unpredictable, chaotic consequences of human desires and corresponding behaviors no longer look unpredictable and chaotic.

What is new in this part of the book from the mainstream literature in economics is that we learn how various problems considered in economics can be seen as problems of (general) systems and spinning yoyos in the light of whole evolution of these systems (Lin, 1999). Thus, it is shown that in a market of free competition, the concept of demand and supply is in fact about how different economic forces mutually restrict each other and mutually react on each other. Because of this, we lay down the theoretical foundation for the introduction of the systemic yoyo model and its methodology into the research of economic entities, organizations, relationships, and evolutions.

As an elementary example, we apply such a yoyo structure to model the evolution of competitions between economic sectors and individual enterprises from the moment when a sector or an enterprise is born to the time when it is gone. By looking at economic competitions as interactions of several or many spinning yoyos in the light of evolution, we are able to readily employ the methods and results discovered and well developed in fluid dynamics to the research of economics and other branches of social science. The materials in Chapters 9 to 11 are based on Lin and Forrest (2008a, 2008b, 2008c).

9.1 Whole Evolution Analysis of Demand and Supply

In a market of free competition, the price P of a consumer good is closely related to the demand D and the supply S. Assume that the changes in P are directly proportional to the difference of the demand and supply:

$$\frac{dP}{dt} = k(D-S), \ k > 0 \tag{9.1}$$

where t stands for time and k is a constant. With all other variables fixed, assume that both the demand D and the supply S are functions of the price P:

$$D = D(P) \text{ and } S = S(P) \tag{9.2}$$

The relationship between the demand D and the price P is generally linear. We can write

$$D(P) = -\lambda P + \beta, \ \lambda, \beta > 0 \tag{9.3}$$

where λ is the rate of change of the demand with respect to the price P and β is the saturation constant of the demand. Because the relationship between the supply S and the price P is generally nonlinear, we can write this symbolically as (at least by means of local approximation)

$$S(P) = \delta + \alpha P + \gamma P^2 \tag{9.4}$$

where $\delta > 0$ is a constant and α and γ are, respectively, the linear and nonlinear intensities of the supply, satisfying $\alpha > 0$ and $\gamma < 0$.

Substituting Equations (9.3) and (9.4) into Equation (9.1) produces

$$\frac{dP}{dt} = AP^2 + BP + C \tag{9.5}$$

where $A = -k\gamma > 0$, $B = -k(\lambda + \alpha) < 0$, and $C = k(\beta - \delta) < 0$.

For Equation (9.1) or (9.5), the majority of the literature focuses on the study of the price stability at the demand-supply equilibrium, while ignoring the whole evolutionary characteristics of the price. Because the price evolution model

(Equation (9.5)) is quadratic, the price in general changes discontinuously and has the characteristic of singular reversal transitions. In particular, let the discriminant of Equation (9.5) be

$$\Delta = B^2 - 4AC = k^2\{\gamma^2 + 4(\lambda + \alpha)(\beta - \delta)\}$$

then we have three possibilities: $\Delta = 0$, $\Delta > 0$, and $\Delta < 0$. In the following, let us analyze the evolution model in each of these cases.

In case 1, when $\Delta = 0$, at the equilibrium there is only one equilibrium price, $P_1 = -\frac{1}{2}\frac{B}{A} > 0$. The characteristics of the whole evolution of the price are described by

$$P = -\frac{1}{2}\frac{B}{A} - \frac{1}{P_0 + At} \qquad (9.6)$$

where P_0 is the integration constant determined by the given initial condition. If $P_0 > 0$, the price P decreases continuously with time and approaches the equilibrium state $(-\frac{1}{2}\frac{B}{A})$. If $P_0 < 0$, then when $t = t_b = -\frac{P_0}{A}$, a discontinuity occurs to the price change. When $t < t_b$, the greater the demand, the higher the price. When $t = t_b$, the price falls abruptly, indicating that either the demand reaches its level of saturation or the supply increases drastically, causing the abrupt drop in the price. When $t > t_b$, the price starts to rebound continuously with time and eventually approaches the demand-supply equilibrium P_1. This process indicates that through the market adjustment, the price is no longer growing as blindly as during the first period of time. Instead, by taking the market situation into consideration, to keep a reasonable equilibrium between the supply and demand, the price eventually stabilizes within a certain range.

This whole evolution analysis of the price is not only more complete than the stability analysis at the equilibrium, but also more practically realistic than studies of continuous evolutions when compared to the objective situations existing in the marketplace of free competitions.

In case 2, when $\Delta > 0$, solving Equation (9.5) leads to

$$\left|\frac{P + \frac{1}{2}\frac{B}{A} - \frac{1}{2}\sqrt{\frac{B^2}{A^2} - \frac{4C}{A}}}{P + \frac{1}{2}\frac{B}{A} + \frac{1}{2}\sqrt{\frac{B^2}{A^2} - \frac{4C}{A}}}\right| = \exp\left(A\sqrt{\frac{B^2}{A^2} - \frac{4C}{A}}\, t + P_{20}\right) \qquad (9.7)$$

where P_{20} is the integration constant. If $\left|P + \frac{1}{2}\frac{B}{A}\right| < \frac{1}{2}\sqrt{\frac{B^2}{A^2} - \frac{4C}{A}}$, then Equation (9.7) indicates that changes in price are continuous. Otherwise, Equation (9.7) indicates that the price evolution contains blown-ups.

In case 3, when $\Delta < 0$, solving Equation (9.5) produces

$$P = \frac{1}{2}\sqrt{\frac{4C}{A} - \frac{B^2}{A^2}}\, \tan\left(\frac{A}{2}\sqrt{\frac{4C}{A} - \frac{B^2}{A^2}}\, t + \frac{1}{2}P_{30}\right) - \frac{1}{2}\frac{B}{A} \qquad (9.8)$$

where P_{30} is the integration constant. This equation implies that at

$$t = t_b = \frac{2}{\sqrt{\frac{4C}{A} - \frac{B^2}{A^2}}} \left(\frac{\pi}{2} + n\pi - P_{30} \right), n = 0, \pm 1, \pm 2, \ldots$$

periodic blown-ups in the price occur. That is, the price of the consumer good shows the behavior of periodic rise and fall.

This simple analysis shows that the concept of demand and supply is in fact about mutual restrictions or mutual reactions of different forces under equal quantitative effects. (The concept of equal quantitative effects was initially introduced by OuYang [1994] in his study of fluid motions. And later, it was used to represent the fundamental and universal characteristics of all materials' movements [Lin, 1998]. By *equal quantitative effects* is meant the eddy effects with nonuniform vortical vectorities existing naturally in systems of equal quantitative movements, due to the unevenness existing in the structures of materials. In this definition, by *equal quantitative movements* is meant such movements that quasi-equal acting and reacting objects are involved, or two or more quasi-equal mutual constraints are concerned. For example, in a principal-agent problem, the principal and the agent can be seen as quasi-equal parties. Each interaction between them can be seen as an equal quantitative movement. The effect of such a movement can be seen as an equal quantitative effect.) Lin (2007) shows that if Einstein's concept of uneven time and space of materials' evolution (Einstein, 1997) is employed, Newton's second law of motion actually indicates that a force, acting on an object, is in fact the attraction or gravitation from the acting object. It is created within the acting object by the unevenness of its internal structure. So, as soon as calculus-based mathematics is employed to the study of such a relationship as that between demand and supply, one has to face the problem of nonlinearity, which is an unsettled mathematics problem. With the introduction of blown-up theory (Wu and Lin, 2002), it has been shown that mathematical nonlinearity is about structures of the entities involved. So, to understand nonlinearity, the quantitative, formal analysis needs to be modified accordingly. To overcome the difficulty encountered by the traditional mathematical methods in the area of weather forecasting, OuYang (Wu and Lin, 2002) proposed the idea of blown-ups and a practical procedure based on the vorticity of materials to resolve evolution problems. In particular, his method is developed on the basis of classifications of materials' structures according to their vectorities and has been proven effective in forecasting zero-probability, disastrous weather conditions. For more details to this end, please consult Part 5 of this book.

On the basis of OuYang's and his students' successes in weather forecasting and the relevant intensive theoretical analysis (Lin, 1998), Wu and Lin (2002) introduced the following systemic yoyo model for each object and every system imaginable. In particular, this model says that each system or object considered in a study is a multidimensional entity that spins about its invisible axis. If we fathom such a spinning entity in our three-dimensional space, we have a structure as shown in Figure 1.1. The black hole side sucks in all things, such as materials, information, investment capital, human resources, etc. After funneling through the short, narrow neck, all things are spit out in the form of a Big Bang. Some of the materials, spit out from the end of the Big Bang, never return to the other side, and some will. Such a structure, as shown in Figure 1.1, is called a (Chinese) yoyo due to its general shape. More specifically, what this model says is that each physical entity in the universe, be it a tangible or intangible object, a living being, an organization, a culture, an economy, etc., can be seen as a kind of realization of a certain multidimensional spinning yoyo with an invisible spin field around it. It stays in a constant spinning motion as depicted in Figure 1.1. If it does stop its spinning, it will no longer exist as an identifiable system and all the orbiting materials will be absorbed by other spinning yoyos.

Experimental evidence for the existence of such a model is provided in Lin (2007). Empirical studies of such a structure are given by Ren et al. (1998). The theoretical justification for this model is OuYang's theory of blown-ups (Wu and Lin, 2002). The mathematical foundation of this yoyo model is Bjerknes' circulation theorem (Hess, 1959). Some initial successful applications of this model in physics, astronomy, and the famous three-body problem are given in Lin (2007) and in Chapters 4 to 6 of this book. For our current purpose, let us see how such a structure can be employed to model the evolution of competitions between economic sectors and individual enterprises.

9.2 The Yoyo Evolution of an Economic Cycle

For the sake of convenience of presentation, we will only look at the general situation for a new economic sector to appear and evolve. To do so, let us see each existing economic sector or an individual company as a spinning yoyo, where the sector's products or services are the materials spit out of the Big Bang side, and the profits and other economic benefits the materials sucked into the yoyo from the black hole side. When several yoyos coexist side by side (Figure 9.1), their mutual interactions lead to the appearance of new products and new services, the subeddies in Figure 9.1. To cash in on the profit opportunities coming along with the new products and services, a new economic sector or new companies are born. At this stage of evolution, the technology of producing the new products is generally quite inefficient, because it is a preliminary combination of the technology and equipment from various existing sectors, where the idea of the new products were initially based. So, the cost of production is quite high. Also, because these new products/services have no share and are relatively unknown in the marketplace, the demand is very small. Consequently, the development in terms of mass production is slow. Even though the spinning of the new yoyo seems to be slow, the gradually increased market awareness and demand indicate that a new technology needs to be introduced. And the increasing spinning strength of the yoyo provides the adequate capital for such a needed technological innovation.

150 ◾ *Systemic Yoyos: Some Impacts of the Second Dimension*

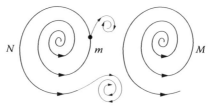

(a) Object *m* is located in a diverging eddy
and pulled by a converging eddy *M*

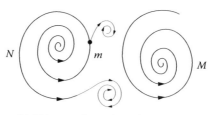

(b) Object *m* is located in a diverging eddy and
pulled or pushed by a diverging eddy *M*

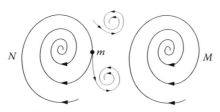

(c) Object *m* is located in a converging eddy
and pulled by a converging eddy *M*

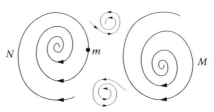

(d) Object *m* is located in a converging eddy and
pulled or pushed by a diverging eddy *M*

Figure 9.1 Some ways small yoyo structures can possibly be created.

In the second stage of evolution, trend-following companies start to pop up at a high speed, similar to the situation of when a calm pond of water is initially disturbed: instead of a big whirlpool being created, many areas of the water start to move. As a consequence, the relevant investment in terms of capital and human resources is drastically increased and the production level is brought up to a higher level. To gain a winning edge in the market competition in terms of market shares (the diverging side or Big Bang side of the yoyo) and revenue (the converging side or black hole side of the yoyo), the technology for mass production is improved and becomes more efficient with lowered costs (the spinning strength of some yoyos becomes strong). So, the market share of this new economic sector, led by a few strong representative companies, opens up quickly. After entering the third stage of evolution, the marketplace is gradually saturated with a more-than-sufficient supply. As the revenue increases, the profit elasticity of the products and services starts to erode. So, due to the weakening profit margins (fewer materials can be sucked in from the black hole side), further advances become slower and more difficult. At the end, the entire sector stops any attempt for further advancement. And the production stabilizes at the level of replacing aged and broken products purchased earlier. As a yoyo's spin becomes weaker, meaning that fewer materials are spit out (from the Big Bang side) and sucked in (from the black hole side), caused by appearances of other different but relevant new yoyos (economic sectors or companies), the yoyo might be absorbed by another faster and stronger spinning structure, or simply destroyed by several adjacent, more energetic yoyos. In fact, each real-life economic entity goes through such a three-staged life span with some minor variations.

One characteristic of this yoyo model is looking at the problem of concern as mutual interactions between several spinning yoyos, that is, from the angle of evolution. It is a dynamic model instead of a static one. Because of this characteristic, one can reasonably expect that such a model can help to bring forward new insights into various existing and new researches. In the following chapters in this part of the book, we will look at several applications of this yoyo model to prove this point.

Chapter 10
The Happy Family

People interact with each other for different purposes. With different individuals and different personalities and preferences arise conflicting interests: How can one keep all these people on the same track, working for a common purpose? To address this problem, Becker (1974), in *A Theory of Social Interactions*, published his rotten kid theorem. This theorem suggests that family members, even those who are selfish, will contribute to efforts to raise family income, if there is a benevolent head that voluntarily bequeaths gifts to other members periodically. The importance of this result has been well described by Bergstrom (1989, p. 1138):

> If it is generally correct, the Rotten Kid theorem must be one of the most remarkable and important results in the theory of incentives. For it tells us that a sufficiently benevolent household head would *automatically* internalize all the external effects that family members have on each other. Benevolent parents of intelligent, though selfish, children can breathe easier. In the family there will be no free riders or principal-agent problems. Elaborate incentive schemes and detection devices are unnecessary. All that is needed is to explain the Rotten Kid theorem to each family member and they will all (except possibly for a few irrational lapses) behave in the common interest. Not only would this be remarkable good news for parents, it would suggest a promising way of avoiding the incentive problems that bedevil firms and other social organizations. Shouldn't it be possible to find group incentive structures similar to those of families with benevolent heads?

In short, the rotten kid theorem can streamline a family, firm, or any organization. To witness this end, various implications and applications of this theorem appear in studies of family altruism and asset transfers (Becker, 1991; Bhalotra, 2004; Altonji, et al., 1997), moral hazard (see Ghaudhuri et al., 2005, and references there), the effects on consumption and savings of public debts, social security, and other governmental transfers among generations (Barro, 1974), patterns of parental bequests (Menchik, 1980; Becker, 1991), child labor and labor supplies (Baland and Robinson, 2000; Bommier and Dubois, 2004; Fernandes, 2000), foreign aids (Federio, 2004), and intergenerational education and culture transmissions (Patacchini and Zenou, 2004), among countless others.

Due to the theorem's wide-ranging importance, Bergstrom (1989) notes that it is worthwhile to find the limits of generality for theorems with strong and interesting conclusions, the rotten kid theorem included. Subsequently, he constructed three examples, each showing where this theorem could go wrong. Along the same idea, after developing a two-time-period model for the study of child labor, which encompasses the human capital dimension, Baland and Robinson (2000) show that the rotten kid theorem does not hold true when parental savings are at a corner. Additionally, Bommier and Dubois (2004) use a similar model to reveal that a child's disutility, a dimension overlooked by previous researchers, considerably affects the validity of this important theorem. Even though 30 years have elapsed since the rotten kid theorem was initially published in 1974, no one has formalized a condition under which the theorem holds true in general (for more details, see Becker, 1991, p. 11).

In this chapter, we will learn how to apply the systemic yoyo model (Wu and Lin, 2002; Lin, 2007) to establish a sufficient and necessary condition under which Becker's rotten kid theorem (the 1974 version; throughout this chapter and book, by Becker's rotten kid theorem, we mean his 1974 version only) holds true in general. In fact, when the yoyo model is applied to the study of the family (Becker, 1991), it is obvious that Becker's rotten kid theorem can only hold true conditionally. By specifying several relevant yoyos and how they interact, we can intuitively see the needed condition under which Becker's theorem holds true in general. Beyond this result, the yoyo model also clearly points to the following two results: the theorem of never-perfect value systems, which shows how any chosen value system would be flawed in practice, and such a theorem that if a tender, loving parent exists, then selfish kids would take advantage of his or tenderness and love.

With the sufficient and necessary condition under which Becker's rotten kid theorem holds true in general shown, Becker's theorem becomes more tangible and easier applied in various theoretical and applied situations. After proving the main theorem, we use examples to show that, in general, even though family members may have conflicts with their individual consumption preferences, the distribution of the benevolent head's income does not have to be affected.

For the first applications of the sufficient and necessary condition for Becker's rotten kid theorem to hold true, we look at the concept of merit goods (Becker, 1991). Interestingly, it is found that in some cases, money just does not buy what the parents want. Like Bergstrom did in 1989, we also go through his examples from a new angle to locate precisely where a conflict of interest occurs, where Becker's rotten kid theorem fails to hold true. Through detailed and rigorous deductions, it is found that many of the claims made by Bergstrom (1989) have to be modified or corrected.

As an astonishing consequence of the newly established result, a so-called theorem of never-perfect value systems is introduced. This result says that in any family setting, regardless how altruistic the parents are, as long as they have a value system in place for children to follow and measure up to, the outcome will unfortunately not be as optimal as expected. When this result is employed to analyze the evolution of the family (Becker, 1991), it is found that the traditional family format, in comparison, might have some advantage over the modern format in terms of passing down family values and value systems. The resourceful-father example shows that to avoid the trap described in the theorem of never-perfect value systems, a divorced but resourceful father tries to stay on good terms with his children by making sure his kids are as happy as they can be, yet equally. Tragically, the consequence entails transforming the kids into potential welfare recipients. At this junction, we cast an extremely important and practical question for our readers to ponder: How do we design a financial system to practically reinforce an established value system?

Based on the sufficient and necessary condition, we afterward look at an elegant and short proof for Bergstrom's results published in 1989 concerning transferable utilities. After that, the established condition is conveniently applied to address the problem of how to avoid the Samaritan's dilemma.

The main contribution this chapter makes to the existing literature is that it shows how to use the new qualitative, figurative analysis method in social science, and how this method can be employed to lay down the foundation for many empirical findings, as a starting point and direction for one to construct traditional models. Beyond these, we show how new and insightful results and understandings can be mapped out analytically on the basis of the figurative analysis.

Throughout this chapter and Chapters 11 to 14, the yoyo model plays the role of a driving force underlying all the thinkings of the works, supplemented with traditional analytic models to produce such results that are more acceptable by the minds of the current, formal education system.

10.1 Becker's Rotten Kid Theorem

Becker (1974) introduced the now-well-known rotten kid theorem, a slightly different version of which was later used in his *Treatise of the Family* (1991) to explain many mysteries and quandaries of household economics. (The wording was only slightly different, but the meaning of the original version was altered greatly in terms of evolution.)

THEOREM 10.1 (Becker's Rotten Kid Theorem)
If a family has a head who cares about all other members so much that he transfers his resources to them automatically, then any redistribution of the head's income among members of the household would not affect the consumption of any member, as long as the head continues to contribute to all. Additionally, other members are also motivated to maximize the family income and consumption, even if their welfare depends on their own consumption alone.

This version of the rotten kid theorem is the original cited from Becker (1974). In Becker (1991), a slightly different version is given. In 1989, Bergstrom established his version of the rotten kid theorem (we will come back to this later, in a sequential part of work). And Dijkstra (2007) generalizes Bergstrom's results to a higher level. He also establishes a condition under which in the Samaritan's dilemma, the altruist reaches her first best when she moves after the parasite. The reason why we base our work on Becker's original result published in 1974 is because this version of the rotten kid theorem provides a sense of evolution of the head's benevolence and the selfish kids' unselfish behaviors and tolerance toward potential redistributions of the head's resources. In particular, if we assume, as in practical situations, that the head transfers his resources to other family members periodically at fixed time intervals, we can depict the implied evolution in Figure 10.1. The scale marks on the timeline represent the moments of individual asset distributions and are given for reference purposes without much practical implication. Mark 0 can be located anywhere on the line as the beginning of our discussion or focus. Negative scale marks represent the moments of distribution of the past, and the positive marks represent the future moments of distribution.

As indicated by Becker (1974), let us assume that at each past moment of distribution, the head transferred $\$R_j$ to selfish member j. If for any reason member k suffers from a temporary financial setback in time period $(-1, 0)$ and the head has to (because of his consideration of others) bail out k by providing k extra monetary support at $t = 0$, and maybe a few following pay periods, then Theorem 10.1 states that the head can adjust his amount of distribution to member j and others to finance his increased support for k. And this adjustment in his transfers will not affect the consumption of any member, as long as the head continues to contribute to all. Intuitively, because of such potential adjustments in transfers, all selfish members feel the sense and existence of the family and are willing to help maximize the family income and consumption, because each of them might one day need such increased financial support.

On the other hand, Bergstrom's (1989) version of the rotten kid theorem focuses on one such a time period as $(k, k + 1)$, and Dijkstra's (2007) work looks at such time periods as $(k, k + 1)$ and $[k, k + 1]$, respectively and individually. Specifically, Bergstrom's result focuses on the case that the head's transfer is made after whatever

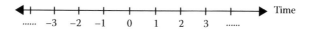

Figure 10.1 Evolution of the head's benevolence and selfish members' behaviors.

the selfish kids do beforehand, and Dijkstra establishes the conditions under which the head reaches his first best when he makes transfers, respectively, either before or after however the selfish kids behave. (Hirshleifer [1977] first pointed out the importance of the order of moves.) Indeed, both Bergstrom's and Dijkstra's results can be employed to successfully analyze practical situations. On the other hand, if we look at the payment cycles of a work environment, where employees get paid periodically at fixed time intervals, as depicted in Figure 10.1, then Becker's original version of the rotten kid theorem is more appropriate. Here, because of potential future transfers, both the benevolent head and the selfish kids will have to adjust their expectations and behaviors from one time period to the next to reach their individual Pareto optimums over time. That is, there are fundamental differences between one-shot deals and repeated businesses. Countless examples have been seen in the business world where people behave in one way for one-shot deals and in completely different manners in potentially repeating business opportunities. Because of this reason, we see the need to look at Becker's 1974 result and see under what general conditions this result holds true.

In 1989, Bergstrom constructed three examples to show that, in general, Becker's rotten kid theorem does not hold true. Then, he introduced the concept of transferable utility as a necessary condition for Becker's rotten kid theorem to hold true. For our purpose here, let us look at Bergstrom's lazy rotten kids with different sets of details added and slightly different outcomes produced. After that, we will develop several important observations based on this example.

Example 10.1: The Lazy Rotten Kids (Bergstrom, 1989)

Let Y_j be an index, $0 \leq Y_j \leq 1, j = 1, \ldots, n$, designed to measure how hard child j works to generate his own income I_j. That is, his own income I_j is a function of Y_j:

$$I_j = I_j(Y_j) \tag{10.1}$$

Assume that each child j's utility U_j is a function of his consumption X_j of goods and leisure, where his leisure is measured by $(1 - Y_j)$.

Suppose that the benevolent head h of the family decides on his distribution of his own resources to the children to maximize his utility function:

$$U_h = U_h(X_h, U_1, \ldots, U_n) \tag{10.2}$$

where X_h represents the amount of consumption by the head h subject to the budget constraint

$$\sum_{j=1}^{n} X_j = \sum_{j=1}^{n} I_j \tag{10.3}$$

where $I_j = I_j(Y_j)$ is given in Equation (10.1).

If the family head h cannot observe the levels of individual efforts Y_j, and the children put in to produce their own incomes $I_j = I_j(Y_j)$, then a lazy and selfish child will not have enough incentive to work. It is because he will not receive any additional pay for his extra bit of effort. Therefore, the family head runs into the problem of incentives with the lazy child. Now, if the child's work is to help the head to make additional income, the lazy child would not automatically put in effort to maximize the family income.

Now, assume that the benevolent head has a system in place to practically measure each child's income I_j and his level of effort Y_j exerted in the production of his income, and that the head's utility depends on the utilities and consumptions of his children. If the benevolent head can only give gifts of money to his children and not be able to direct the distribution of leisure, then, in general, Becker's rotten kid theorem does not hold true.

Specifically, let us consider such a family that the head h has two lazy children named 1 and 2. The utility function of child j is

$$U_j = X_j(1 - Y_j), \quad j = 1, 2 \tag{10.4}$$

and the utility function of the head h is

$$U_h = X_h + \sqrt{U_1} + \sqrt{U_2} = X_h + \sqrt{X_1(1-Y_1)} + \sqrt{X_2(1-Y_2)} \tag{10.5}$$

where X_k is the consumption of person k, $k = 1, 2, h$. The resource transfer plan from the head to each child is defined by

$$h_j = wY_j, \quad j = 1, 2 \tag{10.6}$$

where, to be fair, w is a fixed positive constant and the head h is able to observe the values of Y_j, $j = 1, 2$.

Now, the head's plan of transfer of his resources to his children is carried out so that their consumptions will maximize Equation (10.5) subject to the constraints:

$$X_h + X_1 + X_2 = X_h + I_1 + I_2 + h_1 + h_2 = X_h + I_1(Y_1) + I_2(Y_2) + wY_1 + wY_2 \tag{10.7}$$

and

$$X_j \geq I_j, \quad j = 1, 2 \tag{10.8}$$

Following the assumption of Becker's rotten kid theorem, assume that the head's income $I_h = X_h + h_1 + h_2$ is large enough so that he will always choose to make positive monetary gifts to each child. So, the constraints in Equation (10.8) are not binding and can be ignored. Additionally, because the kids are lazy, we can reasonably assume that $I_1 = 0 = I_2$, meaning that their own incomes produced by their own efforts are zero.

The first-order conditions for maximizing Equation (10.5) subject to Equation (10.7) are given by

$$\begin{bmatrix} \dfrac{\partial U_h}{\partial X_1} \\ \dfrac{\partial U_h}{\partial X_2} \\ \dfrac{\partial U_h}{\partial Y_1} \\ \dfrac{\partial U_h}{\partial Y_2} \end{bmatrix} = \begin{bmatrix} \dfrac{1}{2} U_1^{-1/2}(1-Y_1) \\ \dfrac{1}{2} U_2^{-1/2}(1-Y_2) \\ -\dfrac{1}{2} U_1^{-1/2} X_1 \\ -\dfrac{1}{2} U_2^{-1/2} X_2 \end{bmatrix} = \lambda \begin{bmatrix} -1 \\ -1 \\ w \\ w \end{bmatrix} \qquad (10.9)$$

where λ is the Lagrange multiplier. Dividing the third and fourth row entry equations by the first and second, respectively, leads to

$$w = X_1/(1 - Y_1) \qquad (10.10)$$

and

$$w = X_2/(1 - Y_2) \qquad (10.11)$$

So, Equations (10.10) and (10.11) can be divided and we subsequently obtain

$$X_1/X_2 = (1 - Y_1)/(1 - Y_2) \qquad (10.12)$$

Solving Equation (10.12) for X_1 and X_2 individually and inserting them into Equation (10.7) leads to

$$H_i = C + [(1 - Y_i)/(1 - Y_j) + 1]X_j - h_j, \ i, j = 1, 2, \ i \neq j \qquad (10.13)$$

where $C = X_h - I_h - I_1 - I_2$ is a constant.

What is surprising in Equation (10.13) is that each child's share h_i of the benevolent head's own income is a decreasing function of the effort Y_i he puts in his work. That is, the incentive problem in this case is worse than that in the case where the head does not have a way to observe his children's efforts. Once again, if the children's work is to help the head make additional income, the lazy children might not automatically put in the necessary effort to maximize the family income.

This example shows that Becker's rotten kid theorem does not automatically solve household incentive problems by the presence of a benevolent head. In fact, it shows that by introducing a value system into the decision-making process on how to distribute his own income to his children, the head actually redirects all potential anger and dissatisfaction of the children toward himself.

In Becker's rotten kid theorem, the only action the benevolent head exerts on all the selfish family members is his (voluntary) transfer of his resources. That is, he forces something on other members. Consequently, it is possible that some of the receiving members just do not care about receiving the monetary gifts. Even worse, there might be such a member who could refuse to accept anything from the head. In either of these cases, we say that the head's distribution is in conflict with the consumption preferences of the receiving members. In this regard, Bergstrom (1989) provides a short proof for the rotten kid theorem (Becker's version of 1991). That proof holds true under the assumption, which is not explicitly mentioned there, that the selfish kids' utility is an increasing function of how much they receive from the head. The presentation in this chapter will show that this is exactly where the situation would go wrong.

Based on the analysis of Example 10.1, if we look at Becker's rotten kid theorem of wide-ranging importance in the light of the yoyo model, we can easily see that the correctness of this result must be conditional, not as simply stated as in the theorem. In particular, the family can be seen in two different fashions:

1. The entire family is a spinning yoyo, where the family consumption is seen as the materials spitting out of the Big Bang side and the family income as the materials sucked in from the black hole side.
2. The family is a collection of individually spinning yoyos that interact with each other, where each individual's consumption is seen as the materials spit out of his or her yoyo's Big Bang side and his or her income as the materials sucked in from his or her yoyo's black hole side.

For case 1, if we stand above the yoyo and look at the Big Bang side, similar to the situation where scientists look at the Earth from above its North Pole, then the flows of the materials spit out of the Big Bang side can be simulated by the fluid patterns in the dishpan experiment. (For the convenience of our reader, we will highlight the relevant points of this experiment here.) In particular, Dave Fultz and his colleagues of the University of Chicago (1959) and Raymond Hide (1953) of Cambridge University, England, independently conducted the so-called dishpan experiment. To make the story short, let us describe Fultz's version in some detail. He partially filled a cylindrical vessel with water, placed it on a rotating turntable, and subjected it to heating near the periphery and cooling near the center. The bottom of the container was intended to simulate one hemisphere of the Earth's surface; the water, the air above this hemisphere; the rotation of the turntable, the Earth's rotation; and the heating and cooling, the excess external heating of the atmosphere in low latitudes and the excess cooling in high latitudes.

To observe the pattern of flows at the upper surface of the water, which was intended to simulate atmospheric motion at high elevations, Fultz sprinkled some aluminum powder. A special camera that effectively rotated with the turntable took time exposures so that a moving aluminum particle would appear as a streak and sometimes each exposure ended with a flash, which could add an arrowhead to the forwarded end of each streak. The turntable generally rotated counterclockwise, as does the Earth when viewed from the North Pole.

Even though everything in the experiment was arranged with perfect symmetry about the axis of rotation, such as no impurities added in the water and the bottom of the container was flat, Fultz and his colleagues observed more than they bargained for. First, both expected flow patterns, as shown in Figure 10.2, appeared, and the choice depended on the speed of the turntable's rotation and the intensity of the heating. Briefly, with fixed heating, a transition from circular symmetry (Figure 10.2(a)) would take place as the rotation increased past a critical value. With the sufficiently rapid but fixed rate of rotation, a similar transition would occur when the heating reached a critical strength, while another transition back to the symmetry would occur when the heating reached a still higher critical strength.

What is important about the dishpan experiment is that structures such as jet streams, traveling vortices, and fronts appear to be basic features in rotating heated fluids, and are not peculiar to atmospheres only. With this understanding in place, it is safe for us to conclude that for case 1, where the entire family is seen

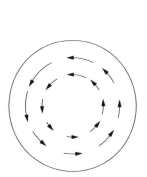
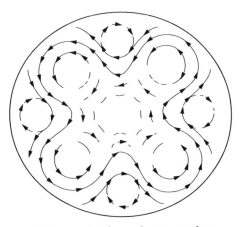

(a) Symmetric flow at the upper surface   (b) Asymmetric flow at the upper surface

Figure 10.2 Patterns observed in Fultz's dishpan experiment.

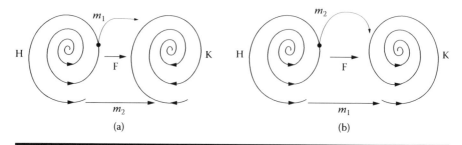

Figure 10.3 Interactions between benevolent head H and a selfish kid K.

as a spinning yoyo, sooner or later the spin field of the family yoyo will have local eddies, which in some locations may not necessarily spin in the same direction of the overall family yoyo. That is, case 1 turns into the situation described in case 2.

For case 2, to make the situation easier to analyze, let us assume that there are only two members, the benevolent head H and a selfish kid K, in the family. In this case, the interactions between the yoyos can be depicted in Figure 10.3, where the benevolence of the head H is represented as a divergent whirlpool and the selfishness of the kid K as a convergent whirlpool. If both m_1 and m_2 stand for the voluntary money transfers from the head, the spin field of K in Figure 10.3 will accept the gift m_1 happily and the transfer m_2 unwillingly (Figure 10.3(b)), or even reject such a monetary gift (Figure 10.3(a)). Here, m_1 is given to the selfish kid K without violating his own preference of consumption, while m_2 is forced on K against his will or personal preferences. That is, we have seen the following result.

THEOREM 10.2
Becker's rotten kid theorem holds true if and only if the distribution of the benevolent head's resources is not in conflict with the consumption preferences of any selfish member.

Proof. ($\Leftarrow$) Assume that Becker's rotten kid theorem holds true. We need to show that the distribution of the benevolent head's resources is not in conflict with the consumption preferences of any selfish member. Let us prove this statement by contradiction. That is, we assume that Becker's rotten kid theorem holds true and that the way the head distributes his resources to other members is in conflict with the consumption preferences of member k.

Let Y be a variable representing one aspect of member k's total consumption preference, satisfying that member k's utility is an increasing function of Y, and that his share h_k of the head's distribution is a decreasing function of Y. To make the situation clear, let us assume that the family has two members only: the benevolent head h and the selfish member k. So, the utility functions can be written as

$$U_k = U_k(X_k, Y), \; U_h = U_h(X_h, U_k) \qquad (10.14)$$

where X_i is the total consumption of member i on various commodities, $i = k, h$, and Y the special variable as given above satisfying

$$\partial U_k / \partial Y > 0 \qquad (10.15)$$

So, the assumption of variable Y means that the distribution of the head's resources to member k is determined by

$$h_k = h_k(Y) \qquad (10.16)$$

satisfying $dh_k/dY < 0$. (The differentiability of h_k is assumed here to take advantage of the methods of calculus.) Without causing any confusion, we assume that no other variables enter into the function h_k.

To the family head, his distribution of his own resources has to maximize his utility function U_h in Equation (10.14) subject to the following budgetary constraint:

$$X_h + X_k = I_h + I_k = X_h + (I_k + h_k) \qquad (10.17)$$

where I_i is the personal income of i, $i = k, h$. So, the first-order conditions for this maximization problem are

$$\begin{bmatrix} \dfrac{\partial U_h}{\partial X_h} \\ \dfrac{\partial U_h}{\partial X_k} \\ \dfrac{\partial U_h}{\partial Y} \end{bmatrix} = \begin{bmatrix} \dfrac{\partial U_h}{\partial X_h} \\ \dfrac{\partial U_h}{\partial U_k} \cdot \dfrac{\partial U_k}{\partial X_k} \\ \dfrac{\partial U_h}{\partial U_k} \cdot \dfrac{\partial U_k}{\partial Y} \end{bmatrix} = \lambda \begin{bmatrix} 1 \\ 1 \\ -\dfrac{dh_k}{dY} \end{bmatrix} \qquad (10.18)$$

Dividing the second row by the third row in Equation (10.18) and simplifying the resultant expression leads to

$$\dfrac{\partial X_k}{\partial Y} = -\dfrac{dh_k}{dY} \qquad (10.19)$$

That is, Equations (10.16) and (10.19) imply that $\partial X_k / \partial Y > 0$. And from Equations (10.16) and (10.17), it follows that $X_k = I_k + h_k$ and $\partial X_k / \partial Y = dh_k / dY < 0$—a contradiction. This contradiction implies that the assumption that the way the head distributes his resources to other members is in conflict with the consumption preferences of member k is incorrect.

($\Leftarrow$) Assume that the distribution of the benevolent head's resources is not in conflict with consumption preferences of any selfish member. We need to prove that

Becker's rotten kid theorem holds true. Once again, we will prove this statement by contradiction. That is, we assume that Becker's rotten kid theorem does not hold true. This means that (1) there is a redistribution of the head's income among members of the household that does affect the consumption of at least one member, say member l, even though the head continues to contribute to all, and (2) there is at least one selfish member k who is not motivated to maximize the head's (the family's) income.

Situation 1 implies that for the one-time redistribution of the head's income, the way of distribution of the head's resources does not agree with how member l prefers to consume his income and monetary gifts. That means that the way the head distributes his income, at least this time, is in conflict with the consumption preferences of member l. This end contradicts the initial assumption that no such conflict exists. That is, situation 1 would not occur.

Now for situation 2, let Y be a variable concerning one aspect of the consumption preferences of member k, representing the reason why member k does not care about maximizing the head's income. If member k's share of the head's distribution is h_k, then it must mean that

$$\left.\frac{\partial h_k}{\partial Y}\right|_{Y=Y_0} < 0$$

where $Y = Y_0$ stands for the special Y-value of the moment of our concern. That is, locally at around $Y = Y_0$, h_k is a decreasing function of Y. In fact, if h_k is not a decreasing function of Y at any local region of $Y = Y_0$, then h_k must satisfy either

$$\left.\frac{\partial h_k}{\partial Y}\right|_{Y=Y_0} = 0 \text{ or } \left.\frac{\partial h_k}{\partial Y}\right|_{Y=Y_0} > 0$$

If h_k satisfies the second condition, then it must be a local increasing function of Y at $Y = Y_0$. In this case, the greater the Y-value (at least locally), the greater share member k receives from the head. Because k's utility is an increasing function of his consumption, he would naturally like to help maximize the head's income. This end contradicts the assumption of Y.

If the first condition above holds true, from the fact that member k is not motivated to maximize the head's income, it follows that h_k must be one of the following three possibilities:

1. It is a local constant function of Y.
2. It reaches a local minimum at $Y = Y_0$.
3. It reaches a local maximum at $Y = Y_0$.

First, h_k cannot be a local constant function or have reached a local minimum at $Y = Y_0$; otherwise, this end would not agree with the fact that member k's utility is an increasing function of his consumption, because if the head's income drops or stays the same, member k's share h_k will have to decrease or stay constant. So, h_k cannot be a local constant function of Y or have reached a local minimum at $Y = Y_0$.

Second, h_k cannot have reached a local maximum at $Y = Y_0$, either; otherwise, it would mean that member k has been helping out the head to maximize his income until $Y = Y_0$. This end contradicts the definition of the variable Y in the first place.

That is, the distribution h_k of the benevolent head's income to member k is in conflict with the consumption preferences of the selfish member k. This end contradicts the initial assumption that no such conflict exists. Therefore, the assumption that Becker's rotten kid theorem does not hold true is incorrect.

10.2 Two Other Mysteries of the Family

First, as a side note, we notice that when two selfish members have conflicting consumption preferences, the benevolent head might get along with both of these selfish members in terms of consumption preferences. Specifically, assume that the utility functions of members i and j are given by

$$U_i = X_i - Y \text{ and } U_j = X_j + Y \qquad (10.20)$$

Then, the benevolent head's utility could be given by

$$U_h = X_h(U_i + U_j) = X_h(X_i + X_j) \qquad (10.21)$$

In Equations (10.20) and (10.21), X_k is the consumption of k, $k = i, j, h$, and Y is a variable of consumption of member i's dislikes and member j's likes. Equation (10.21) indicates that in the face of conflict in terms of consumption preferences existing between i and j, the family head h is completely neutral.

Also, when the benevolent family head h and a selfish member i have conflict in their consumption preferences, it does not necessarily mean that in the overall utility function of the head h, he must have a direct conflict with member i in terms of their consumption preferences. For example, assume that Y stands for a specific consumption of goods that h dislikes while i likes. Let the utility functions of h and i from consumptions be given by

$$U_i = X_i Y \text{ and } U_h = X_h/Y \qquad (10.22)$$

where X_k stands for the total consumption of k besides Y, $k = i, h$. If the overall utility function of h is given by

$$U = U(U_h, U_i) = U_i \cdot U_h = X_i X_h \qquad (10.23)$$

then it implies that individually, h and i have conflicting consumption preferences. However, in his overall utility, the benevolent head h cares about member i as much as if they did not have any conflict in their consumption preferences.

This side note implies that no matter whether or not there exists a conflict in terms of individual consumption preferences between the family members, the benevolent head could care about all members just the same as if there were no conflict at all. That is, in general, the distribution of the head's resources to all members, as seen in Theorem 10.2, is not affected by his personal consumption preferences.

With Theorem 10.2 in place, let us look at how the altruism of parents and the merit goods of children are linked together, assuming that there is no direct bargaining between the parents and children. Here, by *merit goods* we mean particular traits or behavior of children that parents care about: whether they are lazy, studying hard at school, visiting often, drinking excessively, marrying well, or being mean to siblings (Becker, 1991, p. 10).

As studied in Example 10.1, children's laziness can cause a breakdown of Becker's rotten kid theorem as soon as a variable about children's laziness is employed in the distribution of parents' income to the children.

In terms of desiring children to study hard at school, one faces a situation where an index about how hard children study at school needs to be established so that this index can be practically measured. The measurability of this index can lead to significant differences in children's academic achievements, even though no matter if the index is measurable, Theorem 10.2 indicates that Becker's rotten kid theorem is no longer true as soon as the index is used in the determination of how much money each child receives from his or her parents. For example, some parents are happy with their children's schoolwork as long as their grades are B's and A's, especially when these kids are first-generation college-bound. However, if the parents' educational status is relatively high, the index they define to measure how well their children study at school tends to be content-rich. This end could be used to explain why parental education and occupational status have a positive and significant effect on their children's educational achievements (Hill and Duncan, 1987; Norberg-Schonfeldt, 2004).

When a child has grown and is no longer under his parents' roof, a natural desire of the parents will be seeing the adult child coming home to visit often. If the parents are financially successful, it is likely that they will employ monetary means to have the child visit them as often as they hope (Bernheim et al., 1986). However, due to various reasons, obligations, and commitments, the adult child just simply might be too busy to visit his parents as often as his parents want. Now, if the parents apply financial means to make their child visit them often, they will have to face the possible situation, as described in Theorem 10.2, where the distribution of the parents' money is in conflict with the child's consumption preference. For example, the adult child might be working for a promotion he has desired for a long time; he might be dating the woman of his dreams; or he might be working at a second job that brings him considerably more than what his parents could afford to pay him for his visits. Quite frankly, there are millions of possibilities for the child to be too busy to frequently pay his parents visits. However, if the parents are extremely wealthy and the child has no desire to strike it big in life for himself, then Theorem 10.2 implies that Becker's rotten kid theorem will work in this case, and this adult child will pay his parents visits as often as they desire. And maybe the adult child might simply move back into the parents' house together with his family, making the opportunity of taking care of the aging parents his full-time job. These possible scenarios have been well studied and documented by Stanley and Danko (1996) in their over 20 years of research.

For any other merit goods parents want to see their children consume, Theorem 10.2 will apply as long as a variable is introduced to measure how well the children act according to their parents' expectations, and the parents will distribute their monetary rewards to the children according to that specific variable. As long as there is a conflict between how the parents give to the children and the children's consumption preferences, at least one of the children will not be motivated to maximize the parent's income.

Now, let us look at the nightlight controversy, where the conflicting interests between the benevolent head and a selfish member of the family can be seen clearly and vividly. Here, after we add the necessary details, it is found that the conclusion of Bergstrom in this situation is not completely correct. And our modification seems to be more in line with the reality.

Example 10.2: The Nightlight Controversy

This example was initially constructed by Becker in 1974 to show how his rotten kid theorem can be employed in a family setting where the husband has to read at night. The reading light, however, bothers his wife when she tries to sleep. To compensate for her loss or inconvenience, the husband transfers additional monetary contribution to his wife from his increased income, a result of his nightly reading. If Becker's rotten kid theorem holds true for this scenario, then how much and how long both of them will be better off when the selfish wife allows her altruistic husband to read in bed? The reason for asking this question is that the utility of the husband would be raised from his additional reading and thus he would increase his contribution to his wife, more than compensating for the initial harm done by his nightly reading.

To see where Becker's rotten kid theorem could go wrong, Bergstrom (1989) constructed an explicit example using the nightlight usage as the public good. Then he concluded that in order for the wife and husband to be on the same page about the nightlight usage, they have to have an explicit bargain struck. In the following, on the basis of Bergstrom's example, we will show that such a rational bargain cannot be reached for over the long term because the husband is not assured that his nighttime reading will lead to an increased income.

Specifically, let

$$U_H = U_H(U_h(X_h, X_w, Y), U_w(X_w, Y)) = U_h U_w^a, \; 0 < a < 1 \quad (10.24)$$

be the benefit function of the household, consisting of the husband and wife, in two variables U_h and U_w, where U_h and U_w are the utility functions of the husband and wife, respectively, defined by

$$U_h = X_h X_w (Y + 1) \quad (10.25)$$

and

$$U_w = X_w e^{-Y} \quad (10.26)$$

where X_i is the total consumption of goods of member i, $i = h, w$, and Y is the amount of nighttime reading of the husband.

By substituting Equations (10.25) and (10.26) into Equation (10.24), we have

$$U_H = X_h X_w^{1+a}(Y+1)e^{-aY} \quad (10.27)$$

Now, this household wants to maximize U_H subject to the budgetary constraint

$$X_h + X_w = I_h + I_w \quad (10.28)$$

where I_i is the personal income of i, $i = h, w$. The first-order conditions for this optimization problem are

$$\begin{bmatrix} \dfrac{\partial U_H}{\partial X_h} \\[4pt] \dfrac{\partial U_H}{\partial X_w} \\[4pt] \dfrac{\partial U_H}{\partial Y} \end{bmatrix} = \begin{bmatrix} X_w^{1+a}(Y+1)^{-aY} \\ (1+a)X_w^a X_h (Y+1)e^{-aY} \\ X_h X_w^{1+a} e^{-aY}[1 - a(Y+1)] \end{bmatrix} = \lambda \begin{bmatrix} 1 \\ 1 \\ -\dfrac{dI_h}{dY} \end{bmatrix} \quad (10.29)$$

where $\lambda > 0$ is the Lagrange multiplier. The entries in cell (3.1) in Equation (10.29) show that only when

$$Y > 1/a - 1 \quad (10.30)$$

does the husband's income I_h uptrend with an increase of Y. Because the value of a in Equation (10.30) is really not known to the couple, and the wife's consumption

$$X_w = [(1 + a)/(2 + a)](I_h + I_w) \tag{10.31}$$

does not have a clear connection with Y, in a real-life setting, the husband will be in a very inauspicious position to truly strike a deal regarding how much he should read in bed while his wife tries to sleep. It is because Equation (10.30) implies that Y could potentially take a very large value, which to the wife is not realistic at all.

Now, if we turn our attention to Theorem 10.2, the difficulty in this scenario becomes obvious: Becker's rotten kid theorem will not work automatically unless a negotiation takes place. At the same time, the analysis above indicates that the wife might give in temporarily when she imagines a foreseeable a-value. However, if such an imagined a-value were not to materialize in a timely fashion, the final decision would be "absolutely no bedtime reading allowed."

In the same spirit as in the last example, let us look at the following example with our added details and corrections to the earlier versions made. As in the previous case, we will make use of Theorem 10.2 so that the situation can be more easily analyzed and understood than before.

Example 10.3: A Prodigal Son

Based on the work of Lindbeck and Weibull (1988), Bergstrom (1989) constructed this example to show that Becker's rotten kid theorem is not generally true. Specifically, let us consider a selfish kid k and a benevolent parent P. They live in two nonoverlapping time periods 1 and 2. In period 1, the kid k has a certain allotment of wealth to live on. He has the option to spend it all now or spend some now and save the rest for future use. In period 2, he knows he will receive a lump sum of money as a gift from parent k, assuming that the parent cannot make any precommitment to punish k for any of his profligate period 1 behavior. Suppose that the utility functions U_k and U_p of the kid k and parent p are defined by

$$U_k = (c_k^1, c_k^2) = \ln c_k^1 + \ln c_k^2 \tag{10.32}$$

and

$$U_p\left(c_p^1, c_p^2, c_k^1, c_k^2\right) = \ln c_p^1 + \ln c_p^2 + \alpha U_k = \ln c_p^1 + \ln c_p^2 + \alpha \ln c_k^1 + \alpha \ln c_k^2 \tag{10.33}$$

where c_k^i and c_p^i are, respectively, the kid's and parent's consumptions in period i, and $\alpha > 0$ a fixed number. If in period 2, the parent allocates his wealth to k to maximize his utility U_p subject to the budget constraint

$$c_p^2 + c_k^2 = w_p^2 + w_k^2 \tag{10.34}$$

where w_j^2 stands for j's own wealth in period 2, $j = p, k$, then by using the first-order conditions of this optimization problem, we see that to continue to maximize his utility in period 2, the parent will divide the family's total wealth $(w_p^2 + w_k^2)$ in such a way that the fraction $1/(1 + \alpha)$ of the total will go to the parent for his own consumption, and the fraction $\alpha/(1 + \alpha)$ will go to the kid k.

This end implies that if the kid k puts some money away in savings in period 1, he will have a greater sum of wealth in the second period. That is, to obtain more money from the parent in period 2, the kid will have to spend all or more than his period 1 allotment during period 1. In other words, in terms of Pareto efficiency, it will be wise for the kid to squander as much as he can during period 1 so that his altruistic parent will gift more to him during period 2. This end implies that what Becker's rotten kid theorem says is not true—his selfish kid is not automatically motivated to maximize the family wealth and consumption.

Now, Theorem 10.2 explains clearly why Becker's rotten kid theorem fails in this case. First, in Becker's original theorem (Theorem 10.1), "the family income" means the income of the benevolent head instead of the sum of all members' individual incomes. Second, if the kid k saves some money in period 1, he would naturally not want this money being used to figure how much he will be given in period 2. Because in Equation (10.34) the kid's wealth w_k^2 is used against the kid's will, Theorem 10.2 implies that at the very least, the kid k would not be motivated to maximize the parent's income. And this example actually shows a scenario where the kid is motivated to behave much worse.

10.3 Never-Perfect Value Systems and Parasites

The previous result is established based on the fact that when gift m_1 (Figure 10.3) is given to the selfish kid K without violating the kid's preference of consumption, the kid will not fight against the head H. When we look at m_2 in Figure 10.3, the situation is different. Here, no matter whether the kid accepts the transfer (Figure 10.3(b)) or not (Figure 10.3(a)), he will show his unwillingness to cooperate with the head H. This end leads to the following result.

THEOREM 10.3: The Theorem of Never-Perfect Value Systems
In a family of at least two members, one member h is the head who is benevolent and altruistic toward all other members. The head establishes a value system for all members of the family to follow so that in his eyes, every selfish member will be better off now or in the future. If a selfish member k measures up well to the value system, he or she will then be positively rewarded by the head h. Unfortunately, the more effort member k puts in to measure up to the value system, the more he or she will be punished by the reward system.

Before we see the proof, let us illustrate the background on which Theorem 10.3 is composed. As depicted in Figure 10.1, we are not looking at the resource transfers as a one-shot deal. Instead, the benevolent head transfers his resources to other family members in a periodic fashion over time without an end in sight. The idea of value systems implies that the head tells other members at some time moment along the timeline that starting at a certain pay period, each member's behavior will affect how much he or she will receive from the head at the end of that period. And the head will design his response to each member's behaviors in such a way as to maximize his own utility. Once gain, Theorem 10.3 explains a sequential behavioral change and reflection over time between the head and his family members.

Proof. Assume that Y_k is an index, satisfying $0 < Y_k < 1$, established to check how well member k measures up to the value system predetermined by the family head h. This index satisfies that the greater Y_k is, the better member k measures up to the value system. Because k has to put in extra effort to increase the value of Y_k, his utility function $U_k = U_k(X_k, Y_k)$ satisfies

$$\frac{\partial U_k}{\partial X_k} > 0 \text{ and } \frac{\partial U_k}{\partial Y_k} < 0 \tag{10.35}$$

where X_k stands for member k's total consumption of numeraire good. The reason why $\frac{\partial U_k}{\partial Y_k} < 0$ is because all people are lazy to a certain degree.

Let the head h's utility function be

$$U_h = U_h(X_h, U_1, ..., U_n)$$

where it is assumed that other than the head h, the family contains n other members who are all selfish, X_h is the head's own consumption, and U_i is the utility function of family member i, $i = 1, 2, ..., n$.

Assume that the reward from the head h to member k is determined by

$$h_k = h_k(Y_k)$$

such that $\frac{\partial h_k}{\partial Y_k} > 0$. Now, the total consumption of member k is given by

$$X_k = I_k + h_k = I_k + h_k(Y_k)$$

where I_k is member k's own income unrelated to Y_k. To reflect the fact that the index Y_k represents k's performance on the head's value system, which is assumed to make member k better off, we did not explicitly list Y_k as an independent variable in the head's utility function. Instead, Y_k is shown in the total consumption $X_k = I_k + h_k(Y_k)$ of k. That is, an increase in the value of Y_k has a complex effect on member k, both positive and negative. Being positive is in the short term because the head h will help to bring k's consumption to a higher level, and it could be negative because member k has to put in more effort to raise the Y_k-value. So, to make our model more plausible, we can add the assumption that

$$\left| \frac{\partial U_k}{\partial X_k} \cdot \frac{\partial X_k}{\partial Y_k} \right| > \left| \frac{\partial U_k}{\partial Y_k} \right|$$

That is, the increased utility of k, brought forward by the higher-level consumption $X_k = I_k + h_k(Y_k)$, is more than enough to offset the decreased utility of k when he has to put in additional effort to incrementally raise the Y_k-value.

To the head h, his distribution plan of his own income to all other members has to maximize his utility function subject to the following budgetary constraint:

$$X_h + \sum_{j=1}^{n} X_j = X_h + \sum_{j=1}^{n} (I_j + h_j) = I_h + \sum_{j=1}^{n} I_j$$

If we ignore all members except the head h and member k, then the first-order condition for the head's optimization problem is given by

$$\begin{bmatrix} \frac{\partial U_h}{\partial X_h} \\ \frac{\partial U_h}{\partial X_k} \\ \frac{\partial U_h}{\partial Y_k} \end{bmatrix} = \begin{bmatrix} \frac{\partial U_h}{\partial X_h} \\ \frac{\partial U_h}{\partial U_k} \cdot \frac{\partial U_k}{\partial X_k} \\ \frac{\partial U_h}{\partial U_k} \cdot \frac{\partial U_k}{\partial Y_k} \end{bmatrix} = \lambda \begin{bmatrix} 1 \\ 1 \\ \frac{\partial h_k}{\partial Y_k} \end{bmatrix}$$

And member k chooses such a Y_k-value Y_k^* to satisfy his first-order condition

$$\frac{\partial U_k}{\partial Y_k} = \frac{\partial U_k}{\partial Y_k} + \frac{\partial U_k}{\partial X_k} \cdot \frac{\partial h_k(Y_k)}{\partial Y_k} = 0$$

so that he maximizes his utility.

Now, the third equation in the head's first-order condition implies that when the head's expected Y_k-value is greater than Y_k^*, his reward $h_k(Y_k)$ for member k would have a negative rate of increase. That is, we have shown that the more effort member k puts in to measure up to the value system, the more he will be punished by the reward system.

Example 10.4

To see how the result in Theorem 10.3 materially acts out, let us assume that a family consists of two members: the benevolent and altruistic head h and a selfish member k, and Y_k is the index outlined in the proof of Theorem 10.3. Let member k's utility function be given as follows:

$$U_k = U_k(X_k, Y_k) = X_k(1 - Y_k) \tag{10.36}$$

where X_k is member k's total goods consumption and $(1 - Y_k)$ stands for his degree of laziness, and let the head's utility function be

$$U_h = U_h(X_h, U_k) = X_h + \sqrt{U_k} = X_h + \sqrt{X_k(1 - Y_k)} \tag{10.37}$$

Assume that the reward from the head h to member k is determined by

$$h_k = wY_k \tag{10.38}$$

where w is a fixed constant > 0. Then the total consumption of k is

$$X_k = I_k + h_k = I_k + wY_k \tag{10.39}$$

where I_k is member k's own income unrelated to Y_k.

To the family head, he needs to maximize his utility function subject to the budgetary constraint:

$$X_h + X_k = X_h + (I_k + h_k) = I_h + I_k \tag{10.40}$$

The first-order condition for this optimization problem is given by

$$\begin{bmatrix} \dfrac{\partial U_h}{\partial X_h} \\ \dfrac{\partial U_h}{\partial X_k} \\ \dfrac{\partial U_h}{\partial Y_k} \end{bmatrix} = \begin{bmatrix} 1 \\ \dfrac{1 - Y_k}{2\sqrt{X_k(1 - Y_k)}} \\ \dfrac{w(1 - Y_k) - X_k}{2\sqrt{X_k(1 - Y_k)}} \end{bmatrix} = \lambda \begin{bmatrix} 1 \\ 1 \\ -w \end{bmatrix} \tag{10.41}$$

where $\lambda = 1$ the Lagrange multiplier. So, from Equation (10.41), it follows that

$$\sqrt{\frac{1-Y_k}{X_k}} = 2 \text{ and } \frac{X_k - w(1-Y_k)}{\sqrt{X_k(1-Y_k)}} = 2w \qquad (10.42)$$

So, $w = 1/8$ and the selfish member k chooses $Y_k^* = \frac{1}{2} - 4I_k$ to maximize his utility, and when $Y_k \neq Y_k^*$, we have

$$X_k = 1/4(1 - Y_k) \qquad (10.43)$$

Substituting Equation (10.43) into Equation (10.40) and solving for h_k leads to

$$h_k = \frac{1}{4}(1-Y_k) - I_k$$

This equation implies that member k's reward h_k from the family head h is a decreasing function of Y_k, an index that measures how well k is doing in terms of the established value system.

As shown in Theorem 10.3, it can be seen that in our modern society with so many families with divorced parents, if a father wants to stay on good terms with his kids who live with their mother under a different roof, his monetary gifts to the kids should not be permanently tied to a value-based reward system. If they were, the consequence would be infelicitous. To this end, let us consider the following example, where the resourceful father tries to stay out of each of his children's personal affairs so that he can stay on good terms with them.

Example 10.5: The Resourceful Father

Suppose that the family of our concern has a father and two children, named 1 and 2. The utility functions of the children are given by the following in terms of the father's concern:

$$U_i = m_i, \ i = 1, 2 \qquad (10.44)$$

where m_i stands for child i's consumption in the marketplace using his benevolent father's monetary gifts only. The reason why U_i does not involve child i's spending of his own income from his sources is because the father purposely ignores the effect of those incomes so that all the kids would be as happy as they can be when dealing with himself. That is, the children's utility functions in Equation (10.44) are the father's subjective functions. Let the father's utility function be

$$U_h = U_1 U_2 \qquad (10.45)$$

with the budget constraint

$$m_1 + m_2 = I_h \tag{10.46}$$

where I_h is the father's income, and his own consumption, compared to his children's, is roughly zero.

Maximizing the father's utility U_h subject to the constraint in Equation (10.46) implies that the father will distribute his own income I_h to his two children evenly. That is, $U_i = 0.5I_h$, $i = 1, 2$.

To do our comparison, let us now assume that the father also worries how the kids spend their own resources from the marketplace and how much they enjoy their leisure. In this case, the father's subjective utility functions of the children are

$$U_i = m_i + L_i(Y_i), \; i = 1, 2 \tag{10.47}$$

where m_i is kid i's consumption in the marketplace and $L_i(Y_i)$ his spending on leisure, with Y_i being his effort invested in his work to earn his income $I_i(Y_i)$. Assume that the father's utility function is still given in Equation (10.45). So, the family's total income is $I_h + I_1(Y_1) + I_2(Y_2)$. From Equation (10.47), it follows that to maximize U_h in Equation (10.45) subject to the constraint

$$U_1 + U_2 = I_h + I_1(Y_1) + I_2(Y_2) \tag{10.48}$$

the father will distribute his own income in such a way that

$$U_i = 0.5[I_h + I_1(Y_1) + I_2(Y_2)] \tag{10.49}$$

Because $L_i(Y_i)$ is a decreasing function of Y_i—the more effort put into his work, the less opportunity he has for leisure—$I_i(Y_i) - L_i(Y_i)$ is an increasing function of Y_i. That is, the father now faces an incentive problem, similar to the one in Example 10.1.

More specifically, say $I_2(Y_2) = 0$, due to the fact kid 2 did not go to work at all, and $L_1(Y_1) = 0$, because kid 1 worked all his available time so that he did not have any time to enjoy any leisure. Now, Equation (10.49) implies that

$$U_1 = U_2 = \text{total consumption of kids 1 and 2} = 0.5[I_h + I_1(Y_1)]$$

That is, kid 2 enjoyed his leisure to the fullest amount possible, while his total consumption is the same as that of kid 1. If we see how much each of them gets from the father, we find that

Kid 2 gets $0.5[I_h + I_1(Y_1)]$

and

$$\text{Kid 1 gets } m_1 - I_1(Y_1) = 0.5[I_h - I_1(Y_1)]$$

That is, the indolent kid is handsomely rewarded for his indolence, while the assiduous kid is punished for his assiduousness, maybe as severely as giving half of his own income to the lazy sibling and receiving the monetary gift from the father by less than half of his own income.

What went wrong here is that if the father gifts his children using an established value system, none of the kids would be happy with him (Theorem 10.3). Consequently, now the father only cares about how equally his kids would be happy by gifting his money to them accordingly. However, as a consequence, either (1) he creates an environment for the lazy kid to be lazier and the hardworking kid to work harder, or (2) both kids learned their lesson over time and eventually become welfare recipients.

As of now, we only analyzed the situation that the family head H distributes his resources voluntarily to other members. In fact, Figure 10.3 also suggests the possibility that the spin field of the selfish kid K could take its initiative and proactively grab as much of the head's resources as possible, as in the case of the Samaritan dilemma (Buchanan, 1975; Lagerlof, 2004). To this end, we have the following result.

THEOREM 10.4
If a family has such a head who cares about all other members so much that he transfers his resources to them as long as they need to satisfy their own desires of consumption, then each selfish member will devote as little effort to their work as possible to maximize their amounts of transfers from the head.

Proof. Assume that the family head is labeled H, who has n selfish kids 1, 2, ..., n. Because the kids are selfish and only care about satisfying their own consumption desires, their utility functions can be written as

$$U_i = U_i(m_i, Y_i), \quad \frac{\partial U_i}{\partial m_i} > 0 \text{ and } \frac{\partial U_i}{\partial Y_i} < 0, \, i = 1, 2, ..., n \qquad (10.50)$$

where m_i is kid i's commodity consumption in the marketplace and Y_i his effort invested in his work to earn his income $I_i(Y_i)$ satisfying

$$\frac{\partial I_i}{\partial Y_i} > 0, \, i = 1, 2, ..., n \qquad (10.51)$$

Because the head H cares about all other members of his family, his utility can be written as

$$U_H = U_H(m_H, U_1, U_2, ..., U_n) \qquad (10.52)$$

where m_H stands for the head's own consumption of numeraire good. The family's total income is

$$I_H + \sum_{i=1}^{n} I_i(Y_i) = m_H + \sum_{i=1}^{n} m_i \qquad (10.53)$$

To make all the selfish kids as happy as he could help, the head transfers his resources to maximize his utility. So, we have

$$\begin{bmatrix} \dfrac{\partial U_H}{\partial m_i} \\ \dfrac{\partial U_H}{\partial Y_i} \end{bmatrix} = \begin{bmatrix} \dfrac{\partial U_H}{\partial U_i} \dfrac{\partial U_i}{\partial m_i} \\ \dfrac{\partial U_H}{\partial U_i} \dfrac{\partial U_i}{\partial Y_i} \end{bmatrix} = \lambda \begin{bmatrix} 1 \\ -\dfrac{\partial I_i(Y_i)}{\partial Y_i} \end{bmatrix}, i = 1, 2, \ldots, n \qquad (10.54)$$

where λ is the Lagrange multiplier. By dividing the second row entry equation by that of the first row entry equation in Equation (10.54) and simplifying the result, we have

$$\frac{\partial m_i}{\partial Y_i} = -\frac{\partial I_i(Y_i)}{\partial Y_i} < 0, i = 1, 2, \ldots, n \qquad (10.55)$$

Equation (10.55) implies that as long as the selfish kids could obtain financial care from the head, the kids would devote as little effort to their work as possible.

In fact, the result in Theorem 10.4 is empirically supported by countless real-life evidence documented by Stanley and Danko (1996) in their over 20 years of research on financially successful Americans.

Note: The rotten kid theorem is one of the very important results in household economics. Gary Becker's contribution in this area of economics won him the Nobel Prize in 1992. However, applying the concept of conditional transferable utilities, Bergstrom (1989) establishes his version of the rotten kid theorem and constructs three examples to show that Becker's version does not hold true in general. The importance of Theorem 10.2 is not simply about establishing the condition under which Becker's result holds true in general, but also about the fact that on the basis of Theorem 10.2, we can readily show that Bergstrom's rotten kid theorem is a simple corollary of Theorem 10.1 (see below).

According to Becker (1991, p. 278), the family head h is said to be effectively altruistic toward other members of his family, if h's utility function depends positively on the well-being of others and h's behavior is changed by his altruism. In symbols, h's altruism is defined by

$$U_h = U_h(Z_{1h}, \ldots, Z_{mh}, \Psi_1(U_1), \ldots, \Psi_n(U_n)) \qquad (10.56)$$

satisfying

$$\frac{\partial U_h}{\partial U_i} > 0, i = 1,...,n \qquad (10.57)$$

where U_h is the utility function of h, U_i the utility function of member i, Ψ_i a positive function of U_i, $i = 1, ..., n$, and Z_{jh} the jth commodity consumed by h. His altruism is effective if the equilibrium levels of Z_{jh}, for $j = 1, ..., m$, would be different if none of U_i, $i = 1, ..., n$, entered his utility function.

Following this definition of altruism, Becker's rotten kid theorem (Theorem 10.1) implies that in family situations, altruism is more common than in other forms of organizations, and that "even selfish members are induced by the automatic responses of altruistic members to incorporate the interests of altruistic members into their behavior ... even selfish members act as if they were altruistic" (Becker, 1991, p. 344). Our analysis indicates that what's said here means that if altruistic members simply distribute their own monetary income to all other members without any strings attached, then even selfish members would voluntarily help maximize the family income. Now, Theorems 10.2 and 10.3 lead to the following natural question: If in a family, the parents distribute their income to their children with strings attached, how would the children react to the situation? In fact, the proof of Theorem 10.2 implies that some of the selfish children would not automatically help maximize the parents' income. The importance of this question is that in many family settings, the parents are not only obligated to raise healthy children, but also feel the urge to pass the family value systems to the next generation.

So, what is left as an open problem by Theorem 10.3 is how to design a reward system for each given value system the family head wants to impose on or pass down to the next generation.

When studying the "evolution of the family," Becker observed that "parents have fewer children and more is invested in each child in modern than in traditional societies. In traditional societies much of the investment of time and other resources is made by grandparents, aunts, and other kids because of their interest in the children's well-being and behavior. As a result, modern parents are more shocked by the death of a child and generally more concerned about the welfare of each child because of their sizeable commitment of time, money, and energy" (1991, p. 349). With this observation in place, Theorems 10.2 and 10.3 actually explain why in modern societies parents and children are more likely to be at war with each other. It is because when children do not receive their "spending money" regularly without strings attached, they become unhappy and angrier compared to kids from other families, where a different set or not as easily seen set of strings are attached, and sometimes no strings are attached at all.

Because, in modern times, the parents teach a certain set of knowledge and values to children, which are not taught in school, the two sides have to deal with each other more often than in traditional societies. In these traditional societies, children from different but related families were taught by mostly nonparent teachers, and they learned under peer pressure from a much larger environment. This end explains why in traditional societies more family values and trade secrets were kept within families. For a detailed study on the impact of neighborhood quality on children's learning, consult Patacchini and Zenou (2004). For an excellent work on how parents and children are at war at home and how the parents with their effort can be supported by a wider society, see Hao et al. (2000), Annie E. Casey Foundation (1999), Luker (1996), and Gordon (1997).

10.4 Bergstrom's Rotten Kid Theorem and the Samaritan's Dilemma

The original argument in Becker (1974) assumed that the benevolent head transfers his resources to all other members periodically without a terminating date. This explains why during a certain period "any redistribution of the head's income among members of the household would not affect the consumption of any member" (Theorem 10.1). This assumption is also the foundation on which the following observation was derived: "The head's concern about the welfare of other members provides each, including the head, with some insurance against disasters. If a disaster reduced the income of one member alone, k, by say 50 percent, the head would increase his contributions to k, and thereby offset to some extent the decline in k's income. The head would 'finance' his increased contribution to k by reducing his own consumption and his contributions to other members; in effect each member shares k's disaster by consuming less" (Becker, 1974, p. 1076).

By cutting the continuous and infinitely long timeline and by focusing on only one time period where all members live their lives and receive their monetary gifts from the benevolent head at the end of the period, Bergstrom (1989) established his version of the rotten kid theorem on the concept of conditional transferable utilities. Specifically, in this randomly selected single period of time, Bergstrom (1989) used a two-stage game to rewrite the situation described in Becker's rotten kid theorem. Here, the set of players are the family members, including the benevolent head. In the first stage of the game, each player chooses an action. All the individual actions taken by the players in general influence each other's utilities and affect the income of the head available for distribution. In the second stage of the game, the head decides how to distribute his income among the players. Bergstrom calls this two-stage game the one rotten kids play. Now, Becker's rotten kid theorem can be rephrased for this special scenario as follows.

THEOREM 10.5: Becker's Rotten Kid Theorem on One Time Period (Bergstrom, 1989)
The benevolent head's most preferred outcome for this game is the subgame perfect equilibrium of the game, assuming that the head has no control over what action each player, except himself, chooses in the first stage.

Symbolically, assume that the family has n selfish kids. In the first stage of the game rotten kids play, each kid i chooses an action $a_i \in A_i$, a set of possible actions available to him. In the second stage of the game, kid i receives an allotment $t_i = t_i(\vec{a})$ of money from the head. Let $U_i(\vec{a}, t_i)$ be the utility function of kid i, where $\vec{a} = (a_1, ..., a_n) \in \prod_{i=1}^{n} A_i$, satisfying

$$\sum_{j=1}^{n} t_j(\vec{a}) = I(\vec{a}) \tag{10.58}$$

where $I(\vec{a})$ is the total amount of money available for the head to distribute to the kids. After observing the vector $\vec{a}$ of actions, the head decides on the distribution of his income. He surely has many different ways to distribute this sum of money. If the subgame perfect equilibrium for this two-stage game is the same as the Pareto optimum, which is the benevolent head's most preferred outcome, then the rotten kids are said to have behaved well.

THEOREM 10.6: Bergstrom's Rotten Kid Theorem, 1989
If in an n-member family with $(n - 1)$ selfish kids and a benevolent head, named $1, ..., n$, for any $\vec{a} = (a_1, ..., a_n) \in \prod_{i=1}^{N} A_i$, where A_i is the set of possible actions member i can take, and the utility function $U_i(\vec{a}, t_i)$ of member i takes the following form:

$$U_i(\vec{a}, t_i) = A(\vec{a}) t_i + B_i(\vec{a}) \tag{10.59}$$

for some functions $A(\vec{a})$ and $B_i(\vec{a})$, where t_i is i's share of the head's distribution of his own income, $i = 1, ..., n$, then all kids will surely behave well in the game rotten kids play.

Proof. In Bergstrom (1989), an intriguing proof of this result was given using the concepts of utility possibly sets, utility possibility frontiers, transferable utilities, and simplexes. With Theorem 10.2 established in Section 10.1, we can provide another proof of this result as follows.

Because for each vector $\vec{a} = (a_1, ..., a_n) \in \prod_{i=1}^{n} A_i$, the utility function $U_i(\vec{a}, t_i)$ of member i is given by Equation (10.59), it follows that $A(\vec{a}) > 0$, for all a. It is because the t_i-value stands for the level of member i's consumption, and it has been assumed in all literature of relevant works that the more consumption, the higher the utility. That is, member i does not have such a consumption preference that is in conflict with the way the head distributes his own income. So, Theorem 10.2 implies that Becker's rotten kid theorem holds true. Therefore, Theorem 10.5 holds true. That is, all the selfish kids will surely behave well in the first stage of the game rotten kids play.

In fact, the linear format in Equation (10.59), combined with the work of Wu and Lin (2002), was the initial clue for us to search for a result like that in

Theorem 10.2. In particular, studies in Wu and Lin (2002) suggest that as soon as nonlinearity is involved, one has to face uncertainty or chaos. Each uncertainty in general implies eddy motions instead of wave motions. And eddy motions in general indicate conflicts. More detailed results along this line, and complementary to those in Wu and Lin (2002), can be found in Lin (1998).

Example 10.6: The Nightlight Controversy Continued

Because the nightlight in this situation is jointly owned by the two-member family, the husband and wife named 1 and 2, respectively, it can be seen as a public good of the household, and it should enter simultaneously into the utility functions of the husband and wife. Let Y be the amount of nightlight usage negotiated by the husband and wife. No matter how difficult this negotiation can be (see the analysis in Example 10.2), it has to be done in the first stage of the game rotten kids play. If the utility functions of the husband and wife are of the form in Equation (10.59), that is,

$$U_i(Y, t_i) = A(Y)t_i + B_i(Y), i = 1, 2 \qquad (10.60)$$

where A, B_1, and B_2 are fuctions of Y and t_i is the share of the husband's income distributed to i, $i = 1, 2$, then the wife's choice of Y-value will not be greater than the equilibrium Y_{max} value, which makes U_2 reach its Pareto optimality, even though the husband might want Y-value to be as great as possible. Because the t_i-values are determined after the choice of Y-value, the wife would simply employ the t_i-value from the previous distribution. If over time t_2 increases steadily, then the wife would be willing to choose a gradually increased Y-value.

In particular, let

$$U_1 = U_h(Y, t_i) = Yt_1 + Y \text{ and } U_2 = U_w(Y, t_2) = Yt_2 - Y^2 \qquad (10.61)$$

be the utility functions of the husband and wife. Then, the wife would likely choose $Y = t_2/2$, because that is where her utility will be Pareto optimal. This end conclusion seems to be more plausible than that in Example 10.2, because in a loving and caring household, negotiations about the usage of household public goods should possibly end with a win-win conclusion.

Now, with the condition under which Becker's rotten kid theorem holds true being known, the possible application as outlined by Becker (1974) and quoted in the first paragraph of this section actually suggests the following possible abuses of this remarkable theorem: Because of the head h's altruism, there might be a selfish member k who could likely create a disaster for himself as long as h feels it is not k's purposeful wrongdoing. The intention of k is, of course, to receive as much financial transfer from the head h as possible, even though other members' monetary gifts from the head might be greatly reduced. This abuse of Becker's rotten kid theorem is a generalization of Buchanan's (1975) Samaritan's dilemma. This dilemma is about the following paradoxical situation: An altruistic person A is willing to transfer resources to a second person B, if B comes upon hard times. Then, if B today is to decide how much to save for tomorrow, and if B is well aware of A's altruistic concern for him, B will typically save too little compared to what's socially optimal. The dilemma arises because A is unable to commit not to help B out. However, A's willingness to bail B out if he undersaves serves as an implicit tax on B's behavior on savings. For example, if B saves an extra dollar, A will transfer 20 cents less to B than otherwise. This implicit tax distorts B's incentives for a savings plan. For the given equilibrium level of A's support, B would be better off if he consumes less today and more tomorrow. And this would also make A better off given his altruistic concerns for the welfare of B (Bruce and Waldman, 1990, p. 157).

According to Theorem 10.2, the reason why the Samaritan's dilemma appears is because the transfer of the altruistic A is attached to the selfish B's behavior and his consumption preferences.

Lagerlof (2004) fine-tuned the existing literature on the Samaritan's dilemma by changing the assumption of completely known information to that where a substantial degree of incomplete information exists. By employing game theory, he provides a formal analysis on the following intuition: Suppose that B has a piece of private information described by x about some characteristic of himself that is relevant for his gift from A, satisfying that the larger x's magnitude, the lower B's marginal utility of consumption the next day. Because A cares about the welfare of B, A would make a larger transfer to B tomorrow if A believed that x-value were small. The assumptions about x also imply that the smaller x is, the more B wants to save, everything else being equal. Therefore, B has an incentive to make A believe that his x-value is small. And B may try to do so by using his savings as a signaling device (Spence, 1973). In particular, B has an incentive to save more than in the standard setting with complete information. That is, this mechanism can counteract the incentives to undersave in the traditional Samaritan's dilemma model.

Once again, because B's income from A depends on how A perceives B and B's situation, B behaves selfishly without any time left to think about how to help maximize A's income. This situation is similar to the one where an invisible value system is in place and A determines his transfer to B based on the value check of the system. To eliminate B's dependence on A and his selfish game play against A, Theorems 10.2 and 10.3 imply that A should make his monetary transfers to B without looking at B's situation and B's behavior. If B does indeed suffer from a disaster, A should simply do what is suggested in Becker (1974, p. 1076), that "the head (A) would increase his contributions to (B) and thereby offset to some extent the decline in (B)'s income." Here, the key phrase is "offset to some extent," meaning that even though B receives extra help, his marginal utility from the additional help should not be the same or higher than his regular share of contributions from A before B suffers from the disaster. With such a mechanism of reducing the level of utility in place, B's behaviors, such as saving for tomorrow, will be very different from the those described in the Samaritan's dilemma.

10.5 Maximization of Family Income and Child Labor

In the family as described in Theorem 10.1, if the head's distribution of his income is not in conflict with the consumption preferences of any family member, then Theorem 10.2 implies that all members would be motivated to maximize the family income, and consequently the family's consumption and utility. One consequence of this conclusion is that all selfish kids would be willing to work to bring in more monetary income for the family. That is, the problem of child labor appears. The practical problem now is how much the parents would want their children to labor and how much the children would like to labor. Based on Becker's rotten kid theorem, the kids will work as much as their parents demand them to even though (at times) they might wish to work less. To analyze this issue of child labor, let us focus on the concept of child labor efficiency.

Because Theorem 10.2 is established on the observation of Figure 10.3, let us take a look at the other side of the spinning yoyos without standing to the other side of the structures (Figure 10.4). Here, the head's yoyo spins convergently, meaning that he needs to make money to support his family. And the kid K's yoyo rotates divergently, representing the fact that other than taking in his share of the family's income, he is also willing to do something to help out the family. What are described by Theorem 10.2 and Becker's rotten kid theorem are the interactions between the kid K's yoyo and the head H's yoyo in the enclosed areas. In these areas, H's transfers (m_1) are happily accepted by the kid K, and the kid K actively helps out the head H to suck in more resources. At the same time, the

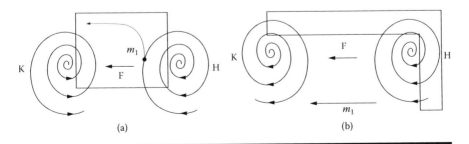

Figure 10.4 Interactions of the other side of the head's and the kid's yoyos.

interactions outside the enclosed areas might not be very cooperating. In fact, in some areas, the kid K's yoyo might be spinning against the spinning direction of the head H's yoyo. This observation is well illustrated by the following example and Proposition 10.1.

Example 10.7

Assume that a family is made up of a parent P and a child C. Similar to what is studied in Baland and Robinson (2000), let the parent's utility function be

$$W_p = c_p + \delta W_c(c_c) \tag{10.62}$$

where c_p stands for the parent's total consumption on various commodities, $W_c(c_c) = c_c^2 + 8$ the child's utility function of consuming as much as c_c on the numeraire of goods, and $\delta \in (0, 1)$ a parameter measuring the degree of the parent's altruism toward the child.

The constraints are given as follows:

$$c_p = 8 + l_c - (l_c + 1) \text{ and } c_c = l_c + 1 \tag{10.63}$$

where 8 is the parent's income from his work, $l_c \in [0, 1]$ the child's income as a child laborer working as much time as l_c out of his endowed unity of available time, and $(l_c + 1)$ the transfer from the parent to the child. Then, from Equations (10.62) and (10.63), the parent's utility function is

$$W_p = 7 + \delta\{(l_c + 1)^2 + 8\}$$

which reaches its maximum at $l_c = 1$, where the child's utility is also optimized. That is, as shown in the enclosed areas in Figure 10.4, the parent's transfers and the kid's willingness to help maximize the family income work out perfectly. However, there is a problem left open here: the child spends his entire endowed unity of time to work as a child laborer. He has no time for school or any other activities, as a kid would enjoy doing. Because of this, the parent should be called a rotten parent.

Similar to the fact that the enclosed areas in Figure 10.4 do not cover the entire spin fields of the parent H's and the kid K's yoyos, the situation constructed in Example 10.7 does not consider the possibility that to some degree, the kid K may not like to spend all his available time to work as a child laborer for income. He may very much like to go out to play with his friends instead of working. To capture this scenario, let us model the situation as follows, based on the idea of Bommier and Dubois (2004). Assume that the parent's and the child's utility functions are given as follows:

$$W_p = U(c_p) + \delta W_c \tag{10.64}$$

and

$$W_c = V_1(1-l_c) + V_2(c_c) \tag{10.65}$$

where $\delta \in [0, 1]$ is the same as in Equation (10.62), $V_2(c_c)$ is the child's utility from consuming c_c amount of money transferred from the parent, and $V_1(1-l_c)$ is the disutility of the child for having to work for the amount of time l_c. Here, we have assumed that the parent controls all the incomes, including both his and the child's. With the setup of our model, it is reasonable to assume that $\frac{\partial c_p}{\partial l_c} > 0$ and $\frac{\partial c_c}{\partial l_c} > 0$. Intuitively, these assumptions mean that by putting the child to work, the family's overall level of consumption increases.

Substituting Equation (10.65) into Equation (10.64) leads to

$$W_p = U(c_p) + \delta V_1(1-l_c) + \delta V_2(c_c) \tag{10.66}$$

The first-order conditions for the parent to maximize his utility W_p with respect to l_c are given by

$$-\delta V_1'(1-l_c) + \delta V_2'(c_c)\frac{\partial c_c}{\partial l_c} = -U'(c_p)\frac{\partial c_p}{\partial l_c} \tag{10.67}$$

At the laissez faire level l_c^* of child labor, we have

$$\left.\frac{\partial W_c}{\partial l_c}\right|_{l_c^*} = \left.-V_1'(1-l_c) + V_2'(c_c)\right|_{l_c^*}$$

$$= -\frac{1}{\delta}U'(c_p)\frac{\partial c_p}{\partial l_c} \text{ (from Equation (10.67))}$$

$$< 0 \text{ (because } U'(c_p) > 0)$$

that is, we have shown the following result.
Proposition 10.1
As long as the child disutility of labor is concerned, the laissez faire level l_c^* of child labor is always inefficiently high.

Intuitively, what Proposition 10.1 says is that the child does not want to work as hard as the parent wants him to in order to maximize the family income. In

186 ■ *Systemic Yoyos: Some Impacts of the Second Dimension*

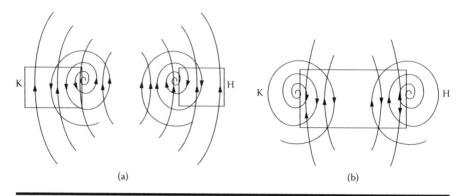

Figure 10.5 **Areas of conflicting desires.**

fact, examples can be constructed to show the possibility that the child does not want to work at all. This end is very well depicted by parts of the interactions of the yoyos in Figure 10.4. In particular, the enclosed areas in Figure 10.5 illustrate the specified interactions, where the parent and the child have contradicting wills and preferences.

After a close analysis on the model established initially by Baland and Robinson (2000), as paraphrased in Example 10.7, we can see a mutually exclusive problem appearing between the child laborer l_c and the child human capital accumulation $(1 - l_c)$. In fact, from our own work experiences or what is recorded by other writers, the child should also be able to accumulate some very important real-life human capital, which can hardly be learned in a school environment (Grootaert and Kanbur, 1995; Canagarajah and Coulombe, 1997).

To not be distracted to the topic of child labor, even though it is very important to investigate for its own sake, from our focus here, the rotten kid theorem, let us stop here with the issue of child labor with the following question open for future research: Is there a theoretical way to balance the kids' motivation to maximize family income and consumption and their needs to have fun as kids?

10.6 Final Words

The work presented in this chapter marks only the first step toward employing the yoyo model, one of the newest results developed in systems science, to the study of economics. As we try to emphasize throughout the chapter, any time a force or a pair of acting and reacting forces exists, the yoyo model can be employed, because the existence of each such pair of forces means an eddy motion as implied in Newton's second law of motion (Lin, 2007). Because of this understanding, we can reasonably expect that we should be able to revisit many or most of the basic concepts in economics in the light of spinning yoyos and their spin fields. In short, the presentation here is only a pebble we throw out there to attract beautiful and more valuable gemstones in the years to come. However, for now, we can recapture what we have done and see what beautiful gemstones we have already attracted:

1. The significance of Becker's rotten kid theorem is that it can be applied to explain many behavioral aspects of household economics and related larger-scale social programs (Becker 1991). Based on the excellent analysis of Bergstrom (1989), we finally have a chance to establish the long-waited sufficient and necessary condition (Theorem 10.2) under which Becker's rotten kid theorem holds true in general.
2. Together with the theorem of never-perfect value systems (Theorem 10.3), this work points to such a paradoxical family situation: To raise healthy kids, the parents are obligated to teach them how to take care of themselves. To achieve this end, they will need to reward the kids for certain desirable habits. That will involve establishing a value system. Theorem 10.3 implies that, consequently, the kids would not be happy with how they are expected to live their lives. To raise socially responsible kids, a similar consequence results. To pass on family value and beliefs to the next generation, Theorem 10.3 indicates that beyond monetary rewards, a lot of convincing and negotiations have to be done. All in all, the parents have to establish reward systems and value systems and at the same time want Becker's rotten kid theorem to hold true for their family. So, an unsettled but extremely important practical question is how they can achieve an operative balance between reinforcing their value systems and keeping the selfish kids motivated to help maximize the family utility.
3. From the discussion about evolution of the family in Section 10.3, one implication of Theorem 10.3 is that to achieve what a parental value system likes to accomplish, the parents need to make use of what's available from outside the family. Such external services and opportunities may easily lessen the burden of the parental reward system. With kids being less angry toward the parents, more agreements in the family can be relatively easily established.
4. Examples in Section 10.1 show that, in general, conflicts among members' consumption preferences may not necessarily create disagreement between how the head's resources are distributed and individual desires of other members. But what can be done when a specific conflict indeed leads to a problematic distribution system? Being unable to resolve this problem can easily lead to a breakup of the family.
5. The previous discussion about the Samaritan's dilemma in fact shows that to raise responsible and independent kids, the parents should never overextend their altruism, even when the kids suffer from setbacks. Otherwise, potentially responsible and independent kids would become welfare recipients. Instead of helping the family to grow positively, they would take advantage of the altruism of the others and their actions would ultimately damage the family harmony.

6. In terms of child labor, Becker's rotten kid theorem implies that all kids would help to maximize the family utility under some conditions (Theorem 10.2). But the real and practical challenge facing parents is how much they should demand the kids to work. How can the parents know with a degree of certainty that the allocation of the kids' endowed time among work, school, sports, and other activities will maximize the kids' potential in life?

Investigations of the open questions posed in this chapter and other related questions are expected to expand the territory of applications of Becker's and Bergstrom's rotten kid theorem, and to bring about brand new understandings on the relevant topics and areas of knowledge.

Chapter 11
Child Labor and Its Efficiency

In this chapter, we present how the systemic yoyo model (Chapter 1, or see Wu and Lin, 2002) and its methodology (Chapter 4, or Lin, 2007) can be meaningfully employed to study problems on child labor in economics. What is new in this chapter from the mainstream literature in economics is that we look at various problems related to child labor as problems of (general) systems and spinning yoyos in the light of whole evolution of these systems (Lin, 1999).

Different from Chapter 10, we now consider the situation that the rotten members of the family are the parents instead of the kids (Bommier and Dubois, 2004), so that the problem of child labor and its efficiency arises. Using the systemic yoyo model as a justification, we introduce three independent variables—child labor, formal schooling, and level of maturity—into the study of child labor. Then, it is shown that as long as children's disutility or utility of labor or formal schooling is concerned, the laissez faire triple of child labor, formal schooling, and level of maturity, which maximizes the parents' utility function, can never be efficient. In particular, such a laissez faire triple is always inefficiently too high. On the other hand, if the efficiency of parent's chosen (for their children) levels of child labor, formal schooling, and level of maturity is determined by the impact on the children's lifetime earnings potential, we can show:

1. The laissez-faire triple of child labor, formal schooling, and level of maturity is efficient either if the parents' savings and bequests are interior when the capital markets are imperfect, or if the parents' bequests are interior and the capital markets are perfect. That is, all externalities of the parents' decision

made for the children would be internalized by their interior transfers in terms of savings (if the capital markets are imperfect) and bequests in order for the children to obtain their optimal lifetime earnings.
2. When the capital markets are perfect, the laissez faire triple of child labor, formal schooling, and level of maturity is efficient if the adult child's transfers to his parents are interior and he lives on a fixed budget. This result implies that when the parents could borrow money in the capital markets, if the child lives below his means when he is an adult, then the parents' decision made for the child, when he was a kid, about the child's labor, formal schooling, and level of maturity, would be efficient in terms of the child's lifetime earnings potential. However, see number 3.
3. When the parents' bequests are at a corner or their savings are at a corner (if the capital markets are imperfect), the laissez faire triple of child labor, formal schooling, and level of maturity is inefficiently high.

To help resolve some of the inefficiencies, which cannot be resolved by individual families, we place family yoyos in a much mightier yoyo, representing the government. In this way, the spin field of the government yoyo forces families and individuals to modify their behaviors accordingly. In particular, if a government regulation is introduced to impose a marginal ban on child labor, then such a ban could be either welfare reducing for both the child and the parents or a Pareto improvement for both the child and the parents, under different sets of specific conditions.

Our presentation below is organized as follows. In Section 11.1, instead of rotten kids, we look at the situation of rotten parents, where children's disutility of work appears. Section 11.2 is devoted to the study of different meanings of the efficiency of child labor and their potentially different outcomes. Because individual families are unlikely to get out of their vicious cycles of child labor on their own, Section 11.3 focuses on the effects of marginal bans on child labor through government regulations.

11.1 Child's Disutility of Work

In this section, we look at the situation in two cases: one contains only one-sided altruism, and the other two-sided altruism.

11.1.1 One-Sided Altruism Model

Similar to the fact that the enclosed areas in Figure 10.4 do not cover the entire spin fields of the parent H's and the kid K's yoyos, the situation constructed in Example 10.4 does not consider the possibility that to some degree, the kid K may not like to spend all his available time to work as a child laborer for income. He may very much like to go out to play with his friends instead of working.

To capture this scenario, going along with Baland and Robinson (2000), Bommier and Dubois (2004), Dessy (2000), and Ranjan (2001), let us assume that there are two time periods, $t = 1, 2$, with no discounting of the future. (Models with only one time period ignore the fact that parental decisions made for children when they are little have lifelong consequences for the children and the parents themselves. See, for example, Basu and Van [1998] and Ravallion and Wodon [2000]). The family of our study consists of parents and children who live through both time periods. In the first period, parents decide how to allocate the children's unit time endowment among work as child labor, formal schooling, and other activities, such as play, sports, and socializing with others. Parents have A efficiency units of labor in each period. In period 1, the children provide $l_c \in [0, 1]$ amount of time at work (child's work includes not just salaried occupations outside the home, but also domestic tasks imposed on children), $e_c \in [0, 1]$ for formal schooling, and $p_c \in [0, 1]$ for other activities, satisfying that $l_c + e_c + p_c = 1$. In this period, parents control all income, including that earned by children. Let $a_c = a_c(l_c, e_c, p_c, m_c) = a_c(l_c, e_c, m_c)$ stand for a child's human capital accumulation acquired through schooling e_c, work l_c, and other activities p_c, where m_c represents the child's level of maturity. (It is our attempt here to model the seemingly controversial empirical literature about the effects of child labor. For example, in their recent study of Ghana, Canagarajah and Coulombe (1997) conclude that "poverty is significantly correlated with the decision to send children to school, and there is a significant negative relationship between going to school and working." To the contrary, Weiner (1991, p. 195) reports, "In India the proprietors of large businesses have not opposed child-labor laws ... one of the complaints of managers of large firms is that their labor force is not sufficiently educated, that too many workers are unable to read manuals or follow the simple instructions written on machines." In terms of social and cognitive skills, Ennew (1982, p. 560) notes that "when relatively young children are forced to care for even younger siblings ... the old child loses the educational opportunity at school, but in addition the younger child may be prevented from developing sufficient verbal and conceptual skills to benefit from formal education.") Assume a_c is strictly increasing and strictly concave upward with respect to each independent variable. Because the sum of l_c, e_c, and p_c is the unity, we can think of a_c as a function of l_c, e_c, and m_c only without having p_c explicitly included. A word about the variable "maturity" m_c: Different children have different levels of maturity no matter what age they are. The level of maturity can be determined partially by a person's ability to interact with others in a socially appropriate manner, and how well he or she takes on his or her responsibilities.

The reason for us to introduce the function a_c for children's accumulation of human capital is that from all activities children acquire knowledge and experience, both positive and negative, that in general will benefit them for the rest of their lives.

In period 2, children are now called adults and working for a living. Their total labor supply in this period is $h(a_c)$, where $h(a_c)$ represents the number of human capital units possessed by an adult who worked for a fraction l_c, and attended formal school for a fraction e_c of his time endowment when a child with a level of maturity m_c. The functions h and a_c are differentiable as needed and strictly increasing and strictly concave upward. In period 2, adults control their own income. We assume that labor markets are all competitive and pay \$1 for each unit of human capital.

We assume that both parents and adult children consume one (aggregate) commodity with unit price 1. For simplicity, we assume that the family has only one child and the parents' and child's utility functions W_p and W_c are related as follows:

$$W_p = U(c_p^1) + U(c_p^2) + \delta W_c(c_c) \tag{11.1}$$

where $\delta \in (0, 1)$ is a parameter measuring the extent to which parents are altruistic toward the child, U and W_c are differentiable as needed, strictly increasing and strictly concave upward, c_p^i is the total consumption of the parents in period $i = 1$, 2, and c_c is the consumption of the adult child in period 2.

Other than choosing the time allocation l_c, e_c, and p_c for the child during period 1, parents in period 2 only decide on monetary transfers to the child, called bequests and denoted $b \geq 0$. Between the periods, they use savings, denoted s, to transfer their income. So, the parents and child face the following budget constraints:

$$c_p^1 = c_p^1(l_c, e_c, m_c) = A + C(a_c) - s(a_c) \tag{11.2}$$

$$c_p^2 = c_p^2(l_c, e_c, m_c) = A - b(s) + s(a_c) = A - b(a_c) + s(a_c) \tag{11.3}$$

$$c_c = c_c(l_c, e_c, m_c) = h(a_c) + b(s) = h(a_c) + b(a_c) \tag{11.4}$$

where we assume that the functions c_p^1, c_p^2, c_c, C, s, and b are all differentiable increasing functions in each of their independent variables; $C = C(a_c)$ is the units of the child's human capital from working as much time as l_c, having formal schooling e_c, and having a level of maturity m_c; $s = s(a_c)$ is the parents' savings in period 1; and $b = b(s) = b(a_c)$ is the parents' bequests in period 2. If we assume that the capital markets are imperfect, then s will have to be nonnegative.

The first-order conditions of maximizing the parents' utility W_p, Equation (11.1), subject to the budgetary constraints in Equations (11.2) to (11.4) with respect to b, l_c, s, e_c, and m_c, are given respectively below:

$$U'(c_p^2) = \delta W_c'(c_c) \text{ if } b > 0, \text{ or } U'(c_p^2) > \delta W_c'(c_c) \text{ if } b = 0 \tag{11.5}$$

$$U'(c_p^1)\left[\frac{\partial C}{\partial l_c} - \frac{\partial s}{\partial l_c}\right] + U'(c_p^2)\left[\frac{\partial s}{\partial l_c} - \frac{\partial b}{\partial l_c}\right] + \delta W_c'(c_c)\left[\frac{\partial h}{\partial l_c} + \frac{\partial b}{\partial l_c}\right] = 0 \tag{11.6}$$

$$U'\left(c_p^1\right) = U'\left(c_p^2\right) \text{ if } s > 0 \text{ or } U'\left(c_p^1\right) > U'\left(c_p^2\right) \text{ if } s = 0 \quad (11.7)$$

$$U'\left(c_p^1\right)\left[\frac{\partial C}{\partial e_c} - \frac{\partial s}{\partial e_c}\right] + U'\left(c_p^2\right)\left[\frac{\partial s}{\partial e_c} - \frac{\partial b}{\partial e_c}\right] + \delta W_c'(c_c)\left[\frac{\partial h}{\partial e_c} + \frac{\partial b}{\partial e_c}\right] = 0 \quad (11.8)$$

$$U'\left(c_p^1\right)\left[\frac{\partial C}{\partial m_c} - \frac{\partial s}{\partial m_c}\right] + U'\left(c_p^2\right)\left[\frac{\partial s}{\partial m_c} - \frac{\partial b}{\partial m_c}\right] + \delta W_c'(c_c)\left[\frac{\partial h}{\partial m_c} + \frac{\partial b}{\partial m_c}\right] = 0 \quad (11.9)$$

The triple (l_c^*, e_c^*, m_c^*) of child labor, formal schooling, and level of the child's maturity is said to be efficient if the triple maximizes the child's lifetime earnings. That is, (l_c^*, e_c^*, m_c^*) maximizes the following function:

$$E_c(l_c, e_c, m_c) = C(a_c) + h(a_c) \quad (11.10)$$

Because work experience and formal schooling are always positive to whomever is involved in terms of human capital accumulation, it is reasonable to assume that $l_c^*, e_c^* > 0$. And because m_c measures a person's level of maturity, we can also assume $m_c^* > 0$.

PROPOSITION 11.1
If the parents' bequests and savings are interior, then the laissez faire triple (l_c^*, e_c^*, m_c^*) of child labor, formal schooling, and level of maturity is efficient.

Proof. Solving Equation (11.6) for $\frac{\partial h(a_c)}{\partial l_c}$ produces

$$\frac{\partial h(a_c)}{\partial l_c} = \frac{-1}{\delta W_c'(c_c)}\left\{U'\left(c_p^1\right)\left[\frac{\partial C}{\partial l_c} - \frac{\partial s}{\partial l_c}\right] + U'\left(c_p^2\right)\left[\frac{\partial s}{\partial l_c} - \frac{\partial b}{\partial l_c}\right] + \delta W_c'(c_c)\frac{\partial b}{\partial l_c}\right\}$$

$$= \frac{-1}{\delta W_c'(c_c)}\left\{U'\left(c_p^2\right)\left[\frac{\partial C}{\partial l_c} - \frac{\partial b}{\partial l_c}\right] + \delta W_c'(c_c)\frac{\partial b}{\partial l_c}\right\} \text{ (from Equation (11.7))}$$

$$= \frac{-1}{\delta W_c'(c_c)}\delta W_c'(c_c)\frac{\partial C}{\partial l_c} \text{ (from Equation (11.5))}$$

$$= -\frac{\partial C(a_c)}{\partial l_c} \quad (11.11)$$

Similarly, solving Equation (11.8) for $\frac{\partial h(a_c)}{\partial e_c}$ and Equation (11.9) for $\frac{\partial h(a_c)}{\partial m_c}$ leads to

$$\frac{\partial h(a_c)}{\partial e_c} = -\frac{\partial C(a_c)}{\partial e_c} \quad (11.12)$$

and

$$\frac{\partial h(a_c)}{\partial m_c} = -\frac{\partial C(a_c)}{\partial m_c} \quad (11.13)$$

Combining Equations (11.11) to (11.13), we have shown that the laissez faire triple (l_c^*, e_c^*, m_c^*), which maximizes the parents' utility, also maximizes the child's lifetime earnings. That is, the triple is efficient.

PROPOSITION 11.2
If the parents' bequests are at a corner, then the laissez faire triple (l_c^*, e_c^*, m_c^*) of child labor, formal schooling, and level of maturity is inefficiently high, meaning that the potential child's lifetime earnings are decreasing at the level of the triple.

Note: It is initially observed by Becker and Murphy (1988) and Nerlove et al. (1988) that nonnegativity constraints on bequests could lead to inefficiencies in the family's resource allocation.

Proof. Because the parents' bequests are at a corner, Equations (11.6) and (11.7) lead to

$$\frac{\partial h(a_c)}{\partial l_c} \leq \frac{-1}{\delta W_c'(c_c)} \left\{ U'(c_p^2) \left[\frac{\partial C}{\partial l_c} - \frac{\partial b}{\partial l_c} \right] + \delta W_c'(c_c) \frac{\partial b}{\partial l_c} \right\}$$

$$< \frac{-1}{\delta W_c'(c_c)} \left[\delta W_c'(c_c) \frac{\partial C}{\partial l_c} \right] \text{ (from Equation (11.5))}$$

$$= -\frac{\partial C(a_c)}{\partial l_c}$$

Similarly, based on Equations (11.8) and (11.9), we have

$$\frac{\partial h(a_c)}{\partial e_c} < -\frac{\partial C(a_c)}{\partial e_c} \quad \text{and} \quad \frac{\partial h(a_c)}{\partial m_c} < -\frac{\partial C(a_c)}{\partial m_c}$$

That is, the laissez faire triple (l_c^*, e_c^*, m_c^*) is inefficiently high.

What is implied in Proposition 11.2 is that when the family is poor, even though the parents are altruistic toward the child, the desired contribution from the child will surely decrease the child's lifetime earnings potential. (This result is similar to what is obtained in Basu and Van [1998] and Eswarran [1996].) The fact that child labor is a facet of poverty has been well established in the empirical literature (see, for example, Rosenzweig, 1981; Labenne, 1997; Chakraborty and Das, 2005; Baland and Robinson [2000] and Bommier and Dubois [2004] establish a clear-up welfare argument that suggests that child labor can be inefficient.) A natural question at this junction is: Would the child be willing to make the wanted contribution to the family? According to the recent work (Chapter 10) on Becker's rotten kid theorem (1974), the answer is both yes and no, depending on how the parents treat the child and what the child's consumption preferences are.

Going back to our case of the poor family, if the child's consumption preference coincides with the desire of the parents, then Becker's rotten kid theorem applies. That is, the child will be willing to meet the parents' demand. Consequently, Proposition 11.2 implies that the child might repeat his parents' lifestyle, being financially poor. On the other hand, if the child's consumption preference is in conflict with how he is treated or demands made on him at home (because in this case, the parents do not have many financial resources to transfer to the child), the child might likely work for the demanded amount of time l_c^* to survive at home. At the same time, he might very well explore other opportunities through formal schooling e_c^* or other activities $p_c^* = 1 - l_c^* - e_c^*$. In this case, the child's level of maturity might rise drastically to have his own consumption preference satisfied.

PROPOSITION 11.3
If the parents' savings are at a corner, then the laissez faire triple (l_c^*, e_c^*, m_c^*) of child labor, formal schooling, and level of maturity is inefficiently high.

Proof. Because the parents' savings are at a corner, Equations (11.6) and (11.7) lead to

$$\frac{\partial h(a_c)}{\partial l_c} < \frac{-1}{\delta W_c'(c_c)}\left\{U'(c_p^2)\left[\frac{\partial C}{\partial l_c} - \frac{\partial b}{\partial l_c}\right] + \delta W_c'(c_c)\frac{\partial b}{\partial l_c}\right\}$$

$$\leq \frac{-1}{\delta W_c'(c_c)}\left[\delta W_c'(c_c)\frac{\partial C}{\partial l_c}\right] \text{ (from Equation (11.5))}$$

$$= -\frac{\partial C(a_c)}{\partial l_c} \qquad (11.14)$$

Similarly, from Equations (11.8) and (11.9), we can produce

$$\frac{\partial h(a_c)}{\partial e_c} < -\frac{\partial C(a_c)}{\partial e_c} \qquad (11.15)$$

and

$$\frac{\partial h(a_c)}{\partial m_c} < -\frac{\partial C(a_c)}{\partial m_c} \qquad (11.16)$$

That is, the laissez faire triple (l_c^*, e_c^*, m_c^*) is inefficiently high.

The inefficiency pointed to in Proposition 11.3 is more likely to arise if the family is poor or lives an extravagant lifestyle in period 1. In the former case, imperfect capital markets prevent the parents from borrowing in the marketplace. So, they borrow from their child's future earnings in period 1. Then, in period 2, they try to internalize the negative effects of child labor on period 2 income through their bequests. In the latter case, where the family lives in extravagant lifestyle in period 1, the capital markets surely will not be perfect enough for the parents to borrow indefinitely. So, they borrow from their child's future earnings. They value period 1 income higher than that of period 2. In such a case, it will be difficult to imagine that the parents will even try to internalize the negative effects of child labor in the second period by bequeathing their child.

Similar to the inefficiency described in Proposition 11.2, if the child's consumption preference in the situation of Proposition 11.3 does not agree with how he is treated at home, but as a child he has to comply with his parents' demand, his level of maturity would potentially rise drastically to have his own consumption preference met while keeping the demanded values of child labor l_c^*, formal schooling e_c^*, and other activities $p_c^* = 1 - l_c^* - e_c^*$.

Both Propositions 11.2 and 11.3 say that when parents leave their children no bequests or transfer no assets from period 1 to period 2 when capital markets are imperfect, the parents fail to internalize the socially efficient trade-off between the levels of child labor, formal schooling, and maturity, and the children's lifetime earning potential.

11.1.2 Adult Child's Altruism toward Parents Model

All assumptions in the previous section are kept the same in this section, except that the child is simultaneously altruistic toward his parents. (Because the inefficiencies of child labor, formal schooling, and level of maturity, as studied in the previous section, are caused partly by the nonnegativity constraint of bequests, we wanted to see if, by introducing filial altruism, the inefficiencies could be eased. Proposition 11.7 shows that they can be if capital markets are perfect. However, when capital markets are imperfect, the same inefficiencies persist.) That is, we assume that the utility function of the child is given as follows:

$$W_c = V(c_c) + \lambda W_p \tag{11.17}$$

where $\lambda \in (0, 1)$ measures the degree of the child's filial altruism and V is differentiable as needed, strictly increasing and strictly concave upward. In this case, we can solve Equations (11.17) and (11.1) for W_p and W_c and obtain:

$$W_p = \frac{1}{1-\delta\lambda}\left[U\left(c_p^1\right) + U\left(c_p^2\right) + \delta V(c_c)\right] \tag{11.18}$$

and

$$W_c = \frac{1}{1-\delta\lambda}\left[V(c_c) + \lambda U\left(c_p^1\right) + \lambda U\left(c_p^2\right)\right] \quad (11.19)$$

subject to the budgetary constraints:

$$c_p^1 = c_p^1(l_c, e_c, m_c) = A + C(a_c) - s(a_c) \quad (11.20)$$

$$c_p^2 = c_p^2(l_c, e_c, m_c) = A - b(a_c) + s(a_c) + \tau(a_c) \quad (11.21)$$

$$c_c = c_c(l_c, e_c, m_c) = h(a_c) + b(a_c) - \tau(a_c) \quad (11.22)$$

where $\tau(a_c) \geq 0$ is a monetary transfer from the child to the parents in period 2. As before, we assume that the functions c_p^1, c_p^2, c_c, C, s, and b are all differentiable increasing in each of their independent variables. Different from Baland and Robinson (2000), we do not assume that the value $\tau(a_c)$ is dependent on the parents' savings s and bequests b.

In this new model, the parents make their choices on l_c, e_c, m_c, s, and b, in both periods to maximize W_p in Equation (11.18). And the child determines his monetary transfers τ to maximize his utility function W_c in Equation (11.19) conditional on l_c, e_c, m_c, and s subject to constraint (11.22). So, the child's first-order condition is given by

$$V'(c_c) = \lambda U'\left(c_p^2\right) \text{ if } \tau > 0, \text{ or } V'(c_c) > \lambda U'\left(c_p^2\right) \text{ if } \tau = 0 \quad (11.23)$$

The first-order conditions of maximizing the parents' utility function W_p (Equation (11.18)) subject to the budgetary constraints in Equations (11.20) to (11.22) with respect to b, l_c, s, e_c, and m_c are given respectively below:

$$U'\left(c_p^2\right) = \delta V'(c_c) \text{ if } b > 0, \text{ or } U'\left(c_p^2\right) > \delta V'(c_c) \text{ if } b = 0 \quad (11.24)$$

$$U'\left(c_p^1\right)\left[\frac{\partial C}{\partial l_c} - \frac{\partial s}{\partial l_c}\right] + U'\left(c_p^2\right)\left[\frac{\partial s}{\partial l_c} - \frac{\partial b}{\partial l_c} + \frac{\partial \tau}{\partial l_c}\right] + \delta V'(c_c)\left[\frac{\partial h}{\partial l_c} + \frac{\partial b}{\partial l_c} - \frac{\partial \tau}{\partial l_c}\right] = 0 \quad (11.25)$$

$$U'\left(c_p^1\right) = U'\left(c_p^2\right) \text{ if } s > 0, \text{ or } U'\left(c_p^1\right) > U'\left(c_p^2\right) \text{ if } s = 0 \quad (11.26)$$

$$U'\left(c_p^1\right)\left[\frac{\partial C}{\partial e_c} - \frac{\partial s}{\partial e_c}\right] + U'\left(c_p^2\right)\left[\frac{\partial s}{\partial e_c} - \frac{\partial b}{\partial e_c} + \frac{\partial \tau}{\partial e_c}\right] + \delta V'(c_c)\left[\frac{\partial h}{\partial e_c} + \frac{\partial b}{\partial e_c} - \frac{\partial \tau}{\partial e_c}\right] = 0 \quad (11.27)$$

$$U'\left(c_p^1\right)\left[\frac{\partial C}{\partial m_c} - \frac{\partial s}{\partial m_c}\right] + U'\left(c_p^2\right)\left[\frac{\partial s}{\partial m_c} - \frac{\partial b}{\partial m_c} + \frac{\partial \tau}{\partial m_c}\right] + \delta V'(c_c)\left[\frac{\partial h}{\partial m_c} + \frac{\partial b}{\partial m_c} - \frac{\partial \tau}{\partial m_c}\right] = 0 \quad (11.28)$$

PROPOSITION 11.4

If the parents' bequests and savings are interior, then the laissez faire triple (l_c^*, e_c^*, m_c^*) of child labor, formal schooling, and level of maturity is efficient.

Proof. From Equation (11.25), it follows that

$$\frac{\partial h}{\partial l_c} = \frac{-1}{\delta V'(c_c)} \left\{ U'(c_p^1) \left[\frac{\partial C}{\partial l_c} - \frac{\partial s}{\partial l_c} \right] + U'(c_p^2) \left[\frac{\partial s}{\partial l_c} - \frac{\partial b}{\partial l_c} + \frac{\partial \tau}{\partial l_c} \right] + \delta V'(c_c) \left[\frac{\partial b}{\partial l_c} - \frac{\partial \tau}{\partial l_c} \right] \right\}$$

$$= \frac{-1}{\delta V'(c_c)} \delta V'(c_c) \frac{\partial C}{\partial l_c} \quad \text{(from Equations (11.26) and (11.24))}$$

$$= -\frac{\partial C}{\partial l_c}$$

Similarly, from Equations (11.27) and (11.28), we get

$$\frac{\partial h}{\partial e_c} = -\frac{\partial C}{\partial e_c} \quad \text{and} \quad \frac{\partial h}{\partial m_c} = -\frac{\partial C}{\partial m_c}.$$

That is, from Equation (11.10), we have shown that the laissez faire triple (l_c^*, e_c^*, m_c^*) is efficient.

PROPOSITION 11.5

If the parents' bequests are at a corner, then the laissez faire triple (l_c^*, e_c^*, m_c^*) of child labor, formal schooling, and level of maturity is inefficiently high.

Proof. Let us assume that the parents' bequests are at a corner. Then, Equation (11.25) implies that

$$\frac{\partial h}{\partial l_c} = \frac{-1}{\delta V'(c_c)} \left\{ U'(c_p^1) \left[\frac{\partial C}{\partial l_c} - \frac{\partial s}{\partial l_c} \right] + U'(c_p^2) \left[\frac{\partial s}{\partial l_c} - \frac{\partial b}{\partial l_c} + \frac{\partial \tau}{\partial l_c} \right] + \delta V'(c_c) \left[\frac{\partial b}{\partial l_c} - \frac{\partial \tau}{\partial l_c} \right] \right\}$$

$$\leq \frac{-1}{\delta V'(c_c)} \left\{ U'(c_p^2) \left[\frac{\partial C}{\partial l_c} - \frac{\partial b}{\partial l_c} + \frac{\partial \tau}{\partial l_c} \right] + \delta V'(c_c) \left[\frac{\partial b}{\partial l_c} - \frac{\partial \tau}{\partial l_c} \right] \right\} \quad \text{(from Equation (11.26))}$$

$$< \frac{-1}{\delta V'(c_c)} \delta V'(c_c) \frac{\partial C}{\partial l_c} = -\frac{\partial C}{\partial l_c} \quad \text{(from Equation (11.24))}$$

Similarly, based on Equations (11.27) and (11.28), we can show that

$$\frac{\partial h}{\partial e_c} < -\frac{\partial C}{\partial e_c} \quad \text{and} \quad \frac{\partial h}{\partial m_c} < -\frac{\partial C}{\partial m_c}$$

From Equation (11.10), it follows that the laissez faire triple (l_c^*, e_c^*, m_c^*) is inefficiently high.

What is implied in Propositions 11.4 and 11.5 is that as long as the filial altruism exists, the child's decision to make monetary transfers to his parents in period 2 has nothing to do with whether the parents' demanded laissez faire triple (l_c^*, e_c^*, m_c^*) in period 1 helps to maximize his lifetime earnings or not. His decision will be purely based on Equation (11.23) to maximize his utility W_c and is determined by his degree of filial altruism. For example, when the condition $V'\bigl(h(a_c^*) + b(a_c^*)\bigr) > \lambda U'\bigl(A - b(a_c^*) + s(a_c^*)\bigr)$, where $a_c^* = a(l_c^*, e_c^*, m_c^*)$, is satisfied, the adult child would not make monetary transfers to his parents even though the parents' bequests $b(a_c^*)$ could be interior.

PROPOSITION 11.6
The adult child's transfers and the parents' bequests cannot both be interior at the laissez faire triple (l_c^*, e_c^*, m_c^*) of child labor, formal schooling, and level of maturity.

Proof. By contradiction, assume that at the laissez faire triple (l_c^*, e_c^*, m_c^*), both the child's transfers and the parents' bequests are interior and the child's utility is maximized. Then, Equations (11.23) and (11.24) imply that

$$V'(c_c) = \lambda U'(c_p^2) = \lambda \delta V'(c_c), \quad 0 < \lambda \delta < 1$$

That is, we must have $V'(c_c) = 0$, which contradicts the hypothesis that V is strictly increasing.

PROPOSITION 11.7
If capital markets are perfect and the adult child's transfers are interior such that the child lives on a fixed budget, then the laissez faire triple (l_c^*, e_c^*, m_c^*) of child labor, formal schooling, and level of maturity is efficient.

Proof. The assumption that capital markets are perfect implies that the equalities in Equations (11.26) and (11.24) hold, even though b might be negative. So, Equation (11.25) leads to

$$\frac{\partial h}{\partial l_c} = \frac{-1}{\delta V'(c_c)} \left\{ U'(c_p^1) \left[\frac{\partial C}{\partial l_c} - \frac{\partial s}{\partial l_c} \right] + U'(c_p^2) \left[\frac{\partial s}{\partial l_c} + \frac{\partial \tau}{\partial l_c} \right] - \delta V'(c_c) \frac{\partial \tau}{\partial l_c} \right\}$$

$$= \frac{-1}{\delta V'(c_c)} \left\{ U'(c_p^2) \left[\frac{\partial C}{\partial l_c} + \frac{\partial \tau}{\partial l_c} \right] - \delta V'(c_c) \left[\frac{\partial C}{\partial l_c} + \frac{\partial \tau}{\partial l_c} \right] + \delta V'(c_c) \frac{\partial C}{\partial l_c} \right\}$$

$$= \frac{-1}{\delta V'(c_c)} \left\{ \left[U'(c_p^2) - \delta V'(c_c) \right] \times \left[\frac{\partial C}{\partial l_c} + \frac{\partial \tau}{\partial l_c} \right] + \delta V'(c_c) \frac{\partial C}{\partial l_c} \right\}$$

where $b(a_c) = 0$ as a consequence of Proposition 11.6. Therefore,

$$\frac{\partial h}{\partial l_c} + \frac{\partial C}{\partial l_c} = \frac{-1}{\delta V'(c_c)}\left[U'\left(c_p^2\right) - \delta V'(c_c)\right] \times \left[\frac{\partial C}{\partial l_c} + \frac{\partial \tau}{\partial l_c}\right] \quad (11.29)$$

By differentiating Equation (11.23) with respect to l_c, we obtain

$$V''\left[\frac{\partial h}{\partial l_c} - \frac{\partial \tau}{\partial l_c}\right] = \lambda U''\left[\frac{\partial s}{\partial l_c} + \frac{\partial \tau}{\partial l_c}\right]$$

and

$$\frac{\partial s}{\partial l_c} + \frac{\partial \tau}{\partial l_c} = \frac{V''}{\lambda U''\left(c_p^2\right)}\left[\frac{\partial h}{\partial l_c} - \frac{\partial \tau}{\partial l_c}\right] \quad (11.30)$$

Based on Equation (11.26), we can rewrite Equation (11.23) as follows:

$$V'(c_c) = \lambda U'\left(c_p^1\right)$$

Differentiating this equation with respect to l_c provides

$$V''\left[\frac{\partial h}{\partial l_c} - \frac{\partial \tau}{\partial l_c}\right] = \lambda U''\left[\frac{\partial C}{\partial l_c} - \frac{\partial s}{\partial l_c}\right]$$

and

$$\frac{\partial C}{\partial l_c} - \frac{\partial s}{\partial l_c} = \frac{V''}{\lambda U''\left(c_p^1\right)}\left[\frac{\partial h}{\partial l_c} - \frac{\partial \tau}{\partial l_c}\right] \quad (11.31)$$

Adding Equations (11.30) and (11.31) leads to

$$\frac{\partial C}{\partial l_c} + \frac{\partial \tau}{\partial l_c} = \frac{V''}{\lambda}\left[\frac{1}{U''\left(c_p^2\right)} + \frac{1}{U''\left(c_p^1\right)}\right]\left[\frac{\partial h}{\partial l_c} - \frac{\partial \tau}{\partial l_c}\right] \quad (11.32)$$

The assumption that the adult child lives on a fixed budget implies that $h(a_c) - \tau(a_c)$ is a constant, his fixed budget. So, we have $\frac{\partial h(a_c)}{\partial l_c} - \frac{\partial \tau(a_c)}{\partial l_c} = 0$. This end means that $\frac{\partial C}{\partial l_c} + \frac{\partial \tau}{\partial l_c} = 0$ (Equation (11.32)) and hence $\frac{\partial h}{\partial l_c} + \frac{\partial C}{\partial l_c} = 0$ (Equation (11.29)).

Similarly, based on Equations (11.27) and (11.28), we can prove

$$\frac{\partial h}{\partial e_c} + \frac{\partial C}{\partial e_c} = 0 \quad \text{and} \quad \frac{\partial h}{\partial m_c} + \frac{\partial C}{\partial m_c} = 0$$

This end implies that the laissez faire triple (l_c^*, e_c^*, m_c^*) is efficient.

In fact, the proof of Proposition 11.7 also implies that if $\frac{d(h-\tau)}{da_c} > 0$, then $\frac{\partial E_c}{\partial l_c} < 0$, $\frac{\partial E_c}{\partial e_c} < 0$, and $\frac{\partial E_c}{\partial m_c} < 0$. That is, the laissez faire triple (l_c^*, e_c^*, m_c^*) is inefficiently high. And, if $\frac{d(h-\tau)}{da_c} < 0$, then $\frac{\partial E_c}{\partial l_c} > 0$, $\frac{\partial E_c}{\partial e_c} > 0$, and $\frac{\partial E_c}{\partial m_c} > 0$. That is, the laissez faire triple (l_c^*, e_c^*, m_c^*) is inefficiently low, meaning that even though the triple (l_c^*, e_c^*, m_c^*) maximizes the parents' and the adult child's utilities, it does not optimize the child's lifetime earnings potential.

PROPOSITION 11.8

When capital markets are imperfect and the parents' savings are at a corner, the laissez faire triple (l_c^*, e_c^*, m_c^*) of child labor, formal schooling, and level of maturity is inefficiently high.

Proof. Both Equations (11.25) and (11.26) imply that

$$\frac{\partial h}{\partial l_c} < \frac{-1}{\delta V'(c_c)} \left\{ U'(c_p^2) \left[\frac{\partial C}{\partial l_c} + \frac{\partial \tau}{\partial l_c} \right] - \delta V'(c_c) \frac{\partial \tau}{\partial l_c} \right\}$$

$$\leq \frac{-1}{\delta V'(c_c)} \delta V'(c_c) \frac{\partial C}{\partial l_c} \quad \text{(from Equation (11.24))}$$

$$= -\frac{\partial C}{\partial l_c}$$

That is, $\frac{\partial E_c}{\partial l_c} < 0$. Similarly, based on Equations (11.27) and (11.28), we can prove that $\frac{\partial E_c}{\partial e_c} < 0$ and $\frac{\partial E_c}{\partial m_c} < 0$. That is, the laissez faire triple (l_c^*, e_c^*, m_c^*) is inefficiently high.

The proof of Proposition 11.8 implies that the laissez faire triple (l_c^*, e_c^*, m_c^*) is inefficiently high even if the adult child's transfers or the parents' bequests are interior.

11.2 Different Efficiencies and Potentially Different Outcomes

Walking along the same path as what we have done so far in this chapter, let us look at Figure 10.2 and consider such a question: When the spin field and the yoyo structure of the kid K is weak, meaning that K is a minor child, the family head H, say a parent, might purposely feed the kid with various gifts or opportunities, such as monetary, educational, or other kinds, in the hope that the kid's yoyo could soon be strong enough to become self-sufficient and independent. As shown in Figure 10.2, not all kinds of gifts the kid wants to accept (the difference between m_1 and m_2), and for a specific kind of gift, the kid's willingness to accept does depend on how it is represented to the kid. So, how can we model the situation that when the parent makes a decision regarding the kid, we know if the decision and its inevitable consequences are good for the kid or not?

As Bommier and Dubois (2004) state, childhood is often related to as the time when one had the largest amount of long-lasting memory of both positive and negative experiences. It is the childhood quality of life that has the greatest impact on the valuation of lifetime utility, even though childhood is only a small fraction of life. Also, during childhood, individuals acquire a large part of their human capital, which is a major determinant of the ability to raise lifetime earnings potential. So, following Bommier and Dubois (2004), let us in this section add a child's disutility from period 1 into our model. In particular, all assumptions in previous sections will be kept the same in this section, except that the child's utility function W_c is given by

$$W_c = V_1(1 - l_c - e_c) + V_2(c_c) + \lambda W_p \tag{11.33}$$

where $\lambda \in (0, 1)$ is defined as before, the child's degree of his filial altruism; V_2 is the child's selfish utility function, dependent on his adult consumption in period 2; and $V_1(1 - l_c - e_c)$ is the child's disutility of working l_c and formal schooling e_c. Assume that V_1 and V_2 are strictly increasing, strictly concave upward, and differentiable as needed. In this section, if the child's utility W_c reaches its maximum at (l_c^*, e_c^*, m_c^*), then we say that the laissez faire triple (l_c^*, e_c^*, m_c^*) of child labor, formal schooling, and level of maturity is efficient.

Combining Equations (11.1) and (11.33), we are able to see that the parents are willing to maximize their utility,

$$W_p = \frac{1}{1 - \lambda \delta} \left[U\left(c_p^1\right) + U\left(c_p^2\right) + \delta V_1(1 - l_c - e_c) + \delta V_2(c_c) \right] \tag{11.34}$$

under their first and second period budgetary constraints in Equations (11.20) and (11.21). And, the child's objective is to maximize his utility,

$$W_c = \frac{1}{1 - \lambda \delta} \left[V_1(1 - l_c - e_c) + V_2(c_c) + \lambda U\left(c_p^1\right) + \lambda U\left(c_p^2\right) \right] \tag{11.35}$$

under the budgetary constraints in Equation (11.22).

The respective first-order conditions for the parents to maximize their utility W_p with respect to l_c, e_c, and m_c are given as follows:

$$\delta V_1'(1 - l_c - e_c) - \delta V_2'(c_c) \frac{\partial c_c}{\partial l_c} = U'\left(c_p^1\right) \frac{\partial c_p^1}{\partial l_c} + U'\left(c_p^2\right) \frac{\partial c_p^2}{\partial l_c} \tag{11.36}$$

$$\delta V_1'(1 - l_c - e_c) - \delta V_2'(c_c) \frac{\partial c_c}{\partial e_c} = U'\left(c_p^1\right) \frac{\partial c_p^1}{\partial e_c} + U'\left(c_p^2\right) \frac{\partial c_p^2}{\partial e_c} \tag{11.37}$$

$$-\delta V_2'(c_c) \frac{\partial C_c}{\partial m_c} = U'\left(c_p^1\right) \frac{\partial c_p^1}{\partial m_c} + U'\left(c_p^2\right) \frac{\partial c_p^2}{\partial m_c} \tag{11.38}$$

PROPOSITION 11.9
The laissez faire triple (l_c^*, e_c^*, m_c^*) of child labor, formal schooling, and level of maturity is inefficiently high.

Proof. From Equation (11.35), it follows that

$$\frac{\partial W_c}{\partial l_c} = \frac{1}{1-\lambda\delta}\left[V_1'(1-l_c-e_c)(-1) + V_2'(c_c)\frac{\partial c_c}{\partial l_c} + \lambda U'(c_p^1)\frac{\partial c_p^1}{\partial l_c} + \lambda U'(c_p^2)\frac{\partial c_p^2}{\partial l_c}\right]$$

$$= \frac{\lambda - \frac{1}{\delta}}{1-\lambda\delta}\left[U'(c_p^1)\frac{\partial c_p^1}{\partial l_c} + U'(c_p^2)\frac{\partial c_p^2}{\partial l_c}\right] < 0 \text{ (from Equation (11.36))}$$

Similarly, from Equations (11.37) and (11.38), it follows that $\frac{\partial W_c}{\partial e_c} < 0$ and $\frac{\partial W_c}{\partial m_c} < 0$. So, we have proven that the laissez faire triple (l_c^*, e_c^*, m_c^*) is inefficiently high.

The inefficiency described in Proposition 11.9 holds true no matter whether capital markets are perfect or not, no matter whether the parents' bequests or savings are interior or at a corner, and no matter whether the child's transfers are interior or not. This result shows that one obtains an inefficiency as soon as children's utility and disutility become the criterion. The inefficiency of a child's time allocation and his level of maturity results only from communications internal to the family between the child and the parents. Driven by two-way altruism, adequate communication between the parents and the child would help to achieve Pareto-improving allocation of child labor l_c^* and formal schooling e_c^* while helping the child to become more mature. In particular, the inefficiency described in Proposition 11.9 is caused by a conflict between the parents' altruism toward the child and the child's consumption preference (Lin and Forrest, 2008). Here, the parents first decide how the child is treated in period 1, and then the child determines if he is willing to help out the family or meet the demand of the parents by deciding whether he should join the workforce as a child laborer, and how he would deal with his formal schooling, and consequently, his maturity would land at an appropriate level.

Also, our analysis above holds true if the child's disutility $V_1(1-l_c-e_c)$ of working and formal schooling is replaced by either the disutility $V_1(1-l_c)$ of working only or the disutility $V_1(1-e_c)$ of formal schooling only, respectively. These individual cases would cover the following situations:

1. The child does not like to work for the purpose of making money, but prefers going to school and any other activities.
2. The child does not like to experience the required formal schooling, but prefers working for money and any other activities.

11.3 Marginal Bans on Child Labor

As analyzed above, especially in Proposition 11.9, it is highly unlikely that individual families are capable of resolving the inefficiencies found in this chapter. So, a natural thinking is to introduce government regulations to mandate some behaviorial changes of all citizens, such as imposing a reduction of child labor. (Historically, trade unions were influential in pressing for banning child labor. One reason is that it is thought to depress wages [Davin, 1982; Zelizer, 1994; Doepke and Zilibotti, 2005].) In terms of the yoyo model, this thinking implies that we place individual family yoyo collections in the influence of the spin field of a much mightier yoyo, which is the government. The reason this government yoyo is much mightier than the individual family yoyo or an individual person's yoyo is because the government yoyo has the court system, legal system, police force, etc., at its disposal. Figure 11.1 shows how such a mightier spin field could affect individuals differently. Briefly, it flattens all small yoyos in the same direction. No matter whether the spin field influence of the government yoyo goes upward or downward, the effect of a partial ban on child labor will be mixed, depending on the specific circumstances that individuals are in. For example, in the enclosed area in Figure 11.1(a), if the government's spin field goes downward, a partial ban on child labor will be welfare reducing for both the child K and the parent H. This is a case where we assume that a linear technology is in place, meaning that the field of H spins at a constant speed.

Let us first analyze the partial equilibrium (or linear technology) effect of a marginal ban on child labor. Here, the partial equilibrium (or linear technology) is defined to mean that any variation of child labor supply has no effect on child and adult wages and firms' profits. Then, we have the following results.

PROPOSITION 11.10
For all three models studied above, the model of one-sided altruism, that of two-sided altruism, and that with a child's disutility or utility, at partial equilibrium (or with linear technology), a marginal ban on child labor is welfare enhancing for the child but not for the parents if, for the first two models, the parents' bequests or savings are at a corner.

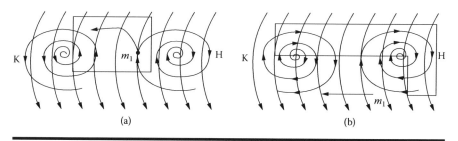

Figure 11.1 Government yoyo's influence on individual's behavior and preferences.

Proof. For both the one-sided and two-sided altruism models, Propositions 11.2, 11.3, 11.5, and 11.8 imply that if the parents' bequests or savings are at a corner, the laissez faire triple (l_c^*, e_c^*, m_c^*) of child labor, formal schooling, and level of maturity is inefficiently high. So, when the child labor l_c^* is marginally reduced, the parents' utility W_p will be short from its maximum, while the efficiency of child labor will be accordingly improved.

For the model with child's disutility or utility from period 1, Proposition 11.9 and the discussions earlier imply that the laissez faire triple (l_c^*, e_c^*, m_c^*) is inefficiently high. So, the same reasoning as given in the previous paragraph applies here, which ends the proof.

PROPOSITION 11.11
If either of the following conditions holds,

(i) capital markets are perfect and $\frac{d(h-\tau)}{da_c} > 0$;
(ii) capital markets are imperfect and parents' savings are at a corner,

then a marginal ban on child labor is welfare enhancing for the child but not for the parents.

Proof. The result follows from the remarks beneath Proposition 11.7 and Proposition 11.8.

PROPOSITION 11.12
If capital markets are perfect and $\frac{d(h-\tau)}{da_c} < 0$, then a marginal ban on child labor is welfare reducing for both the child and the parents.

Proof. The proof of Proposition 11.7 implies that when capital markets are perfect and $\frac{d(h-\tau)}{da_c} < 0$, the laissez faire triple (l_c^*, e_c^*, m_c^*) maximizes the parents' utility, but not enough to optimize the child's lifetime earnings potential. So, in this case, a marginal ban on child labor will not only make the parents' utility W_p short from its maximum, but also pull the child farther away from reaching his maximum lifetime earnings potential.

PROPOSITION 11.13
If one of the following conditions holds true,

(i) for both one-sided and two-sided altruism models, the parents' bequests and savings are interior;
(ii) for the two-sided altruism model, capital markets are perfect, the child's transfers are interior, and the child lives on a fixed budget,

then a marginal ban on child labor is welfare reducing for both the child and the parents.

Proof. The conclusion follows from Propositions 11.1, 11.4, and 11.7.

Second, suppose that a marginal ban on child labor is accompanied by a government legislation requiring an additional amount of formal schooling for each child. Then the child's human capital accumulation $a_c = a_c(l_c, e_c, m_c)$ might not necessarily drop with the marginal ban on child labor. For example, let $a_c = a_c(l_c, e_c, m_c) = l_c + (1+e_c)^2 + m_c^2$, where we assume that the level of child's maturity m_c is an index ≥ 2. If the marginal ban on child labor reduces the total available child labor supply by Δl_c (> 0) and the government legislation requires an increasing amount of formal schooling by this available time slot Δl_c, then we have

$$a_c(l_c - \Delta l_c, e_c + \Delta l_c, m_c) = (l_c - \Delta l_c) + (1 + e_c + \Delta l_c)^2 + m_c^2$$
$$= a_c(l_c, e_c, m_c) + 2e_c \Delta l_c + \Delta l_c + \Delta l_c^2$$
$$> a_c(l_c, e_c, m_c)$$

In this case, both the total child income $C(a_c)$ and the adult child income $h(a_c)$ will in fact be increased.

At the same time, when the child labor drops by Δl_c (> 0) and formal schooling increases by Δl_c, if the child's level of maturity drops as much as $2\Delta l_c$ as a consequence of the marginal ban on child labor, then we have

$$a_c(l_c - \Delta l_c, e_c + \Delta l_c, m_c - 2\Delta l_c) = a_c(l_c, e_c, m_c) + \Delta l_c(2e_c + 1 + 5\Delta l_c - 4m_c)$$
$$< a_c(l_c, e_c, m_c)$$

That is, in this case, both child income $C(a_c)$ and adult income $h(a_c)$ will drop. And both the parents and child are worse off.

Therefore, to guarantee success of any marginal ban on child labor, the corresponding levels of formal schooling and maturity should be kept up. Additionally, this analysis also suggests that a complete ban on child labor might not be a good idea. Here, by *complete ban* we mean that children are not allowed to work for income in any possible manner. It is because working experience, both positive and negative, surely helps to increase the value of m_c, the level of maturity.

Third, let us look at the general equilibrium (or nonlinear technology) effects of a marginal ban on child labor, where a ban on child labor reduces the supply of child labor in period 1 and increases the supply of adult labor in period 2. As a result, period 1 wages might rise and period 2 wages might fall. (If we position this situation on an infinite timeline, it can be seen that all individuals are first children and then adults. So, the transition from a situation where there is no ban on child labor to a situation where there is a ban on child labor will be costly for the generation that spend their childhood before the transition and their adulthood after the transition. So, in a partial equilibrium approach or with linear technology, the introduction of a ban on child labor does not induce Pareto improvement, even though such a ban will benefit all the subsequent generations, because it is costly only to one generation. This remark also implies that when such government legislation as a ban on child labor is introduced, it should be implemented gradually to reduce the cost of the generation that lives across the transition.) To develop our analysis, assume that there are agents in the economy who run firms and use labor to produce the numeraire good. For simplicity, we assume that only one such agent (a representative firm) exists, who lives through both time periods, has no children, and is endowed with nonlinear technologies for converting efficiency units of labor into commercial good.

The firm's profit is given by

$$\pi = \sum_{t=1}^{2} \sum_{i=p,c} [f_i(L_{it}) - w_{it} L_{it}] \quad (11.39)$$

where L_{it} represents the firm's demand of i ($= p, c$) kind of labor, $p =$ parents and $c =$ children, in period t ($= 1, 2$); w_{it} is the wage rate of i kind of labor in period t; and f_i, $i = p, c$, is the firm's economic output of the parental or child labor so that $f_p(AL_p)$ is the output produced by parents in each time period, and $f_c(C(a_c)L_p)$ and $f_c(h(a_c)L_p)$ are the outputs produced by children in periods 1 and 2, respectively. Assume that the production functions f_i, $i = p, c$, are differentiable as needed, strictly increasing and strictly concave upward. The firm's objective is to choose L_{it} to maximize its profit π given in Equation (11.39). So, the optimal labor demand L_{it} satisfies

$$f'_i(L_{it}) - w_{it} = 0 \quad (11.40)$$

Considering available labor supplies in periods 1 and 2 (in efficiency units), equilibrium wages w^*_{it} should satisfy the following market-clearing conditions:

$$\left. \begin{array}{ll} L_{p1}(w_{p1}) = AL_p, & L_{c1}(w_{c1}) = L_p C\left(a^*_c\right) \\ L_{p2}(w_{p2}) = AL_p, & L_{c2}(w_{c2}) = L_p h\left(a^*_c\right) \end{array} \right\} \quad (11.41)$$

where $a^*_c = a_c\left(l^*_c, e^*_c, m^*_c\right)$ for the laissez faire triple (l^*_c, e^*_c, m^*_c) of child labor, formal schooling, and level of maturity.

We now study when a marginal ban of child labor will be Pareto improving to the parents, the child, and the representative firm in the economy under the general equilibrium effects.

First, for the model of one-sided altruism, the parents maximize their utility W_p in Equation (11.1) subject to the following budgetary constraints:

$$\left. \begin{array}{l} c^1_p = c^1_p(l_c, e_c, m_c) = w_{p1} A + w_{c1} C(a_c) - s(a_c) \\ c^2_p = c^2_p(l_c, e_c, m_c) = w_{p2} A - b(a_c) + s(a_c) \\ c_c = c_c(l_c, e_c, m_c) = w_{c2} h(a_c) + b(a_c) \end{array} \right\} \quad (11.42)$$

where as before, c_p^1, c_p^2, c_c, C, s, and b are all differentiable increasing functions in each of their independent variables. So, the first-order conditions in Equations (11.5) and (11.7) still hold true.

For the firm, from Equation (11.39) it follows that the effects of a small ban of child labor on its profits are given by

$$\frac{\partial \pi}{\partial l_c} = -\frac{\partial w_{c1}}{\partial l_c} L_{c1} - \frac{\partial w_{c2}}{\partial l_c} L_{c2} \qquad (11.43)$$

Differentiating the market-clearing conditions in Equation (11.41) provides

$$L'_{c1}(w_{c1}) \frac{\partial w_{c1}}{\partial l_c} = L_p \frac{\partial C}{\partial l_c} \quad \text{and} \quad L'_{c2}(w_{c2}) \frac{\partial w_{c2}}{\partial l_c} = L_p \frac{\partial h}{\partial l_c} \qquad (11.44)$$

So, Equation (11.43) can be rewritten as

$$\frac{\partial \pi}{\partial l_c} = -L_p \left(\frac{\partial C}{\partial l_c} w_{c1} \varepsilon_{c1} + \frac{\partial h}{\partial l_c} w_{c2} \varepsilon_{c2} \right) \qquad (11.45)$$

where $\varepsilon_{it} = \frac{\partial w_{it}}{\partial L_{it}} \frac{L_{it}}{w_{it}}$ is the elasticity of the wage rate to the amount of labor demanded, $i = p, c, t = 1, 2$. So, $\frac{\partial \pi}{\partial l_c}$ is nonpositive if

$$\frac{\partial C}{\partial l_c} w_{c1} \varepsilon_{c1} + \frac{\partial h}{\partial l_c} w_{c2} \varepsilon_{c2} \geq 0 \qquad (11.46)$$

For parents, the effects of the ban are given by

$$\frac{\partial W_p}{\partial l_c} = U'(c_p^1) \left[w_{c1} \frac{\partial C}{\partial l_c} (\varepsilon_{c1} + 1) + w_{c2} \frac{\partial h}{\partial l_c} (\varepsilon_{c2} + 1) \right]$$

assuming that the parents' bequests and savings are interior. That is, if we have the following, then the effects on the parental welfare are nonpositive:

$$w_{c1} \frac{\partial C}{\partial l_c} (\varepsilon_{c1} + 1) + w_{c2} \frac{\partial h}{\partial l_c} (\varepsilon_{c2} + 1) \leq 0 \qquad (11.47)$$

For the child, the effects of the ban are

$$\frac{\partial W_c}{\partial l_c} = W'(c_c) \left[w_{c2} \frac{\partial h}{\partial l_c} (\varepsilon_{c2} + 1) + \frac{\partial b}{\partial l_c} \right]$$

which are nonpositive if

$$w_{c2} \frac{\partial h}{\partial l_c} (\varepsilon_{c2} + 1) + \frac{\partial b}{\partial l_c} \leq 0 \qquad (11.48)$$

PROPOSITION 11.14

For the one-sided altruism model, no ban on child labor can be a Pareto improvement.

Proof. If the parents' bequests and savings are interior, then a ban on child labor is a Pareto improvement if Equations (11.46) to (11.48) hold true simultaneously. But Equations (11.46) and (11.47) cannot hold true at the same time.

If the parents' bequests or savings are not interior, then the first-order conditions (11.5) and (11.7) imply that

$$\frac{\partial W_p}{\partial l_c} \geq \delta W'_c(c_c) \left[w_{c1} \frac{\partial C}{\partial l_c} (\varepsilon_{c1} + 1) + w_{c2} \frac{\partial h}{\partial l_c} (\varepsilon_{c2} + 1) \right]$$

$$= \delta W'_c(c_c) \left[\left(w_{c1} \varepsilon_{c1} \frac{\partial C}{\partial l_c} + w_{c2} \varepsilon_{c2} \frac{\partial h}{\partial l_c} \right) + \left(w_{c1} \frac{\partial C}{\partial l_c} + w_{c2} \frac{\partial h}{\partial l_c} \right) \right]$$

So, if Equation (11.45) holds true, we must have $\frac{\partial W_p}{\partial l_c} > 0$.

PROPOSITION 11.15

For the one-sided altruism model, a marginal increase on formal schooling is a Pareto improvement, if

$$E = \frac{\partial C}{\partial e_c} w_{c1} \varepsilon_{c1} + \frac{\partial h}{\partial e_c} w_{c2} \varepsilon_{c2} \leq 0 \tag{11.49}$$

$$E + \left(w_{c1} \frac{\partial C}{\partial e_c} + w_{c2} \frac{\partial h}{\partial e_c} \right) \geq 0 \tag{11.50}$$

and

$$w_{c2} \frac{\partial h}{\partial e_c} (\varepsilon_{c2} + 1) + \frac{\partial b}{\partial e_c} \geq 0$$

Proof. The details are similar to those used in the proof of the previous proposition and are omitted.

Second, for the model of two-sided altruism, the parents' and child's utilities are given in Equations (11.18) and (11.19) subject to the following constraints:

$$\left. \begin{aligned} c^1_p &= c^1_p(l_c, e_c, m_c) = w_{p1} A + w_{c1} C(a_c) - s(a_c) \\ c^2_p &= c^2_p(l_c, e_c, m_c) = w_{p2} A - b(a_c) + s(a_c) + \tau(a_c) \\ c_c &= c_c(l_c, e_c, m_c) = w_{c2} h(a_c) + b(a_c) - \tau(a_c) \end{aligned} \right\} \tag{11.51}$$

where c_p^1, c_p^2, c_c, C, s, b, and τ are the same as before with the first-order conditions in Equations (11.24) and (11.26) holding true for the parents to maximize their utility W_p.

PROPOSITION 11.16
For the two-sided altruism model, no ban on child labor can be a Pareto improvement.

Proof. If a ban on child labor is a Pareto improvement, then its effects on the firm's profit π, the parents' utility W_p, and the child's utility W_c must satisfy

$$\frac{\partial \pi}{\partial l_c} \leq 0, \quad \frac{\partial W_p}{\partial l_c} \leq 0, \quad \frac{\partial W_c}{\partial l_c} \leq 0$$

However, based on Equations (11.18), (11.24), and (11.26), it can be shown that

$$\frac{\partial W_p}{\partial l_c} \geq \frac{\delta V'(c_c)}{1-\delta\lambda}\left[w_{c1}\frac{\partial C}{\partial l_c}(\varepsilon_{c1}+1)+w_{c2}\frac{\partial b}{\partial l_c}(\varepsilon_{c2}+1)\right]$$

Therefore, if $\frac{\partial \pi}{\partial l_c} \leq 0$ or Equation (11.46) holds true, we have $\frac{\partial W_p}{\partial l_c} > 0$.

PROPOSITION 11.17
For the two-sided altruism model, a marginal increase on formal schooling is a Pareto improvement, if Equations (11.49), (11.50), and the following hold true:

$$w_{c2}\frac{\partial b}{\partial e_c}(\varepsilon_{c2}+1)+\delta\lambda w_{c1}\frac{\partial C}{\partial e_c}(\varepsilon_{c1}+1) \geq (1-\delta\lambda)\left(\frac{\partial \tau}{\partial e_c}-\frac{\partial b}{\partial e_c}\right) \quad (11.52)$$

Proof. Both Equations (11.49) and (11.50) imply that $\frac{\partial \pi}{\partial e_c} \geq 0$ and $\frac{\partial W_p}{\partial e_c} \geq 0$. The effects of a marginal increase of formal schooling on the child are given by

$$\frac{\partial W_c}{\partial e_c} = \frac{1}{1-\delta\lambda}\left[V'(c_c)\frac{\partial c_c}{\partial e_c}+\lambda U'(c_p^1)\frac{\partial c_p^1}{\partial e_c}+\lambda U'(c_p^2)\frac{\partial c_p^2}{\partial e_c}\right]$$

$$\geq \frac{V'(c_c)}{1-\delta\lambda}\left[\frac{\partial c_c}{\partial e_c}+\lambda\delta\frac{\partial c_p^1}{\partial e_c}+\lambda\delta\frac{\partial c_p^2}{\partial e_c}\right]$$

$$= \frac{V'(c_c)}{1-\delta\lambda}\left[w_{c2}\frac{\partial b}{\partial e_c}(\varepsilon_{c2}+1)+\delta\lambda w_{c1}\frac{\partial C}{\partial e_c}(\varepsilon_{c1}+1)+(1-\delta\lambda)\left(\frac{\partial b}{\partial e_c}-\frac{\partial \tau}{\partial e_c}\right)\right]$$

So, Equation (11.52) implies $\frac{\partial W_c}{\partial e_c} \geq 0$. That is, when the required conditions are satisfied, any marginal increase on formal schooling is a Pareto improvement.

Third, for the model with a child's disutility from period 1, the parents' and child's utilities are given in Equations (11.34) and (11.35) subject to the budgetary constraints in Equation (11.51). So, Equations (11.26), (11.36), and (11.37) hold true as the first-order conditions of the parents' utility maximization with respect to s, l_c, and e_c.

The effects of a small ban of child labor on parental welfare are given by

$$\frac{\partial W_p}{\partial l_c} = \frac{1}{1-\delta\lambda}\left\{U'\left(c_p^1\right)\frac{\partial c_p^1}{\partial l_c} + U'\left(c_p^2\right)\frac{\partial c_p^2}{\partial l_c} - \left[\delta V_1'(1-l_c-e_c) - \delta V_2'(c_c)\frac{\partial c_c}{\partial l_c}\right]\right\}$$

$$\leq \frac{1}{1-\delta\lambda}\left\{U'\left(c_p^1\right)\left[\frac{\partial c_p^1}{\partial l_c} + \frac{\partial c_p^2}{\partial l_c}\right] - \left[U'\left(c_p^1\right)\frac{\partial c_p^1}{\partial l_c} + U'\left(c_p^2\right)\frac{\partial c_p^2}{\partial l_c}\right]\right\}$$

($\because$ Equations (11.26) and (11.36))

$$\leq \frac{1}{1-\delta\lambda}\left[U'\left(c_p^1\right) - U'\left(c_p^2\right)\right]\cdot\left[\frac{\partial c_p^1}{\partial l_c} + \frac{\partial c_p^2}{\partial l_c}\right]$$

where the equality holds true when parents' savings are interior or capital markets are perfect.

The effects of the ban on the child are

$$\frac{\partial W_c}{\partial l_c} = \frac{\delta\lambda-1}{(1-\delta\lambda)\delta}\left[U'\left(c_p^1\right)\frac{\partial c_p^1}{\partial l_c} + U'\left(c_p^2\right)\frac{\partial c_p^2}{\partial l_c}\right] \text{ (from Equation (11.36))}$$

< 0 (because $\delta\lambda - 1 < 0$)

Therefore, we have partially shown the following result.

PROPOSITION 11.18
For the model with a child's disutility or utility from period 1, if the parents' savings are interior or capital markets are perfect and Equation (11.46) holds true, then a ban on child labor is a Pareto improvement.

Note

Results in Propositions 11.10 to 11.14 and 11.16 agree with Basu and Van (1998), who also find that a ban on child labor is not Pareto improving. The result in Proposition 11.18 roughly agrees with what's obtained by Bommier and Dubois (2004).

Proof. For the case where the child's utility in period 1 is considered, discussions in Section 11.2 beneath Proposition 11.9 apply here.

Similar to this analysis, Equation (11.37) implies the following.

PROPOSITION 11.19
For the model with a child's disutility or utility from period 1, no marginal increase on formal schooling is a Pareto improvement.

Proof. The result follows from the fact that based on Equation (11.37), we have

$$\frac{\partial W_c}{\partial e_c} = \frac{\delta\lambda - 1}{(1-\delta\lambda)\delta}\left[U'(c_p^1)\frac{\partial c_p^1}{\partial e_c} + U'(c_p^2)\frac{\partial c_p^2}{\partial e_c}\right] < 0$$

11.4 Conclusion

In this chapter, we emphasize the importance of all three variables—child labor, formal schooling, and level of maturity—in the context of a child's accumulation of human capital. Because these variables may well interact with each other, we produce some interesting results differing significantly from all those works that do not consider such possible interactions.

The results in this chapter show that when the parents are altruistic and well disciplined in their lives, for example, they save in period 1 and bequeath in period 2, or if the adult children are altruistic and transfer to their parents, then efficiencies of child labor, formal schooling, and level of maturity can be achieved, assuming that the children have no feelings about their childhood experiences. However, if the children's period 1 feelings toward labor or formal schooling matter, the inefficiencies of child labor, formal schooling, and level of maturity become unavoidable. This is where the situation becomes interesting. In particular, if the laissez faire levels of child labor l_c^* or formal schooling e_c^* are not zero, then it leads to the rotten parents effect, as so named by Bommier and Dubois (2004), where the parents rationally sacrifice some of the children's childhood utility by making them work and study too much with the anticipation that this will result in higher future earnings of the children, and consequently larger transfers from the children.

One possible method to reduce the size of inefficiencies is to introduce government legislations for a marginal ban on child labor and a marginal increase in child's formal schooling. By doing so, there is a possible chance to achieve a Pareto improvement for all parties involved: the parents, children, and firms. This is because with more formal education, the children's level of maturity can be expected to rise and the human capital accumulation to accelerate. So, the earnings power as a child laborer or as an adult will accordingly increase.

The analysis in this chapter, in fact, also indirectly shows the reason why it is difficult or impossible to abolish child labor in reality. (The phenomenon of child labor is age old. In the contemporary world, the International Labor Organization estimates that worldwide 120 million children between the ages of 5 to 14 work fulltime. See, for example, ILO [1996]. And in 2005, Krueger and Donohue quantified the effects of child labor legislation on human capital accumulation and distribution of wealth and welfare. They found that households with significant financial asset holdings unambiguously lose from any government intervention, high-wage workers benefit most from a ban on child labor, while low-wage workers benefit the most from free education, and that a child labor ban induces welfare losses because it reduces income opportunities for poor families without being effective in stimulating education attainment.) This is because under different circumstances, the laissez faire triple (l_c^*, e_c^*, m_c^*) of child labor, formal schooling, and level of maturity can be efficient, inefficiently high, or inefficiently low. And in reality, these circumstances can easily evolve from a state where the laissez faire triple is efficient to one where the triple becomes inefficient. And not even government legislations on a ban of child labor can be guaranteed to be Pareto improving.

Chapter 12

Economic Eddies and Existence of Different Industry Sizes

In this chapter, we turn our attention to the study of interindustry wage structures. We first model each commercial firm as a specific spinning yoyo, called an economic yoyo. Then, on the basis of the evolution of a flow of such yoyos, we can see how these economic yoyos interact with each other through combinations and breakups. This end shows why there exist companies of different sizes in each economic sector or industry. Similar analysis shows why at any chosen moment of time in history, there are economic sectors and industries of various scales. With such a dynamic systemic analysis, which points to certain key structures for each existing commercial entity, such as a firm, economic sector, or industry, in place, we establish a simple profit maximization model to study large and small firms and their differences in areas of production, determination of product selling prices, and the cost basis of their products. Among what we find is that when a firm has limited resources, its business has a glass ceiling for its potential maximum level of profits.

By employing this simple profit maximization model to the study of interindustry wage differentials, it is found that financially resourceful companies have a large array of advantages over those companies that are limited by their resources. One example of such advantage is that the former companies bring in handsome profits in two dimensions, the products and their personnel, while the latter companies could only produce their profits from a single dimension, the products, with an invisible glass ceiling. To maximize their profit in the human resource dimension, financially resourceful companies need to spend extra money on their employees. This end provides the long-sought-after explanation for the stable wage differentials existing interindustrially over time and across national borders (Thaler, 1989). Beyond this wonderful outcome, our model also provides plausible economic reasons for other relevant questions, including, but not limited to, why high-wage industries tend to have low quit rates, why accounting profits and market power are reliable predictors of industry wages, why the association between wages and labor's share of costs in an industry is negative, why industries with high capital-labor ratios tend to pay higher wages, why unionization rate increases wages for both union and nonunion members in a firm, etc. What is worth mentioning here is that in the labor market, the law of one price does not hold true for job opportunities that look identical in their job descriptions.

Note: In the past half a century, many first-class economists, including Nobel laureates George Stigler of 1982 (Stigler, 1958), Robert Solow of 1987 (Solow, 1979), George Akerlof of 2001, and Daniel Kahneman of 2002 (Kahneman et al., 1986), have contributed to the understanding of interindustry wage pattern. However, as Thaler (1989) shows, none of the established attempts seem to explain the existing pattern in a satisfactory manner without assuming something difficult to accept. To this end, our work in this chapter, thanks to the yoyo model and methodology, provides the most intuitive and plausible explanation without assuming anything difficult to swallow.

More specifically, over a half century ago, economists began to notice stable wage differentials existing interindustrially over time and across national borders. Some industries pay their workers more than other industries (Slichter, 1950). And the differentials apply across all occupations. That is, if one occupation in an industry is high paid, then all other occupations tend to be so too, and over time appear to be so internationally.

To comprehend this wage structure and the mechanism behind it, over the past decades many scholars have looked at the problem from various angles and made much important progress. Even so, the interindustry wage differential still cannot be fathomed in its entirety without assuming some hypotheses that are difficult to accept by the community of economists. These hypotheses include, but are not limited to:

1. Firms are choosing not to maximize their profits (Krueger and Summers, 1987).
2. Firms pay attention to perceived equity in setting wages (Akerlof and Yellen, 1990).
3. High-wage firms find that lowering wages would decrease their profits (Krueger and Summers, 1987).

The idea that firm managers would choose not to maximize profits and instead have highly paid employees, including the blue-color workers far removed from the managers' milieu, is a real enigma. Due to this reason, there has not been an attempt to explain interindustry wage differentials using any agency model.

On the other hand, several attempts (for example, Stiglitz, 1976; Akerlof and Yellen, 1990; Berlinski, 2000) have been made to apply the concept of norms of internal equity in setting wages as constraints. They lead to the appearance of persistent interindustry wage differentials and uniformity across occupations of the interindustry wage structure. However, such an assumption seems to be controversial to economists (Thaler, 1989).

For hypothesis 3 to hold, one has to assume either that higher wages can increase production output, which is the foundation of the efficiency wage models (Yellen, 1984), or that higher wages are offered as a natural response to the threat of possible collective actions (Dickens, 1986).

Thaler (1989) clearly shows that none of the current attempts seem to explain the interindustry wage differentials in a satisfactory manner without assuming something difficult to accept. To meet this challenge, let us turn to the yoyo model and see what we can obtain beyond what is known now in this area of research.

12.1 Economic Yoyos and Their Flows

Let us model a commercial company as a spinning yoyo, as shown in Figure 12.1, where the black hole side sucks in all the basic supplies to sustain the vitality of the company, such as all needed raw materials, utilities, human resources, various services, and, most importantly, the profit. The Big Bang side emits the company's products. Through marketing effort, the company establishes its reputation

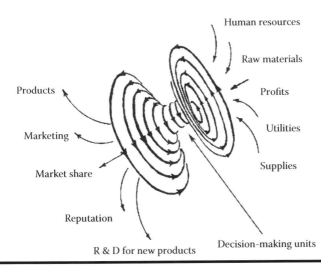

Figure 12.1   The yoyo model for a viable commercial company.

for quality and occupies a certain percentage of the consumers' market. With the help of its R&D efforts, the company continually eyes expansions in new areas of business. Various internal decision-making units, including the board of directors, the CEO and other executives, etc., connect the two sides of the spinning yoyo (Gillan, 2006).

The force necessary for such a yoyo to exist is the continued inflow of profits at and above the minimum sustainable level. Without the profits or the ability to generate the adequate profits, the company will fail and its economic yoyo will disappear. When the opportunity for earning adequate profits exists in the marketplace, many entrepreneurs will enter the same sector of the economic actions to fight for their share of the potential pool of profits. Therefore, an industry or economic sector starts to emerge. If we imagine all the possible spinning yoyos of an industry, each of which stands for a company, floating against each other, we then can see the fact that some yoyos will have the tendency to combine into greater yoyos, and other yoyos may break into smaller yoyos or simply disappear, depending on the yoyos' spinning directions, speeds, and angles of tilt. For details on how spinning yoyos could interact with each other, consult Chapters 4 to 6 or Lin (2007). As long as the flow of demands is unstable (see analysis in Section 9.2), the economic yoyos will continue to fight for their own:

1. Existence
2. Survival
3. Possible expansion

When the picture of demands begins to crystallize, that is, when the acting and reacting forces start to stabilize, the intensive struggles for market shares between the economic yoyos will decline to a minimum and the entire economic sector begins to stabilize. This description is based on studies of fluid motions (OuYang, 1994) without using any of the terminology.

Now, if each economic yoyo, representing an individual company, in the previous imagined picture is replaced by such a bigger economic yoyo that represents an economic sector or industry, then a similar evolution of struggles for market share or the available resources and profit opportunities between industries or economic sectors would exist. This evolution of struggle would be dynamic if we place it in the flow of time. In particular, some yoyos, representing economic sectors or industries, will have the tendency to combine into greater yoyos, and others may break apart into smaller yoyos or simply disappear. This evolution process is similar to that of a current rushing down a high land. If the boundary conditions are complicated, the state of the current flow pattern will be difficult to model using the traditional mathematics. It is because the overall state of motion is moving downward mixed with local variations, such as jet streams, whirlpools of different sizes, strengths, and directions, etc. So, if we take a snapshot for the evolution of economic power struggle between the economic yoyos of various industries, in the still momentary picture, we can expect to see spinning yoyos of different sizes, meaning some industries (yoyos) suck in more profits and human resources than others. It is reasonable to expect that industries whose yoyo structures suck in more profits and other available resources behave differently than other industries.

The qualitative analysis of economic yoyos as above can continue along a line parallel to the evolution process of a fluid motion, and a corresponding mathematical analysis can be introduced to support such a qualitative analysis. In the following, we will establish a simple analytic model based on the premise of the yoyo model to see that the analysis above actually works out mathematically. Also, the previous analytic model provides a platform for us to study the interindustry wage differentials without introducing any hard-to-accept hypothesis.

12.2 Simple Model for Perfect Capital Markets

Assume that a retailer sells a specific product for $\$p_s$ each unit. The total cost for the entire process of acquiring the product, shipping and handling, insurance, storage, salesperson's salary, etc., is $\$p_p$ each unit. Let $n = n(p_s)$ be the total number of units sold at the price $\$p_s$ per unit. Then, for this particular retailer, his profit from this single line of product is given by

$$P = \text{profit} = n(p_s)(p_s - p_p) \tag{12.1}$$

If the capital markets are perfect, meaning that the retailer can borrow as much funds to purchase or produce as many units of the product as he needs to at any desirable time, then he would determine such a selling price $\$p_s$ so that his profit P in Equation (12.1) will be maximized. The first-order condition for this maximization problem is given as follows:

$$\frac{\partial P}{\partial p_s} = n'(p_s)(p_s - p_p) + n(p_s) = 0 \tag{12.2}$$

For each fixed p_s- and p_p-value, we have

$$n'(p_s) = -\frac{n(p_s)}{p_s - p_p}$$

This is a separable differential equation; its solution is

$$n(p_s) = \frac{C}{p_s - p_p} \tag{12.3}$$

where C is the integration constant. If at the price level p_{s0} the initial market demand for the product is $n(p_{s0})$ units, then Equation (12.3) implies that $C = n(p_{s0})(p_{s0} - p_p)$. Substituting this C-value and Equation (12.3) into Equation (12.1) leads to

$$P = \text{profit} = n(p_{s0})(p_{s0} - p_p) \quad (12.4)$$

Equation (12.3) implies that if the retailer plans to make a profit on each unit of his product, meaning $p_s > p_p$, then the constant C will be positive and $n(p_s)$ increases indefinitely as the selling price $p_s \rightarrow (p_p)^+$. For reasons like competition or clearing out the specific product line, the retailer could choose to get rid of the product by selling it at a price p_s below the cost p_p. In this case, the constant C will be less than 0. If the retailer is involved in a competition to occupy a greater market share for his product, then Equation (12.3) also implies that he has to keep the unit price p_s as close to the cost basis p_p as possible to maximize the marker demand $n(p_s)$.

To maximize his profit P, Equation (12.4) indicates that if the difference $(p_{s0} - p_s)$ stays constant, the lower the p_{s0}-value, the greater the demand $n(p_{s0})$ and the greater the total profit P. So, to generate as much profit as possible, the retailer has to keep his per-unit cost p_p as low as possible. Because costs like shipping and handling, insurance, storage, salesperson's salaries, etc., are exogenous to the retailer and are quite robust, the retailer's efficient strategy to drastically reduce the cost basis p_p is to locate a manufacturer who can massively produce the needed product at a price below all other competitors. There are many ways to achieve this goal. For example, the manufacturer can hire a labor force at a below-the-market price, purchase his raw materials at extremely low prices, or introduce a revolutionary technology to drastically increase the labor's productivity.

Because raw materials and energies are international commodities mostly traded on global markets, and due to the advances of information technology, all competing manufacturers can quite easily catch up with technological innovations, to purchase products at below-the-market prices, the retailer would naturally go to places where the labor quality is adequate and the labor costs are as low as possible. When many retailers look for such ideal labor markets, manufacturing businesses naturally start to relocate in different geographic locations with as close to ideal labor bases as possible. That is, the sea of economic yoyos starts to flow, caused by the uneven force of competition. This end explains why the current trend of moving manufacturing operations from industrialized nations to third world countries does not seem reversible in the foreseeable future, as long as international transportation costs stay low and the global economic system stays open and competitive. In fact, this end is simulated well by the fluid patterns found in the dishpan experiment.

Assume in Equation (12.1) that the selling price p_s drops an increment $\Delta p_s > 0$ per unit. If the retailer does not pick up any additional market demand for his product at this reduced price $(p_s - \Delta p_s)$, then he would have a loss in the amount of $n(p_s)\Delta p_s$. So, if at the lower unit price level, the retailer expands the market demand

$n(p_s)$ to that of $n(p_s - \Delta p_s)$, and if the profit increment $P(p_s - \Delta p_s) - P(p_s)$ is greater than the theoretical loss $n(p_s)\Delta p_s$, that is,

$$n(p_s - \Delta p_s)(p_s - \Delta p_s - p_p) - n(p_s)(p_s - p_p) > n(p_s)\Delta p_s \qquad (12.5)$$

then the retailer would rather reduce his unit selling price p_s to maximize his profit. Equation (12.5) is equivalent to the following:

$$\frac{dP(p_s)}{dp_s} = \lim_{\Delta p_s \to 0^+} \frac{P(p_s - \Delta p_s) - P(p_s)}{-\Delta p_s} < -n(p_s) \qquad (12.6)$$

Now, instead of reducing the per-unit selling price p_s by an increment $\Delta p_s > 0$, the unit cost basis p_p is increased by an increment $\Delta p_p > 0$. If the retailer does not pick up any additional demand for his product with his increased cost basis, then he would experience a loss in the amount of $n(p_p)\Delta p_p$. However, if such an increase in his cost basis produces a new market demand $n(p_p + \Delta p_p)$ and the change in profit P satisfies

$$P(p_p + \Delta p_p) - P(p_p) > n(p_p)\Delta p_p \qquad (12.7)$$

then the retailer would rather increase his unit cost basis by as much as necessary to maximize his profit. Now, Equation (12.7) is equivalent to the following:

$$\frac{dP(p_p)}{dp_p} = \lim_{\Delta p_p \to 0^+} \frac{P(p_p + \Delta p_p) - P(p_p)}{\Delta p_p} > n(p_p) \qquad (12.8)$$

Combining Equations (12.6) and (12.8) provides the following conclusions.
Proposition 12.1
Assume that the capital markets are perfect. If at a unit selling price $p_s = p_{s0}$ and a unit cost basis $p_p = p_{p0}$, the total profit $P(p_s, p_p)$ from selling the product satisfies:

(i) $\left.\frac{\partial P}{\partial p_s}\right|_{p_s = p_{s0}} < -n(p_{s0})$, then the retailer can reduce his unit selling price p_{s0} to increase his total profit; or

(ii) $\left.\frac{\partial P}{\partial p_p}\right|_{p_p = p_{p0}} > n(p_{p0})$, then the retailer can increase his unit cost basis p_{p0} to reap in additional profits; or

(iii) $\left.\frac{\partial P}{\partial p_s}\right|_{p_s = p_{s0}} < -n(p_{s0})$ and $\left.\frac{\partial P}{\partial p_p}\right|_{p_p = p_{p0}} > n(p_{p0})$, then the retailer can both reduce his unit selling price p_{s0} and increase his unit cost basis p_{p0} to boost his total profit from his specific line of product.

What is implied in the previous analysis includes the fact that if by reducing his unit profit $(p_s - p_p)$ the retailer can greatly increase his total profit by gaining additional market demand, he will try all he can to accomplish that goal. More specifically, he might:

1. Launch an aggressive commercial campaign where p_s is increased to expand the market demand.
2. Acquire a larger number of units of the product at a much lower price, where p_p is decreased so that p_s can be accordingly lowered to attract more customers.
3. Lower per-unit shipping and handling costs with increased volume of business.
4. Lower per-unit insurance costs with increased volume of business so that the savings can be passed on to the customers to create a healthier market demand.
5. Increase the salesperson's wages. With such a potential of drastically increasing take-home pays, the salesperson would work harder and smarter so that the market demand is consequently pushed to new extremes.

Example 12.1

Assume that for a special product, the unit cost basis $p_p = \$5.00$ and its market demand is determined by $n(p_s) = 1{,}000 \exp(5 - p_s)$, where p_s is the unit selling price. So, the total profit from selling $n(p_s)$ units at $\$p_s$ each is

$$P = n(p_s)(p_s - 5) = 1{,}000 e^{5-p_s}(p_s - 5)$$

If $p_s > 7$, then we have $\frac{\partial P}{\partial p_s} < -n(p_s)$. Proposition 12.1(i) implies that if the retail company is well funded, it can increase its profit by reducing the unit selling price as close to \$7+ as possible.

Similar examples can be constructed for the scenarios in Proposition 12.1(ii) and (iii).

12.3 Simple Model for Imperfect Capital Markets

In this section, we analyze the simple model established in the previous section for the case of imperfect capital markets. That is, the profit P in Equation (12.1) is subject to the following constraint:

$$n(p_s)p_p = I \qquad (12.9)$$

where I is the total available funds for the retailer to invest in his line of product and $n(p_s)$ is seen as the size of the inventory. The first-order conditions for maximizing the profit P subject to the constraint in Equation (12.9) are given by

$$\frac{\partial P}{\partial p_s} = n'(p_s)(p_s - p_p) + n(p_s) = \lambda n'(p_s) p_p \tag{12.10}$$

and

$$\frac{\partial P}{\partial p_p} = n'(p_s) \frac{dp_s}{dp_p}(p_s - p_p) + n(p_s)\left(\frac{dp_s}{dp_p} - 1\right) = \lambda\left[n'(p_s)\frac{dp_s}{dp_p} p_p + n(p_s)\right] \tag{12.11}$$

where λ is the Lagrange multiplier. Substituting Equation (12.10) into Equation (12.11) leads to

$$(1+\lambda)n(p_s) = 0$$

Equation (12.9) implies that $n(p_s) \neq 0$, which means $\lambda = -1$. So, Equation (12.10) becomes

$$n'(p_s)p_s = -n(p_s)$$

Solving this equation for $n(p_s)$ gives us

$$n(p_s) = \frac{n(p_{s0})p_{s0}}{p_s} \tag{12.12}$$

where $n(p_{s0})$ is the initial market demand when the product is sold for $\$p_{s0}$ per unit. So, the retailer's profit P is given by

$$P = \text{profit} = \frac{n(p_{s0})p_{s0}}{p_s}(p_s - p_p) \tag{12.13a}$$

$$= n(p_{s0})p_{s0}\left(1 - \frac{p_p}{p_s}\right) \tag{12.13b}$$

Equation (12.12) indicates that to expand the market demand, the retailer has to decrease his unit selling price. Because the capital markets are imperfect, this means that the retailer has a limited resource to invest in his product. That is, he cannot afford to compete with a retailer who has unlimited resources. Similar to the situation of a financially powerful retailer, our small retailer can also increase his profit by reducing his selling price p_s, if he can keep the unit profit $(p_s - p_p)$ constant. However, unlike the powerful retailer in Section 12.2, our small retailer has a cap, $n(p_{s0})p_{s0}$ (Equation (12.13b)), on how much he can expand his potential of total profit. This comparison tells us the following two facts:

1. When financial resources are limited, any venture will have a glass ceiling for its maximum level of profits.
2. Because of their limited resources, small retailers or poorly funded ventures do not have many opportunities to locate extremely low-priced manufacturers. One reason is that they do not have the ability to place large orders, and also, they do not have the financial strength to create their own low-priced manufacturing operations to strengthen their ability to compete.

By comparing Equations (12.13a) and (12.13b) to conclusions 1 to 5 in the previous section, we can see the following facts:

1. While financially powerful companies are promoting their product to expand their market share and appearance, companies with limited resources cannot afford to devote much of their scarce resources to do so. One reason is that they do not have much money to allocate for the purpose of promotion. Another reason is that, as Equation (12.13b) indicates, an excessive amount of spending will keep their unit selling price p_s high. To increase their profit potential, companies with limited resources have to control their spending so that their profit can be maximized by lowering their unit selling price p_s.
2. While financially powerful companies are placing large orders at much reduced wholesale prices, companies with limited resources just cannot take such opportunities. Similarly, other volume-related savings are not available to ventures with limited resources.

Chapter 13
A Fresh Look at Interindustry Wage Differentials

Because the systemic yoyo model can be employed to analyze nonlinear interactions, as described in the previous chapters, in this chapter we apply this method to provide a fresh look at interindustry wage differentials.

Over the years, numerous scholars have tried to provide various reasons for the existence of the interindustry wage pattern. It has been found that high wages can be explained in part by:

1. Unpleasant and unsafe working environment
2. The purpose of hiring better workers for their both measurable and immeasurable labor quality
3. Compensation differences naturally existing between industries
4. Each employee's effort $e(w)$ being an increasing function of his wage rate w
5. Firms' engagement in monitoring their workers' performance so that those who are caught shirking will be fired
6. The purpose of reducing the rate of employees quitting, because hiring and training workers can be expensive (Hamermesh, 1993)
7. Making employees feel that they are paid fairly
8. The firms' ability to pay
9. Labor union density, etc.

The uniformity of wage differentials across occupations works against both explanations 1 and 2 (Thaler, 1989). And although explanation 3 is undoubtedly true to a degree (Rosen, 1986), it still cannot explain why such compensation differences exist from one industry to another. Murphy and Topal (1987) support explanation 2, believing that the unexplained variance is due to employees' unobserved abilities. By identifying "the unobserved abilities" as intelligence, in particular as IQ test scores, Blackburn and Neumark (1988) find that there is a negative relationship between an industry's wage and the average IQ scores of its employees.

Explanations 4 to 7 are the fundamental premises of the so-called efficiency wage models using the mathematical rigor (Yellen, 1984). These models have attracted a great deal of attention and tend to prove that higher than competitive wages can be profitable.

By specifying the positive effort-wage relationship, as described in explanation 4, four main classes of efficiency wage models have been established. More specifically, shirking models (Shapiro and Stiglitz, 1984) are developed to describe how hard employees work at their jobs where piece rates are impractical due to either the difficulty of counting "pieces" or high costs of monitoring, or both. By paying above market wages, firms engage in monitoring their workers and fire those who are caught shirking. As expected, these models indicate that high-wage industries are those with high monitoring costs or those bearing high costs of employee shirking. About shirking models, Thaler (1989) casts the following natural questions: Do employees work harder when they think they are in danger of losing a high-paying job? Do employees work enough harder to justify the higher wages? Are the firms that pay high wages those who would gain the most from an increase in workers' effort? The analysis in this chapter will address these important questions from a new angle completely and definitely.

The so-called turnover models are established on the assumption that explanation 6 holds true (Salop, 1979; Stiglitz, 1974). Consequently, these models predict that high-wage industries are those with the highest costs associated with quitting. Because data on quit rates are published, it has been confirmed that paying high wages does decrease quit rates. Adverse selection models (Stiglitz, 1976; Weiss, 1980) are established based on explanations 2 and 4 by assuming that the average quality of the applicant pool increases with the wage rate. These models imply that industries that are sensitive to labor quality or spend a lot on measuring quality offer high wages. Evidently, these models do not provide a convincing explanation for the interindustry wage patterns.

On explanation 7, fair-wage models are created (Akerlof and Yellen, 1990; Solow, 1979) by assuming that employees will exert more effort if they think they are being paid fairly. Under this premise, firms pay wages above competitive levels whenever their workers' perceived fair wage exceeds the competitive wage. As a result, these models lead to such conclusions as industries with high profits or where team work and employee cooperation are particularly important will pay high wages (Kahneman et al., 1986). Even though both common sense and social psychological research on equity theory suggest that workers are more productive if the morale is high, the following practical problem remains open: Is the true efficiency wage that sets the marginal gains from increased morale equal to marginal costs (Thaler, 1989)? To this end, Raff and Summers (1987) evaluated Ford's decision in 1913 to double wages.

Evidently, explanation 8 does not make sense, because each dollar wage increase means a dollar dividend less for the stockholders. As for explanation 9, our analysis below will show that labor unions are attracted to high-wage industries instead of the other way around, even though studies find that industry wage rates are corrected to union density (the percentage of workers in an industry who belong to a union), and that the unionization rate increases wages for both union and nonunion members in an industry.

With this brief historical note in place, we are now ready to apply the simple model established in the previous chapter to explain the interindustry wage differentials easily and clearly. Assume that a company produces and sells a line of special product. Then, the company has two sources of income: (1) producing and selling the product and (2) hiring each worker. In particular, from each unit of the product produced and sold, the company makes as much profit as $p_s^p - p_p^p$, where p_s^p is the selling price per unit and p_p^p the unit cost. And for each worker it hires, the company generates as much profit as $p_s^W - p_p^W$, where p_s^W is the average revenue the worker is expected to make and p_p^W the average cost associated with the worker.

To help us uncover the underlying mechanism of the interindustry wage differentials, we will analyze the company in two situations:

1. The company has strong financial backings. That is, the model in Section 12.2 applies to this company.
2. The company's financial network and resources are limited. That is, the company does not have a perfect capital market available to it. So, the model in Section 12.3 applies here.

13.1 Financially Resourceful Companies

For situation 1, according to Equation (12.3), for the company's product, the market demand is given by

$$n_p\left(p_s^p\right) = n_p\left(p_{s0}^p\right) \frac{p_{s0}^p - p_p^p}{p_s^p - p_p^p} \text{ (units)} \quad (13.1)$$

and the total profit from selling its product, based on Equation (12.4), is

$$P^p = \text{profit of its product} = n_p\left(p_{s0}^p\right)\left(p_{s0}^p - p_p^p\right) \quad (13.2)$$

where $n_p(p_{s0}^p)$ is the initial market demand for the product selling at $\$p_{s0}^p$ each unit. Equation (13.1) implies that when $p_s^p \to (p_p^p)^+$, the demand will approach infinity. Equation (13.2) indicates that if the unit profit $(p_{s0}^p - p_p^p)$ stays constant, the lower the p_{s0}^p-value, the greater the demand $n_p(p_s^p)$ and the greater the total profit P^p. For details, see the analysis of Equations (12.3) and (12.4).

Similarly, the company's staffing need is given by

$$n_W(p_s^W) = n_W(p_{s0}^W) \frac{p_{s0}^W - p_p^W}{p_s^W - p_p^W} \quad \text{(persons)} \tag{13.3}$$

and the total profit from hiring $n_W(p_s^W)$ employees is given by

$$P^W = \text{profit of personnel} = n_W(p_{s0}^W)(p_{s0}^W - p_p^W) \tag{13.4}$$

where $n_W(p_{s0}^W)$ stands for the company's initial need for additional personnel, hired at the initial expected revenue $\$p_{s0}^W$ per worker.

Different from Equation (13.1), in Equation (13.3) the higher the initial p_{s0}^W-value, the more workers the company would like to hire initially. After then, the smaller the difference $(p_s^W - p_p^W)$ is and the less p_p^W is controlled, the greater number $n_W(p_s^W)$ of workers will be needed. When Equations (13.3) and (13.4) are combined, it can be seen that if the difference $(p_{s0}^W - p_p^W)$ can stay constant, the closer $p_s^W \to (p_p^W)^+$, the greater the $n_W(p_s^W)$-value will be and the greater the total profit P^W will become. Now, $p_s^W \to (p_p^W)^+$ means that when the p_s^W-value is relatively stable, p_p^W should be increased as much as possible. That is, workers' wages can go up so that the total per-employee cost can approach the expected per-employee revenue p_s^W as much as possible.

When the p_p^W-value is increasing, the per-unit product cost p_p^p will also increase accordingly. But Equations (13.1) and (13.2) indicate that as long as the difference $(p_{s0}^p - p_p^p)$ does not decrease and $(p_s^p - p_p^p)$ drops, the total profit from the line of product will still go higher. That is, our analysis leads to the following result.

PROPOSITION 13.1

If a business venture is well funded, assuming all other aspects of the operation stay the same, then:

(i) The market demand for the product increases as the unit-selling price drops close to the unit cost basis, while the total profit increases drastically.
(ii) The company taking on the venture will hire additional employees at higher than competitive wage rates with the total profit soaring.

Corresponding to Proposition 12.1 in terms of hiring employees, we have the following:

PROPOSITION 13.2

Assume that the company has all necessary funds for its operation. If at the expected revenue level $p_s^p = p_{s0}^p$ a new hire would generate for the company and at a total cost $p_p^W = p_{p0}^W$, the total profit $P^W(p_s^W, p_p^W)$ all employees together are expected to generate satisfies

(i) $$\left.\frac{\partial P^W}{\partial p_s^W}\right|_{p_{s0}^W} < -n_W\left(p_{s0}^W\right)$$

then the company can reduce its expected per-employee revenue p_{s0}^p to increase its total profit; or

(ii) $$\left.\frac{\partial P^W}{\partial p_p^W}\right|_{p_{p0}^W} > n_W\left(p_{p0}^W\right)$$

then the company can raise its expected per-employee cost basis p_{p0}^p to reap in additional profit; or

(iii) $$\left.\frac{\partial P^W}{\partial p_s^W}\right|_{p_{s0}^W} < -n_W\left(p_{s0}^W\right) \text{ and } \left.\frac{\partial P^W}{\partial p_p^W}\right|_{p_{p0}^W} > n_W\left(p_{p0}^W\right)$$

then the company can both reduce its expected per-employee revenue p_{s0}^p and raise its per-employee cost basis p_{p0}^p to maximize its overall profit.

13.2 Companies with Limited Resources

For situation 2, where the company's resources are limited, our model in Section 12.3 applies. The total profit is given by

$$P_{total} = P^p + P^W = n_p\left(p_s^p\right)\left(p_s^p - p_p^p\right) + n_W\left(p_s^W\right)\left(p_s^W - p_p^W\right) \quad (13.5)$$

subject to the budget constraint

$$n_p\left(p_s^p\right)p_p^p + n_W\left(p_s^W\right)p_p^W = I \quad (13.6)$$

where *I* stands for the total funds available to the company. By solving the maximization problem of Equation (13.5) subject to Equation (13.6), we establish the following results:

$$n_p(p_s^p) = \frac{n_p(p_{s0}^p) p_{s0}^p}{p_s^p} \tag{13.7}$$

$$n_W(p_s^W) = \frac{n_W(p_{s0}^W) p_{s0}^W}{p_s^W} \tag{13.8}$$

and

$$P_{total} = P^p + P^W = \frac{n_p(p_{s0}^p) p_{s0}^p}{p_s^p}(p_s^p - p_p^p) + \frac{n_W(p_{s0}^W) p_{s0}^W}{p_s^W}(p_s^W - p_p^W) \tag{13.9}$$

$$= n_p(p_{s0}^p) p_{s0}^p \left(1 - \frac{p_p^p}{p_s^p}\right) + n_W(p_{s0}^W) p_{s0}^W \left(1 - \frac{p_p^W}{p_s^W}\right) \tag{13.10}$$

where $p_{s0}^p n_p(p_{s0}^p)$ and $p_{s0}^W n_W(p_{s0}^W)$ are defined the same as in the analysis of situation 1.

Equation (13.7) is the same as Equation (12.12). So, for its analysis, refer to that of Equation (12.12). Equation (13.8) indicates that to hire more employees, the company has to lower the average expected per-worker revenue. Because the capital markets are not perfect for the company, this result implies that the company has to limit how many workers it can afford to hire. To maximize its profit in the dimension of human resources, Equation (13.10) implies that $p_s^W \gg p_p^W$. So, in such a company with limited resources, it can either hire a relatively large number of employees at low wage rates or hire a relatively small number of high-quality workers at relatively higher wage rates. For the latter case to occur, the workers' productivity must be very high, which in general means that the company needs to invest a great deal in technology. And this end might not be possible. Together with the company's weak financial standing, the high-quality worker option may never be practically possible for the company to take.

When Equation (13.10) is compared to Equation (13.4), it can be seen that

1. Financially resourceful companies can spend extra money on workers' retraining programs to lower the average per-worker cost basis p_p^W, while firms with limited resources cannot. This is because in the latter case, extra spending on workers' training programs increases both p_s^W-and p_p^W-values. So, the ratio p_p^W/p_s^W may not change in the favorable direction to the firms.

2. Similar reasoning indicates that financially resourceful companies can afford to invest in programs to make their workers feel good and to raise their morale, while firms with limited resources just cannot afford such luxuries. Consequently, workers in financially strong firms produce more and are less likely to quit when compared to those hired by financially strained companies.
3. While financially resourceful companies hire large numbers of workers so that these companies can easily reduce their per-employee benefit costs, companies with few workers have to pay the inflated market prices for the same benefit packages. That is, volume savings are not available to firms with limited resources.

This analysis explains why in a firm, if an occupation is high paid, so are all other occupations. In particular, if one occupation is paid at a level above the competitive wage rates, then all other occupations within the firm must play supporting roles. The increased productivity from the central occupation will be more than enough to finance the supporting occupations so that their respective wages are higher than those occupations' competitive market rates. For more detailed discussion about this end, please refer to the section below.

13.3 Look Back at Some of the Existing Literature

Our analysis, together with our yoyo model, in fact provides a plausible economic explanation for:

1. Why high-wage industries tend to have low quit rates (Katz and Summers, 1989; Akerlof et al., 1988). It is because of a sense of job security (financially strong firms), being treated with respect (because each of them brings profit to the company), and the feeling of being paid in excess of their opportunity costs. In terms of our yoyo model, high-wage industries are those yoyos that spin powerfully so that fewer objects can escape their spin fields.
2. Why firm sizes are an indicator of the levels of compensation (Brown and Medoff, 1989; Groshen, 1988). It is because large firm sizes reflect the market share the firms occupy and how much they make from the head count of their workers. In terms of our yoyo model, the large size of an industry shows its volume of materials, including profits, being sucked in from the black hole side.
3. Why accounting profits and market power are reliable predictors of industry wages (Thaler, 1989). It is because they indicate how fast the firms' products and their workers generate revenue. From the angle of spinning yoyos, they are indicators of the speed of materials being sucked into the yoyo of the industry.

4. Why the association between wages and labor's share of costs in an industry is negative (Slichter, 1950). It is because in Equation (13.3), to reduce the difference ($p_s^W - p_p^W$), the firm may well spend a good amount of money to improve productivity, while raising its workers' wages disproportionably slowly. In the economic yoyo of an industry, the overall spinning strength has to be distributed among all aspects of the Big Bang side. And workers' wages are only one aspect of the many.

5. Why industries with high capital-labor ratios tend to pay higher wages (Lawrence and Lawrence, 1985; Dickens and Katz, 1987). Our model indicates that if a firm requires a high level of capital investment for its operation, the unit product cost of the firm must be high compared to that of firms operating on relatively much lower budgets. Equation (13.2) implies that for such a capital-intensive firm, the initial unit profit ($p_{s0}^p - p_p^p$) must be enough high for the investors to put up the necessary venture capital. After achieving the firm's initial business success, continued high level of capital investment will be needed to reduce the unit cost basis p_p^p by heavily investing in advanced and innovative technology to increase the workers' productivity. For a long period of time, the delicate technology may require large amounts of capital to maintain its working conditions.

6. Why the unionization rate increases wages for both union and nonunion members in a firm. Based on our model in Equations (13.1) to (13.4), the firm will have a lot of weight in both its product market and the labor market. The large number of employees is good for the firm in the sense that the firm actually generates profits in the dimension of human resources. But this same crowd of workers is also, or could also be, a problem and a source of troubles. The first reason is that when the firm is small, its workers are few in numbers and low in quality (relatively speaking). So, they can be replaced relatively easily. Using our yoyo model, the small group of workers could not form a strong enough whirlpool to overthrow or severely interrupt the operation of the economic yoyo of the firm. When the number of workers reaches a critical mass, and when these employees discover the lucrative cash flows in and out of the firm's bank accounts because of their work, they will see the need to take collective actions to share a greater proportion of the firm's profits. This end is simulated well by the dishpan experiment, where only with enough water, asymmetric and nonuniform flow patterns will appear. Our analysis here explains why Dickens (1986) predicts that industries with high wages suffer from the highest threat of union actions.

Similar to the situation of making a snowball on a cold, snowy winter day (a kind of yoyo action), firms or industries that pay wages above the competitive rates can reap in profits with added advantage and convenience: high workers' morale, desirable work ethics, low quit rates, better-quality people, etc. To see this end, let us identify our firm, which pays its workers wages above the competitive rates, with the benevolent head in Becker's rotten kid theorem and Theorem 10.2, and all employees with those selfish members of the family. Then, Theorem 10.2 says that as long as the distribution of the firm's resources to all employees is not in conflict with the consumption preferences of any (selfish) employees, then all the employees will be motivated to maximize the firm's income. Because the employees can tell by comparison that their wages are above the competitive market rates, their consumption preferences—getting paid more than others in the same occupation—are not in conflict with the need for the firm to pay its employees as much as possible (Equations (13.3) and (13.4)). So, Becker's rotten kid theorem implies that of course none of the workers would quit, except for those who have different consumption needs and preferences beyond wages, and of course, the workers would have high morale toward their work and company, because they are motivated to maximize the firm's income.

Beyond what is said above, what Becker's rotten kid theorem says also provides an answer to Thaler's (1989) question:

> Consider the case of two large firms with plants located in the same community. The firms have clerical staffs that perform virtually identical services. Firm H is in a high-wage industry and pays its clerical staff W_H, while Firm L is in a low-wage industry and pays its clerical staff only $W_L < W_H$. Suppose that Firm H decides to save money by cutting the wage of its clerical workers to W_L. Is this action profitable?

As what Becker's rotten kid theorem and Theorem 10.2 describe, if the cut is temporary and is for savings that will be used to finance some crisis related to the firm's operation, then the wage cut will still be profitable to the firm. Our analysis based on the yoyo model and that developed in Chapter 12 indicates that if Firm H plans to hang on to its competitive edge, meaning that it will continue to occupy similar market share for its products and reap similar profits from its human resources, it will naturally not make the cut permanent, because the firm also expects to generate profit from each and every employee it hires.

What we have obtained in this section shows that what Krueger and Summers (1987) said is correct: for some reason, high-wage firms find that lowering wages would decrease profits. What is different, though, is that our analysis shows that wages above opportunity costs for the workers are consistent with profit maximization without any prerequisite like:

1. Higher wages can increase output as assumed in efficiency wage models.
2. Higher wages are a rational response to the threat of collective actions.

At the same time, our analysis provides a theoretical explanation of why when a high-wage firm contracts an activity out, it would not pay the same high wages to the outside contractor (Berlinski, 2000), because the contractor is not in the human resource equation of the firm's profit maximization.

13.4 The Law of One Price

Even though we provided a plausible explanation for the existence of interindustry wage differentials, this phenomenon still seems to violate the law of one price, a fundamental component of the theory of competitive markets. (Pratt et al. [1979] show that the law of one price may not hold true in product market. And Lamond and Thaler [2003] cited sources that indicate that this law might be violated in financial markets, too.) To see what has gone wrong with this law in our situation at hand, let us first recite the law from a standard textbook and look at its background reasoning.

> **Law of one price** (Bodies and Merton, 2000): In a competitive (and efficient) market, if two assets are equivalent, they will have the same market price.

The intuition for this law is that all sellers will flock to the highest prevailing price and all buyers to the lowest current market price. So, if the market is efficient and competitive, the convergence to one price is instant.

The long-lasting and stable existence of interindustry wage differentials indicates that this law does not apply to the situation of wage structures. In particular, when, say, janitor positions with identical job descriptions are for sale by various companies, each selling price of these positions is the unseen (to the buyers, the job applicants) expected revenue each prospective janitor will bring to the specific hiring company. For example, a janitor working for a small dentist's office is expected to save the owner dentist time from cleaning the clinic complex and related works. So, the expected revenue the janitor creates is the additional revenue the dentist actually produces from his relieved cleaning responsibilities. If this value were $100/day on the average, then the selling price for this janitor position would be $3,000/month. So, if the dentist hires a janitor for $1,500/month, he still pockets $1,500/month more than before hiring the janitor. On the other hand, if the same janitor works for a large and extremely profitable law firm, his work might help the firm produce an additional $3,000/day, even though he would carry exactly the same responsibilities as if he worked for the dentist. In this case, this janitor position's selling price would be tagged at $90,000/month. If the law firm paid a janitor $10,000/month, a purposely exaggerated number for such an occupation to make a point here, the firm would still make $80,000/month extra compared to before the firm hired any janitor.

Because the selling prices in our job market situation are hidden either purposely or unconsciously, there is no way for all sellers, the hiring firms, to flock to the highest prevailing price. On the other hand, even when these selling prices are known to everyone, the sellers still will not flock to the highest prevailing price due to varied productivities and price tags being involved. Because of this reason, those companies that are most profitable can easily offer the highest market wages to occupations like janitor. At the same time, all job applicants flock to the advertised positions with the highest wages and best benefits, assuming that such information of advertisement can reach all potential job applicants in exactly the same fashion.

Also, our analysis indicates that the highest advertised wages could well be the lowest expenses the hiring companies would pay to the advertised positions compared to other positions with identical job descriptions available in the marketplace, considering how much revenue the workers generate for their companies.

Comparing to studies from the past in this area, the main contribution of our models and analysis to the existing literature is that we made no assumptions that are difficult to accept. And the maximization of our models is achieved using the simple Lagrange method and some knowledge of differential equations. A second contribution is how we expand the understanding of the purpose of hiring workers: a new dimension of profit making. By analyzing the major differences between perfect and imperfect capital markets, we uncovered some of the invisible differences between firms, and consequently between industries. These subtle differences lead to differences in business opportunities, causing the firms to behave differently in terms of wage setting and other practical aspects of daily operations.

Chapter 14

Dynamics between Long-Term and Short-Term Projects

Because the yoyo model can be employed to analyze nonlinear interactions, as described in the previous chapters, we will in this chapter apply this method to look at how the systemic yoyo model can be employed to study the problem of corporate governance, where corporations need capital to fund their ventures or provide the founders with cash-out opportunities, and suppliers of finance want to be assured that they get a return on their investment. We will study the mutual restrictions and interactions between a firm's board of directors, which is dominated by long-term, large shareholders, and the CEO, who is not a long-term, large shareholder. Before establishing a concrete analytic model, we first develop the systemic yoyo model foundation for some of the relevant empirical discoveries. On such a qualitative foundation, we provide an explanation for:

1. Why an elaborate legal system is needed for venture capital to flow smoothly and why certain governments around the world are willing to go extra miles to establish such as legal system
2. Why the agency problem existing between the large shareholders and the CEO cannot be resolved completely and the best one can do about this problem is to reduce its severity
3. Why outside financers still leave their money to managers in all market economies from around the world even with the knowledge of the unsolvable agency problem
4. Why there are boards of directors in the first place, among others

237

With these new and deepened understandings about the mutual interactions and restrictions between the board and the CEO, we establish a simple analytic model to study the price behaviors of different investment projects, the dynamics between long-term and short-term projects and assets, and the power struggle between the board and the CEO. Our quantitative analysis indicates that when assets are undervalued, the mispricing of the long-term asset in equilibrium is worse than that of the short-term asset. When the assets are overpriced, the mispricing of the short-term asset in equilibrium is worse than that of the long-term asset. Based on this result, we provide an explanation for why CEOs would prefer undervalued short-term investment projects.

In terms of CEOs' choices of long-term and short-term projects, it is found that even though the boards, dominated by long-term, large shareholders, prefer long-term, value-building projects, CEOs successes on short-term projects have a consequence favorable to the CEOs: they gain additional trust from and control over the boards for the CEOs. More interestingly, we can prove that the number of short-term projects a CEO likes to take on is less than the number the board would like him to take on. That is, in practice, no matter how much the large shareholders worry about the possibilities that the CEO could expropriate their investment funds or spend the funds on his or her private benefits of control, and how much they do to prevent their worries from becoming true, the CEO still has substantial control right (gained over time) and discretion to allocate funds as he or she chooses.

Similar to what is done in Chapters 12 and 13, let us imagine each corporation and each supplier of finance as an economic yoyo, where the black hole side sucks in funds from whatever available sources, while the Big Bang side spends the money in various ways. More specifically, if a yoyo models a corporation, then its black hole side absorbs funds from various sources, such as banks, mutual funds, pension plans, insurance companies, individual investors, etc. And part of the materials spit out of the Big Bang side contains such a portion of the corporation's profits that it is returned to the investors as capital gains or dividends. If a yoyo represents an investor, either large or small, then the black hole side stands for ways and amounts he could get his return on his investments, and the Big Bang side contains opportunities of future finance provided to each particular corporation. Hopefully, our modeling can help to shed new lights on the understanding of corporate governance so that rich economies can be marginally improved and major institutional changes can be stimulated in places where they need to be made.

14.1 The Yoyo Model Foundation for Empirical Discoveries

According to Alchian (1950) and Stigler (1958), we have the following evolutionary theory: no one should worry about corporate governance reform, because in the long run market competition would force firms to minimize costs. And as part of this cost minimization to adopt rules, firms would be able to raise external capital at the lowest cost. If we imagine that all the yoyos, representing firms and investors, coexist side by side, interacting with each other, then it is ready to see that due to differences in spinning directions and angles, some yoyos will be destroyed and some will combine into greater (yoyo) structures with enhanced ability to absorb more investors' funds. To prevent abusive competitions and maximize the cash flows (so that the governments could potentially collect the most taxes), regulatory entities must be involved to help those tiny (yoyos) investors from being crushed so that when necessary, firms can still attract capital for their venture needs from all potential sources, including these investors. That is, even though market competition is a powerful force toward economic efficiency, it alone is not sufficient to solve the problem of corporate governance. That is, the aforementioned evolutionary theory holds true conditionally in such a sense that the field of competition must be level for all competitors (yoyos) to participate (spin). To this end, countries like the United States, Britain, Germany, and Japan have well introduced sophisticated laws to protect investors.

For a publicly traded firm, because financers generally are either too small or not qualified or not informed enough to make detailed operational decisions, they rely on the manager to run the firm. Consequently, the manager ends up with substantial control rights and discretion to allocate funds as he chooses. He can expropriate the funds (Zingales, 1994) or spend the funds on his private benefits of control, such as building an empire for himself (Owen, 1991), spending on consumption perquisites (Burrough and Helyar, 1990), expanding the firm irrationally, pursuing pet projects that do not benefit the investors (Jensen, 1986), etc. And worst of all, managers can expropriate investors by staying on the job even if they are no longer competent enough to run the firm (Shleifer and Vishny, 1989). To prevent any of these possibilities of how the manager could misuse the investors' investments—an agency problem—the systemic yoyo model suggests that one possible solution would be make sure to place the manager's personal yoyo in the same spin field of the yoyos of all of the investors—earn a return for his finance or effort in the firm. That is, tie the financial interests of the manager with those of the financers (Jensen and Meckling, 1976). (One method of tying the manager's interest with those of the investors is to employ incentive contracts, which have been in use since at least the time of Berle and Means [1932]. However, evidence [Yermack, 1997] exists showing that incentive contracts create enormous opportunities for self-dealing for the managers.) However, as suggested by the enclosed areas in Figure 14.1, no solution would work to solve this agency problem completely; as long as the manager is not the same as the numeraire investor, an abstract investor representing the common characteristics and desires of all financers together, there will be conflicts. At the same time, the dishpan experiment shows that as long as the manager is not the same as the only investor, the agency problem will appear in one form or another. In particular, individual investors have a single goal: earn a return on their investment. However, the manager in general has more goals to achieve and more

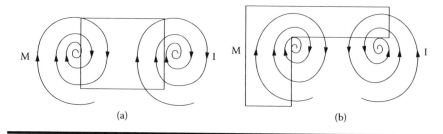

Figure 14.1 Conflicts of interests between the manager M and the numeraire investor I.

benefits, such as any of the aforementioned private benefits of control, to enjoy beyond simply making an earning or return on his labor effort and talent for being the manager. Our analysis implies that in terms of corporate governance, as long as there exist investors who are not the manager, there will be an unsolvable agency problem. The best one can do is to reduce its severity.

The reason why the yoyos M and I are convergent in Figure 14.1 is that the manager M and the investor I both want to grab more of the financial return and many other privileges. In Figure 14.1(a), yoyos M and I have a tendency to combine into a greater yoyo due to their harmonic spinning direction. In Figure 14.1(b), the yoyos M and I will try to destroy each other without the slightest chance to combine due to their opposite spinning directions.

If we walk to the other side of the yoyos in Figure 14.1, we observe the spin fields M and I as shown in Figure 14.2. Because the yoyos are all divergent in this angle, they do not tend to combine into a greater yoyo in the case of Figure 14.1(a), unless M (or I) is much more powerful than I (or M) (Figure 14.2(a)). Once again, the yoyo interactions in Figure 14.2(b) will never combine relatively peacefully due to their opposite spinning directions unless one of them is destroyed completely by the other.

That is, if the relationship between the board of directors and the CEO is as depicted in Figures 14.1(a) and 14.2(a), the firm has a chance to grow into a successful business venture. If the relationship between the board and the CEO turns

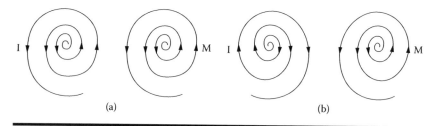

Figure 14.2 Conflicting ideas of how to manage the firm between the manager M and investor I.

out like the one shown in Figures 14.1(b) and 14.2(b), then the firm is heading to its eventual destruction, unless the CEO is replaced or some of the main shareholders or creditors give up their controlling influence.

Our conclusion above points directly to the following question: Why do investors leave their money to managers who have enormous discretion on how to expropriate much of it for their personal gain (Shleifer and Vishny, 1997)? What is puzzling is that outside finance occurs in all market economies and on an enormous scale in the developed ones, such as those in the United States, Britain, Germany, and Japan. To this end, the systemic yoyo model provides the answer. That is, based on the fact that eddy motions exist in pairs, one can see that when there is an opportunity, there will be risk takers. Among the risk takers, some are prudent, others are not and are gullible and get taken. The reason why in developed market economies, enormous outside finance exists (also explaining why these economies are more developed than others) is because the societal structures (yoyos), such as the local and national governments, underlying the economic yoyos of the developed economies spin in the same directions and angles as the yoyos of these economies. In particular, the fiscal welfare of the governments is closely tied to the cash flows in the economies through taxation. So, to make all the yoyos of large scales, such as different levels of the governments and firms, do well financially, the governments introduced laws to protect investors. Because of these laws, many small investors, as the tiny particles evolving about in the spin fields of the underlying yoyos, become gullible and get taken either sometimes or most of the time. On the other hand, the laws established are not simply for protecting investors whose financial gains in general do not contribute much to the fiscal welfare of the governments, but also for playing the magic role of monitoring the behaviors of firms. Every time investors get taken, some noise will be made and heard, which very likely brings additional monetary awards to the governments through the use of the legal system.

Another outcome of the laws and regulations is that based on the prevalent existence of governing boards from all over the world throughout recorded, recent history, most firms are required to have a board that meets a list of requirements: it must contain at least a certain number of directors, meet with some specified regularity, have various committees, contain a proportion of the membership independent from the management, etc. To this end, Hermalin and Weisbach (2003) question why there are boards of directors in the first place? The systemic yoyo model suggests that the board of a firm exists as the concentrated area of interests of all involved parties, including the management of the firm, investors, and creditors, specifically those large shareholders and significant creditors. The board serves as the narrow neck of the economic yoyo of the firm. It connects the activities of the firm with all the parties whose interests are tied to the success of the firm. Without the neck (or the board), the firm would not exist both financially and physically. The existence of a functioning board induces investors to trust the manager with their money.

Through the use of law protections, large investors provide external financing in exchange for their control rights of the firm (Hart, 1995). A natural question at this junction is: If these investors want to control a profitable or potentially profitable firm through their financing, why don't they simply go out and start their own firms because they have the financial resources? The systemic yoyo model suggests that starting a new firm (a yoyo) and succeeding in the competition with other existing firms in the same economic sector through absorbing other firms (yoyos) and destroying those which cannot be absorbed, while causing fierce competition, is more difficult than simply controlling the firm that has been most or potentially most successful in its economic sector.

The most important legal right investors, as shareholders of a firm, have is that of voting on important corporate matters and in elections of the board of directors, which in turn has monitoring rights over the manager (Manne, 1965). As expected from the systemic yoyo model, the manager could effectively interfere with the voting process in his favor (Pound, 1988) if control rights, such as votes, are split among many shareholders (yoyos that are too small compared to that of the manager). So, to break the imbalance in sizes, some investors with a collectively large cash flow at stake gain more effective control rights by being large. There are several forms that concentration can take, such as large shareholders, takeovers, and large creditors. When some investors own a substantial percentage of shares, they would have the incentive and strength to monitor, put pressure on, or oust the management through proxy fight or a takeover, or gain the outright control of the firm (Shleifer and Vishny, 1986) with 51 percent or more ownership. When large shareholders are less common, hostile takeovers, as a mechanism for consolidating ownership, emerge (Jensen and Ruback, 1983). In a hostile takeover, a bidder makes a tender offer to the dispersed shareholders of the target firm. If the shareholders accept the offer, the bidder acquires control of the target firm. Takeovers are considered a mechanism in the United States used to effectively control managerial discretion (Easterbrook and Fischel, 1991). Similar to large shareholders, significant creditors are also potentially active investors, who want the returns on their investments materialized. These creditors' power comes from the control rights when firms default or violate debt covenants (Smith and Warner, 1979) and future lending needs of the firms. So, the creditors control substantial cash flow rights and the ability to interfere with the firms' major decision making.

Even if shareholders elect the board, directors do not necessarily represent their interests. In the United States, boards, especially those dominated by outside directors, sometimes remove top managers after poor performance (Weisbach, 1988). Using the systemic yoyo model, this end is quite clear. When the board of a firm is dominated by outside directors, the board would likely be controlled by large shareholders or creditors. Because the large shareholders and creditors may not qualify or be informed enough to judge and monitor the performance of the manager, they are more likely to place on the board their spokesperson and representatives from outside the firm. These outside directors are employed to counter the power and influence of the manager. Otherwise, more directors from inside the firm or interlocking directors, which means the managers of firms serving on each other's boards, would dominate the board. This result explains:

1. Why the board composition, as measured by the insider-outsider ratio, is not correlated with firm's performance (MacAvoy et al., 1983)
2. Why the size of a board is negatively related to the firm's financial performance (Lipton and Lorsch, 1992; Jensen, 1993; Yermack, 1996), because the dishpan experiment indicates that the larger a spin field, the more likely for chaotic movements within the field to exist. That is, the greater the board sizes, the more less effective the board.
3. With higher proportions of outside directors, smaller boards tend to make arguably better decisions concerning acquisitions (Shivdasani, 1993), poison pills (Brickley et al., 1994), executive compensation (Core et al., 1999), and CEO replacement (Denis and Denis, 1995)

14.2 CEO's Choices of Projects

Based on our analysis in Section 14.1, it should be logical to expect that for most companies from around the world, the ownership should not be completely dispersed, and instead it should be concentrated in the hands of families and wealthy investors. This end is verified by Eisenberg (1976) for developed economies. In the United States, large shareholders, especially over 50 percent majority ownership, are relatively uncommon, probably due to legal restrictions on high concentrations of ownership and exercise of control by banks, mutual funds, insurance companies, and other institutions (Roe, 1994). Even so, ownership is not completely dispersed and also concentrated by families and wealthy investors (Eisenberg, 1976; Shleifer and Vishny, 1986). (DeLong [1991] points to a significant governance role played by J. P. Morgan Partners in the companies J. P. Morgan invested in in the early 20th century. More recently, U.S. banks play a major governance role in bankruptcies, where they change managers and directors [Gilson, 1990].) In Germany, large commercial banks through proxy voting arrangements often control over a quarter of the votes in major companies, and also have smaller but significant cash flow stakes as direct shareholders or creditors (Frank and Mayer, 1994). It is estimated that about 80 percent of the large German companies have an over 25 percent nonbank large shareholder (Gorton and Schmid, 1996), where bank and nonbank block holders improve the performance of German companies. In smaller German companies, the norm is family control through majority ownership or pyramids (Frank and Mayer, 1994). In Japan, although ownership is not nearly as concentrated as in Germany, large cross-holdings as well as shareholdings by major banks are the norm (Prowse, 1990). (Yafeh and Yosha [1996] find that large shareholders reduce discretionary spending by Japanese managers.) In France, cross-ownership and so-called core investors are common (OECD, 1995). In most of Europe, Latin America, East Asia, and Africa, corporations typically have controlling owners, who are often founders or their offspring. The effectiveness of large shareholders is closely tied to their ability to defend their rights and put a lighter burden on the legal system than smaller investors might, if they tried to enforce their rights.

Large shareholders and investors, to protect their investment, guarantee their return, and prevent the managers from expropriating funds and spending on their private benefits of control, would monitor the managers' moves and decision making closely. To study analytically how the board of a publicly traded company and the CEO interact, let us focus our attention on such a problem as how the CEO chooses between long-term and short-term projects (assets). (In this chapter, we use projects and assets interchangeably, because for investors, they mainly care about when mispricings of their holdings disappear so that they can close their positions with profits.) By short-term projects, we mean such projects that they cannot stay mispriced for long. Otherwise, these projects would be called long-term.

Assume that the board is effectively controlled by large, long-term shareholders. The controlling interest wants long-term growth, even though it could mean that they sometimes have to suffer from some occasional temporary losses. And assume that the CEO is not a long-term, large shareholder of the company. That is, there is a conflict of interest here between the board and the CEO. In particular, the board desires the CEO to devote all his or her talent and labor effort to work on the existing long-term projects and initiate new ones so that the market value of the company would grow steadily over time. On the other hand, the CEO needs short-term equity performance, because poor equity performance raises the likelihood of replacement by the board or through a hostile takeover. In either case, the CEO loses his or her job and all the privilege of control. (Kaplan and Minton [1994] show that companies with large shareholders or a principal banking relationship are more likely to replace managers in response to poor performance than companies without them. Shivdasani [1993] shows that large outside shareholders increases the likelihood that a company is taken over, whereas Denis and Serrano [1996] show that if a takeover is defeated, management turnover is higher in poorly performing companies that have block holders.)

14.2.1 Price Behavior of Different Investment Projects

To address our problem of the CEO's choices between long-term and short-term projects, let us first look at the mispricings of both of these different investment projects. Throughout this chapter, we assume that there is no discount of the future over time.

Assume that a short-term asset is traded at $p_s < v_s$ a share in time period 1, where v_s is the fundamental share-value of the asset. In this period, an investor buys $n(p_s)$ shares at the market price p_s a share. His total cost of the investment is $I_s = n(p_s)p_s$. Because this asset is a short-term investment, the mispricing disappears in period 2 and the trading price now equals the fundamental value v_s. So, the profit from this short-term asset is

$$n(p_s)v_s - I_s R = \frac{I_s v_s}{p_s} - I_s R = I_s\left(\frac{v_s}{p_s} - R\right) \tag{14.1}$$

where $R > 1$ is the gross interest spent on the total investment amount I_s.

Assume that for a long-term asset, the same investor buys in period 1 $n(p_L)$ shares at p_L a share, satisfying $p_L < v_L$, where v_L is the fundamental share-value of the asset. So, the total cost to the investor is $I_L = n(p_L)p_L$. Because the asset is a long-term investment, its mispricing does not disappear until period 3, when the total profit is

$$n(p_L)v_L - I_L R^2$$

Discounting this amount to period 2 provides

$$\frac{1}{R}\left[n(p_L)v_L - I_L R^2\right] = \frac{1}{R}\left[\frac{I_L v_L}{p_L} - I_L R^2\right] = I_L\left[\frac{v_L}{p_L R} - R\right] \tag{14.2}$$

If the investor wants to produce the same return on his investments from the short-term and long-term assets, we have from Equations (14.1) and (14.2) that

$$I_s\left[\frac{v_s}{p_s} - R\right] = I_L\left[\frac{v_L}{p_L R} - R\right]$$

From the assumption that $I_s = I_L$, meaning that the investor allocated the same amount of funds to each opportunity, we have

$$\frac{v_s}{p_s} = \frac{v_L}{p_L} \cdot \frac{1}{R}, \quad R > 1 \tag{14.3}$$

That is, for the investor to put his money with the long-term asset in period 1, this asset must be more mispriced than the short-term asset. (This result is the same as that obtained by Shleifer and Vishny [1990], where they establish a different model. Our results in the following paragraphs are opposite of those predicted by these authors.)

Similar to the analysis above, we can compare the mispricings between short-term and long-term assets in three additional different cases:

1. Both the long-term and short-term assets are traded in period 1 at a per-share price $p_i > v_i$, $i = s, L$.

2. The investor is risk averse and the long-term asset is more risky than the short-term investment, and the trading prices are $p_i < v_i$, $i = s, L$.
3. The same situation as in case 2 holds true, except that the trading prices are $p_i > v_i$, $i = s, L$.

For case 1, an equation similar to Equation (14.3) holds. However, this new equation means something completely opposite of that of Equation (14.3). In particular, for this current situation, Equation (14.3) implies that for the investor to put his money in the short-term asset, the asset must be more mispriced than the long-term asset. This end confirms the practical experience that going short can make a quicker and more handsome profit in a relatively short period of time than going long (Soros, 1998).

For case 2, we obtain

$$\frac{v_s}{p_s} < \frac{v_L}{p_L} \cdot \frac{1}{R_L}, R_L > 1 \qquad (14.4)$$

where R_L is the long-term gross interest rate. For case 3, we have

$$\frac{v_L}{p_L} \cdot \frac{1}{R_L} = \frac{v_s}{p_s} + (R_L - R_s) > \frac{v_s}{p_s}, R_s > 1 \qquad (14.5)$$

where $R_s < R_L$ stands for the short-term gross interest rate.

Summarizing what we have obtained, it can be seen that when assets are undervalued, the mispricing of the long-term asset in equilibrium is worse than that of the short-term asset. When the assets are overvalued, the mispricing of the short-term asset in equilibrium is worse than that of the long-term asset. And when a risky long-term asset is involved, the mispricings of the long-term asset is even worse if both assets are undervalued; and the mispricing of the safer short-term asset is worse if both assets are overvalued, than when risk-free assets are considered.

Because it takes a longer time for fundamental uncertainties to resolve, CEOs are typically averse to severely underpricing their long-term equity. Considering their job security, they also try to avoid overvalued short-term projects and all long-term projects, even though some of these projects could be detrimentally important to the long-term health and growth of their companies. Combining this reasoning with the fact that CEOs' compensation depends typically in part on short-term equity performance, CEOs should prefer undervalued short-term projects, if the CEOs are not large, long-term shareholders. So, to this end, how do CEOs and their boards, dominated by large, long-term shareholders, agree on which projects, long-term or short-term, to devote their limited energy and labor efforts to? In the next subsection, we will establish a theoretical model to study this problem.

14.2.2 Dynamics of Projects

To study the interactions between the CEO and the board of directors of a publicly traded company, we have assumed that the controlling shareholder directors care about near-term earnings of the company for the stability of their investment portfolio. However, their main focus is more about how to improve the long-term, healthy growth, even if they have to temporarily suffer some short-term financial setbacks. At the same time, the board newly hired a CEO who is not a large, long-term shareholder of the company.

14.2.2.1 The Model

Because of the somewhat contradicting desires of the board, the shareholder directors implicitly or explicitly allow the CEO to embark on both long-term and short-term projects for the company. In the following, we will establish a simple model for the effect of the CEO taking on short-term projects on the value of the company in the eyes of the shareholder directors. Suppose that there are two time periods, $t = 1, 2$. In period 1, the CEO puts in his labor and time to work on various projects, both long-term and short-term. As in Conyon and Read (2006), in this period, the value of the company is

$$y = a + e + \varepsilon \tag{14.6}$$

where a stands for the CEO's level of ability, e his labor effort, and ε a normally distributed random error with mean zero. That is, investors in the (stock) market value the company in terms of the CEO's ability, effort, and a random noise that is not influenced by the variables a and e.

Because the CEO does not want to lose his job, he has to work on his company's short-term equity performance, even though he is aware of the fact that the shareholder directors like him to focus as much of his labor effort as possible on the long-term growth of the company. So to the board, the real cost for the CEO to take on short-term projects is the opportunity costs of his effort, assuming that the capital markets are perfect. That is, investment capital is not a problem. When needed, the company can always take out loans. Time and labor effort not spent on long-term projects may well result in lost long-term growth opportunities. That is, the CEO's total labor effort e is split into two parts: e_s and e_L, where the subscript s stands for the amount the CEO spends on short-term projects and the subscript L, on long-term projects. So, the value of the company as seen in the eyes of the shareholder directors becomes

$$y = a + e_L + \varepsilon \tag{14.7}$$

where $e_L = e - e_s$ and $(-e_s)$ represents the opportunity costs for the CEO's working on short-term projects.

One of the reasons why the board allows the CEO to work on short-term projects is that such projects generally can realize their value in a short period. So, they provide an opportunity for the board to observe the CEO further, and such projects give the CEO a chance to harness his ability through working with people in the company new to him, exploring new knowledge and possibilities, etc. In general, through taking on numerous short-term projects, the CEO is expected to become more effective in his work and decision making. Let the change in the CEO's ability in period 1 gained solely from taking on short-term projects be

$$h_1 = f(a_0)n, \text{ satisfying } f(a_0) > 0 \text{ and } f'(a_0) > 0$$

where n stands for the number of short-term projects the CEO initiates and manages in period 1, and a_0 his endowed initial level of ability, which the board knows from records and references. Because h_1 is an increasing function in n, we have assumed that taking on various short-term projects can only enrich the CEO's knowledge base and make him more able to manage and initiate long-term projects. Therefore, the ability of the CEO in period 1 is given by $a_0 + h_1$, and the company's value in period 1 is given, as seen by the shareholder directors, as follows:

$$y = a_0 + h_1 + e_L + \varepsilon \qquad (14.8)$$

That is, in period 1, the CEO decides on how many short-term projects n to initiate and manage and how much labor effort $e_s = e - e_L$ to devote to these short-term projects so that he maximizes his personal utility function:

$$U = E(w) + \beta(E(y) + E(b)) - e_s n \qquad (14.9)$$

where $E(w)$ is the CEO's expected wage from the company, which is determined by the board in period 2, and $E(b)$ (≥ 0) his expected nonpecuniary benefits gained from these short-term projects, such as recognition locally, reputation within the industry of the company, etc., defined by

$$E(b) = b(a_0)e_s n, \text{ satisfying } 1 > b(a_0) > 0 \text{ and } b'(a_0) > 0$$

$E(y)$ stands for the expected value of his company in period 1, as seen by the shareholder directors, and $\beta \in (0, 1)$ is a parameter measuring the degree the CEO cares about how well his company does under his leadership. Here, $(-e_s n)$ stands for the CEO's disutility of initiating and managing short-term projects.

In period 2, based on the observed value of the company,

$$y^* = y(n^*, e_s^*) = a_0 + h_1^* + e_L^*$$

where the y^*-value is uniquely determined by n^* and e_s^*; the board decides on how much compensation w to pay the CEO. Here * stands for observed values. Based on the idea of Nash bargaining, the board and the CEO negotiate a specific amount w of compensation to pay the CEO by solving

$$V = (y^* - w) \times w^\alpha = \left\{ \left(a_0 + h_1^* + e_L^* \right) - w \right\} \times w^\alpha \quad (14.10)$$

where $\alpha \in (0, +\infty)$ represents the bargaining power of the CEO over the board. For instance, when $\alpha \approx 0$, the board has almost 100 percent bargaining power over the CEO. On the other hand, when $\alpha \to \infty$, the CEO has a dominating influence over the board.

14.2.2.2 The Analysis

To solve our theoretical model, let us begin by considering the negotiation between the board and the CEO in period 2. The negotiation results in the maximization of V in Equation (14.10) with respect to w subject to the constraint $w \leq a_0 + h_1^* + e_L^*$. Denote the solution to this problem by $\hat{w}$. Then, we have

$$\hat{w} = \frac{\alpha}{1+\alpha} y^* \quad (14.11)$$

At the extremes, we have $\alpha \approx 0$ and $\alpha \to \infty$. When $\alpha \approx 0$, that is, when the board has nearly total control over the CEO, the board pays the CEO almost none of his contributions to the value of the company. According to Becker's rotten kid theorem and Theorem 3.1, at such an extreme, the CEO would not voluntarily try to maximize the value of the company. This is, of course, not what the board wants to happen. On the other hand, if the CEO has a dominating influence over the board, $\alpha \to \infty$, the board would pay the CEO nearly the entirety of his contributions to the value of the company. Because this circumstance does not fit well with the setup of our situation, where large, long-term shareholders control the board, this extreme is unlikely to occur. So, what is left open is interesting: Equation (14.11) implies that the CEO, other than trying to create value, is motivated to gain as much control over the board as possible, and the board wants to stay as independent from the CEO as possible.

Now, in period 1, anticipating the compensation $\hat{w}$ in Equation (14.11), the CEO chooses n- and e_s-values. The CEO's indirect utility is given by

$$U = U(n, e_s) = \left(\frac{\alpha}{1+\alpha} + \beta \right) E(y) + \beta E(b) - e_s n \quad (14.12)$$

Assume that n and e_s (or e) are variables independent of each other. Then, the first-order conditions with respect to n and e are given respectively by

$$\left(\frac{\alpha}{1+\alpha}+\beta\right)f(a_0)+\beta b(a_0)e_s - e_s = 0 \tag{14.13}$$

and

$$\left(\frac{\alpha}{1+\alpha}+\beta\right)+\beta b(a_0)\left(1-\frac{\partial e_L}{\partial e}\right)n - \left(1-\frac{\partial e_L}{\partial e}\right)n = 0 \tag{14.14}$$

That is, in period 1, the CEO chooses the total number n^* of short-term projects to initiate and to manage, and his overall level $e^* = e_L^* + e_s^*$ of labor effort to work on his job as the CEO of his company, with e_L^* amount spent on the long-term projects and e_s^* on the short-term projects, where

$$n^* = \frac{\frac{\alpha}{1+\alpha}+\beta}{\left(1-\frac{\partial e_{L^*}}{\partial e}\right)(1-\beta b(a_0))} \tag{14.15}$$

and

$$e_s^* = \frac{\left(\frac{\alpha}{1+\alpha}+\beta\right)f(a_0)}{1-\beta b(a_0)} \tag{14.16}$$

If the CEO exerts roughly a fixed amount of labor effort e_L on the management and initiation of long-term projects, then he would initiate and manage as many as

$$n^* = \frac{\frac{\alpha}{1+\alpha}+\beta}{1-\beta b(a_0)} \tag{14.17}$$

short-term projects.

Both Equations (14.15) and (14.17) indicate that the higher the level the CEO's initial ability a_0 is, the more short-term projects he would initiate and manage, and according to Equation (14.16), the more labor effort, correspondingly, he will devote to these projects. His additional effort spent on the short-term projects helps him gain those nonpecuniary benefits $b = b(a_0)e_s n$, which he can use to gain more bargaining power over the board. At the same time, the CEO's increased number of successful short-term projects raises his level of ability by as much as $h_1 = f(a_0)n$. So, he becomes more effective in his managerial duties, and consequently, the company's value increases instead of decreases due to the CEO's increased commitment to short-term projects.

Additionally, Equation (14.15) implies that only when the marginal increase in the total labor effort e from working on all projects is greater than the marginal increase in the labor effort e_L spent on the long-term projects (meaning the CEO tried to use some of his personal time to work on short-term projects) does the CEO take on short-term projects. Because taking on additional short-term projects generally means that the amount of time spent on the management of existing long-term projects decreases, that is, $\frac{\partial e_L}{\partial e} < 1$, Equation (14.15) implies that in general, the CEO would take on short-term projects.

14.2.2.3 Power Struggle between the Board and the CEO

The fact that the CEO has been hired by the board means that the majority of the directors feel that the CEO could do a good job to satisfy their needs. That is, starting on the first day when the CEO begins to work for the company as the executive officer, he has a certain amount of bargaining power over the board. Equation (14.11) implies that it is natural for the CEO to try to gain more bargaining power through whatever is necessary and workable, as long as he is able to make a positive contribution to increase the value of the company. To achieve this end, Equation (14.8) indicates that the CEO would take on as many short-term projects as possible while keeping his effort level e_L on long-term projects as steady as possible.

From Equations (14.15) and (14.16), we obtain

$$\text{sign}\left(\frac{\partial n^*}{\partial \alpha}\right) = \text{sign}\left(1 - \frac{\partial e_L}{\partial e}\right) \qquad (14.18)$$

and

$$\frac{\partial e_s^*}{\partial \alpha} = \frac{f(a_0)}{(1 - \beta b(a_0))(1 + \alpha)^2} > 0 \qquad (14.19)$$

Equation (14.18) implies that the more bargaining power the CEO has over the board, the more his ideas on short-term projects will be implemented. Equation (14.19) indicates that in this case, the CEO would also spend more of his available time and possibly high-quality service to manage the short-term projects.

On the other hand, from Equations (14.15) and (14.16), we obtain

$$\frac{\partial \alpha}{\partial n^*} = \frac{A}{(An^* - \beta - 1)^2} \qquad (14.20)$$

and

$$\frac{\partial \alpha}{\partial e_s^*} = \frac{B}{(Be_s^* - \beta - 1)^2} \qquad (14.21)$$

where $A = \left(1 - \frac{\partial e_L}{\partial e}\right)(1 - \beta b(a_0))$ and $B = (1 - \beta b(a_0))/f(a_0)$. That is, Equations (14.20) and (14.21) imply that taking on short-term projects and spending extra labor effort on them could potentially increase the CEO's bargaining power over the board.

Because labor effort spent on short-term projects has a direct impact on how much bargaining power the CEO has over the board, we can now assume that the effort level e_L (respectively, e and e_s) is a function of n. It is reasonable for us to assume that $\frac{\partial e_L}{\partial n} < 0$ and $\frac{\partial^2 e_L}{\partial \alpha^2} > 0$. Intuitively, this assumption means that the more short-term projects the CEO takes on, the less time he has available for his long-term projects. And when he takes on even more short-term projects, his time spent on long-term projects will asymptotically approach a minimally required level to take care of long-term projects.

Now, let us look at the relationship between the optimal value n of short-term projects taken by the CEO from the standpoint of the board, and the value from the standpoint of the CEO. The board wants to maximize its expected value in Equation (14.8). So, the board's optimal n value satisfies

$$\frac{\partial E(y)}{\partial n} = f(a_0) + \frac{\partial e_L}{\partial n} = 0 \tag{14.22}$$

And at this n-value,

$$\frac{\partial U}{\partial n} = \left(\frac{\alpha}{1+\alpha}\beta\right)\left(f(a_0) + \frac{\partial e_L}{\partial n}\right) + (\beta b(a_0) - 1)\left(e_s + n\frac{\partial e_s}{\partial n}\right)$$

$$= (\beta b(a_0) - 1)\left(e_s + n\frac{\partial e_s}{\partial n}\right) \tag{14.23}$$

Based on the same reasoning for the assumption that $\frac{\partial e_L}{\partial n} < 0$ applies here, we have $\frac{\partial e_s}{\partial n} > 0$. So, Equation (14.23) implies that $\frac{\partial U}{\partial n} < 0$. That is, the CEO's personal optimal n-value is less than the board's optimal n-value. Here, we have shown the following result.

PROPOSITION 14.1
The number of short-term projects the CEO likes to take on is less than the number the board would like him to take on.

This result is comforting to both the CEO and the board. It means that the CEO can simply work on as many short-term projects as he pleases without the need to worry that his attempt and hard work might offend some of the large, long-term shareholder directors. At the same time, Equations (14.20) and (14.21) spell out the fact that as long as the CEO is successful with his short-term projects while keeping all long-term projects on track, the board would put more trust in him and automatically give him more bargaining power.

14.3 Conclusion

All the work, as presented in this chapter and Chapters 9 to 13, marks only one of the first steps toward employing the yoyo model, one of the newest results developed in systems science, in the study of problems in economics and corporate finance. As we try to emphasize throughout these chapters, any time a force or a pair of acting and reacting forces exists, the yoyo model can be employed, because the existence of each such pair of forces means an eddy motion as implied in Newton's second law of motion (Chapter 4 or Lin, 2007). Because of this understanding, one can reasonably expect that he or she should be able to revisit many or most of the basic concepts in economics in the light of spinning yoyos and their spin fields. Consequently, he or she could explain the results, at least some of them, obtained in behavioral economics and behavioral finance. More specifically, for example, we should now be more equipped than before to analyze the anomaly existing between utility maximization, as economists do all the time to produce their often surprising inferences on people's desires and consumption behaviors, and experienced utility, as each one of us, as consumers and decision makers, goes through on a daily basis (Kahneman and Thaler, 2006). As another example, we could look at the connection between economics and corporate culture (Hermalin, 2001) from a different angle with nonlinearity involved.

In short, the work presented in Chapters 9 to 14 is only a pebble we throw out there to attract beautiful and more valuable gemstones in the years to come.

STRUCTURE OF HUMAN THOUGHTS AND INFINITY PROBLEMS IN MODERN MATHEMATICS

Chapter 15

A Quick Glance at the History of Mathematics

With an invitation from Ronald Mickens in 1987, Yi Lin had the honor to join some of the very well-known scholars from around the globe, such as Wendell Holladay (Vanderbilt University), Saunders Mac Lane (University of Chicago), John Polkinghorne (Cambridge, UK), Robert Rosen (Dalhousie University, Canada), and others, to express his opinion from the angle of systems research on Nobel laureate Eugene P. Wigner's assertion about "the unreasonable effectiveness of mathematics":

> The miracle of the appropriateness of the language of mathematics for the formulation of the laws of physics is a wonderful gift which we neither understand nor deserve. We should be grateful for it and hope that it will remain valid in future research and that it will extend, for better or for worse, to our pleasure even though perhaps also to our bafflement, to wide branches of learning. (Wigner, 1960)

His paper, entitled "A Few Systems-Colored Views of the World," was eventually published in 1990 (Mickens, 1990). Continuing this train of thought, based on the observation that when faced with practical problems, new abstract mathematical theories are developed to investigate them, predictions made based on the newly developed theories are very "accurate." In 1997, with Hu and Li, Yi Lin posted the following question: Does the human way of thinking have the same structure as that of the material world?

Our thinking behind this question is that the accuracy in our predictions, derived from the newly developed theories, is an indication that the human way of thinking is roughly the same as how the material world is constructed. In other words, our way of thinking and the material world have the same structure. But how can this end be proven? The difficulty here is: What is meant by *structure*? And how can he get started to uncover this structure? With these questions in mind, starting in 1995, together with colleagues, Yi Lin began his search in two directions, one in theory and the other in applications. Along the line of theory, he and colleagues analyzed what makes some theories in history successful and long lasting. Evidently, mathematics, as a system of theories, has been representative of such a theory. Its appearance can be traced back to the very beginning of human existence. And as a form of life itself, mathematics still goes on very strongly. Along the line of applications, Yi Lin notices that some human endeavors have been under way since the dawn of our recorded history. So, what is the connection here? What keeps mathematics young and thriving? And what underlies the human motivation to pursue those difficult and long-lasting endeavors?

Over the years since 1995, Yi Lin and his colleagues have published in theory, in applications, and about interactions between theories and applications. Finally, in 2002, they discovered such a structure that underlies all physical and imaginary systems (Wu and Lin, 2002). This structure is given the name of systemic (Chinese) yoyo model in Lin (2007). In that publication, the yoyo model is applied to generalize Newton's laws of motion, and to shed new insights on Kepler's laws on planetary motion, Newton's law of universal gravitation, and the study of the three-body problem. And in the following papers (Lin and Forrest, 2008b, c), this yoyo model is successfully employed to study the rotten kid theorem, initially established by Nobel laureate Gary Becker (1974), in household economics, child labor in labor economics, interindustrial wage differentials in the economics of wage structure, and the interactions between the CEO and the board of directors of corporate governance. Now, in this chapter, we want to use the history of mathematics to uncover the systemic yoyo structure in human thoughts. Here, we assume that mathematics is an organic system of thoughts created on the bases of experience, intuition, and logic thinking.

By sorting through the ideas appearing in and concepts studied in the history of mathematics, we discover that some ideas and concepts, such as those of infinitesimals, infinities, limits, etc., underlie each crisis in the foundations of mathematics. After introducing the concepts of actual and potential infinities, we are able to uncover some hidden contradictions in the system of modern mathematics and see the return of the Berkeley paradox in the foundations of mathematics. By formally mapping the history of mathematics onto our yoyo model, it can be shown using available statistics that the periodic transitions between uniform and symmetric flows and chaotic currents existing in rotating fluids, the dishpan experiment, imply that the fourth crisis in the foundations of mathematics has occurred.

So, all in all, through our systemic analysis of the history of mathematics, it is found that the systemic yoyo structure existing in human thoughts is indeed the same as that of the physical systems studied in Lin (2007) and that of economic systems studied in Lin and Forrest (2008b, c). For the completeness of our presentation, let us look at some other scholars who have done works in recent years on related concepts, such as actual and potential infinities. Limited by space, we will only mention a few relevant works. Interested readers can either consult the papers cited here and the references listed therein, or do a more thorough literature search along all the lines outlined here.

In 1992, T. Hailperin of Lehigh University established a formal characterization of a potential infinite sequence as a rule-generatable sequence. In 1997, in his study of ontologically neutral (ON) language Q and the concept of models for an ON language, Hailperin employed the concept of actually infinite sets in the definition of verity functions. On this basis, he showed that any assignment of truth-values to the atomic sentences could be extended so as to be a verity function on Q^{cl}, the closed formulas of Q. And based on this infinitistic assumption and validity defined to be all possible assignments of truth-values to the atomic sentences, he provided a completeness proof along traditional lines. To improve his completeness proof for ON logic, Hailperin (2001) introduced the concept of potential infinite verity functions on potentially infinite sequences, leading to the notion of potential infinite models, and proved the following fundamental theorem of potential infinite semantics along with several other important results.

THEOREM 15.1.
(Hailperin, 2001) Let M be a potential infinite model with potential infinite limit L for an ON language $Q(p_1, ..., p_r)$. Then,

(i) there is a unique function V_M that extends M from atomic sentences so as to be a potential infinite verity function on closed formulas of Q;
(ii) for any closed formula $(\Sigma i)\varphi$ (or $(\Pi i)\varphi$) the sequence of values $V_M(\varphi\, 0), ..., V_M(\varphi\,(v)), ...$ ends with a tail of all 0s or all 1s;
(iii) for any closed formula ϕ, its V_M value is equaled to that of its L-finitization, that is, $V_M(\phi) = V_M(S_L(\phi))$.

J. Engelfriet and T. Gelsema (2004) of Leiden University (The Netherlands) studied the middle congruence, a notion of structural equivalence of processes, in which replication of a process is viewed as a potential infinite number of copies of the process in the sense that copies are spawned at need rather than produced all at once. Their work is based on the standard congruence studied in Milner (1992) and the extended congruence first introduced in Engelfriet (1996). They introduced additional laws to model !P as an actual infinite number of copies of P, and their notion of middle congruence shares with standard congruence the suggestion that replication is only potentially infinite. They proved that middle congruence has the same desirable properties as extended congruence: it is decidable and has a concrete multiset semantics, leading to the conclusion that these properties do not depend on the distinction between potential and actual replication.

Our purpose here is not to briefly and simply walk through the milestones in the history of mathematics. Instead, we like to look at this history as a system of human thoughts in light of whole systemic evolution. We will focus on several threads of thoughts that have existed since the beginning of recorded history. By walking through history along these threads, we can clearly see how the underlying systemic yoyo structure exists and how the spin field of this structure goes through the pattern changes as vividly described in the dishpan experiment. By doing so, we expect to show that in the thought system called mathematics, another crisis is overdue, and in fact, such a crisis has arrived and appeared in the foundations of the system. For this end, please consult Chapter 16.

This chapter is organized as follows. In Section 15.1, we look at how mathematics got started in the beginning of time. In Sections 15.2 to 15.4, we walk respectively through the first, second, and third crises in the foundations of mathematics with some relevant details.

A note about references: All the information regarding the history of mathematics below without mentioning the source is from either Eves (1992) or Kline (1972). And the presentation in this and the next chapter is mainly based on Lin et al. (2008).

15.1 The Beginning

Archeological evidence suggests that man employed counting as far back as 50,000 years ago. Since our recorded history is much shorter than that, we have to apply our imagination to fill in a huge gap here. Due to the need for survival, humans started out with a sense of numbers and recognition of more and less. With the gradual evolution of society, counting became imperative. Probably the earliest way of keeping a count was by some simple tally method, employing the principle of one-to-one correspondence. An assortment of vocal sounds was developed as a word tally against the number of objects in a group. Then, an assortment of symbols was derived to stand for these numbers. After developing simple grouping systems for numbers, multiplicative groupings and positional numeral systems were established. Many of the modern-day computing patterns in elementary arithmetic appeared only in the 15th century due to a lack of a plentiful and convenient supply of suitable writing media in the earlier days.

Early mathematics was developed to meet practical needs and the evolution of more advanced forms of society. So, usable calendars, systems of weights and measures, and methods of survey were invented. A special craft came into being for the cultivation, application, and instruction of this practical science. In such a circumstance, tendencies toward abstraction were bound to develop, and the science was then studied for its own sake. It was in this way that algebra grew out of arithmetic and geometry out of measuration.

In the last centuries of the second millennium BC, along with economic and political changes, some civilizations disappeared or waned, and new people, including the Greeks, came to the fore. The alphabet was invented and coins were introduced. The new civilization made its appearance in towns along the coast of Asia Minor and later on the mainland of Greece, Sicily, and the Italian shore.

For the first time, men started to ask *why* in fundamental ways. The past empirical processes, sufficient for the question *how*, were no longer sufficient to answer these more scientific inquiries of *why*. Thus, mathematics, in the modern sense of the word, was born in this atmosphere of rationalism. Thales is the first-known individual with whom mathematical discoveries are associated. He used logical reasoning instead of intuition and experiment.

The next outstanding Greek mathematician is Pythagoras. His philosophy rested on the assumption that whole numbers are the cause of the various qualities of man and matter. This led to an exaltation and study of number properties and arithmetic, along with geometry, music, and spherics. Among many important works, Pythagoras independently discovered the Pythagorean theorem: the square of the hypotenuse of a right triangle is equal to the sum of the squares of the two legs.

15.2 First Crisis in the Foundations of Mathematics

Historically speaking, the concept of integers arose from the need to count finite collections of objects. The need to measure quantities, such as length, weight, and time, gave rise to the concept of fractions, because it is very seldom that a length, as an example, is an exact integral number of linear units. Thus, a rational number is defined as the quotient of two integers p/q, where $q \neq 0$. The system of rational numbers seemed to be sufficient for practical purposes at the time.

The rational numbers can be understood with a simple geometrical interpretation. Imagine that we are given a horizontal straight line and we mark two distinct points, O and I, on the line, with I located to the right of O. Let us use the segment OI as a unit of length. If we understand the mark O as 0 and I as 1, then all the positive and negative integers can be presented by a set of points on the line spaced at unit intervals apart, with the positive integers marked to the right of O and the negative integers to the left of O. The fractions with denominator q may be then represented by the points that divide each of the unit intervals into q equal parts. So, for each rational number, there is a corresponding point on the line. Due to the fact that in early applications of mathematics, the system of rational numbers was closed with respect to the arithmetic operations, +, −, ×, ÷, and met the need of all practical purposes, it was natural for the early scholars to believe that all the points on the line were used up in the way just described.

It was not until the fifth century BC that the unexpected discovery of irrational numbers, called the Hippasus paradox, together with other paradoxes, such as those constructed by Zeno, gave rise to the first crisis to the foundations of mathematics. This discovery is one of the greatest achievements of the Pythagoreans.

This new finding implied that there are still points on the line not corresponding to any rational number. In particular, the Pythagoreans found that the diagonal of a unit square does not correspond to a rational number point on the line.

The rigorous proof for the existence of irrational numbers was surprising and disturbing to the Pythagoreans, because it was a mortal blow to the Pythagorean philosophy that all in the world depend on whole numbers. Geometrically, this discovery was startling, because who could doubt that for any two given line segments, one is able to find some third line segment that can be marked off a whole number of times into each of the two given segments? Let the two segments be s and d, such that s is the side of a square and d the diagonal. If there were a third segment t that could be marked off a whole number of times into s and d, one would have $s = bt$ and $d = at$, where a and b are positive integers. Because $d = s\sqrt{2}$, we have $d = at = s\sqrt{2} = bt\sqrt{2}$ or $\sqrt{2} = a/b$, a rational number. This contradiction implies that there exist incommensurable line segments, meaning segments without a common unit of measure.

This discovery that like magnitudes may be incommensurable proved to be extremely devastating. It caused some consternation in the Pythagorean ranks. Not only did it appear to upset the basic assumption that all in the world depend on whole numbers, but because the Pythagorean definition of proportion assumed that any two like magnitudes are commensurable, all the propositions in the Pythagorean theory of proportion had to be limited to commensurable magnitudes, making the theory unsound, and their general theory of similar figures became invalid.

For quite a long time, $\sqrt{2}$ was the only known irrational number. Later Theodorus of Cyrene (ca. 425 BC) showed that $\sqrt{3}, \sqrt{5}, \sqrt{6}, \sqrt{7}, \sqrt{8}, \sqrt{10}, \sqrt{11}, \sqrt{12}, \sqrt{13}, \sqrt{14}, \sqrt{15}$, and $\sqrt{17}$ are also irrational numbers. The final resolution of this crisis in the foundations of mathematics was achieved in about 370 BC by Eudoxus.

To get a sense on how Eudoxus resolved the crisis, let us look at the following.

PROPOSITION 15.1

(i) Triangles having equal bases and equal altitudes have equal areas
(ii) Of any two triangles having the same altitude, the one that has the greater area has the greater base.

Evidently, statement 2 above is a corollary of statement 1.

Assume that we have two triangles ABC and ADE, where the bases BC and DE lie on the same straight line MN, as shown in Figure 15.1. Without knowing the existence of irrational numbers, the Pythagoreans assumed implicitly that any two line segments are commensurable. So, BC and DE have some common unit of measure u so that BC = pu and DE = qu. That is, the unit u goes into BC p times and DE q times. Mark off these points of division on BC and DE and connect each of them with vertex A. So, the triangles ABC and ADE are divided, respectively, into p and q smaller triangles. From Proposition 15.1(i), it follows that all these

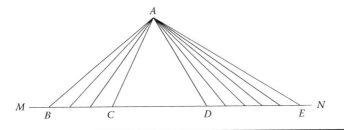

Figure 15.1 Triangles ABC and ADE.

smaller triangles have the same area. So, $\triangle ABC:\triangle ADE = p:q = BC:DE$. That is, the Pythagoreans had shown the following result.

PROPOSITION 15.2
The areas of triangles having the same altitude are to one another as their bases.

With the later discovery that two line segments do not have to be commensurable, this Pythagorean proof, along with others, became incorrect. This is why the discovery of irrational numbers was extremely devastating to the Pythagoreans.

Here is how Eudoxus resolved the crisis. He first redefined proportion as follows: magnitudes are said to be in the same ratio, the first to the second, and the third to the fourth, when, if any equi-multiples whatever taken of the first and third, and any equi-multiples whatever taken of the second and fourth, the former equi-multiples alike exceed, are alike equal to, or are alike less than the later equi-multiples taken in corresponding order. In symbols, if A, B, C, and D are any four unsigned magnitudes, A and B being of the same kind (both line segments, or angles, or areas, or volumes), and C and D being of the same kind, then the ratio of A to B is equal to that of C to D when, for arbitrary positive integers m and n, $mA \geq, =,$ or $\leq nB$ according to $mC \geq, =,$ or $\leq nD$.

Going back to the previous example, on CB produce (mark off) successively from B, $m - 1$ segments equal to CB, and connect the points of division, B_2, B_3, ..., B_m, with vertex A, shown in Figure 15.2. Similarly, on DE produce (mark off) successively from E, $n - 1$ segments equal to DE, and connect the points of

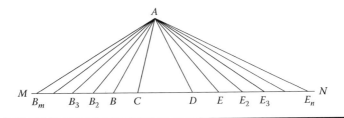

Figure 15.2 The Eudoxus construction.

division, $E_2, E_3, \ldots, E_n$, with vertex A. Then, $B_mC = m(BC)$, $\triangle Ab_mC = m(\triangle ABC)$, $DE_n = n(DE)$, and $\triangle ADE_n = n(\triangle ADE)$. Now, Proposition 15.1 implies that $\triangle Ab_mC \geq, =,$ or $\leq \triangle ADE_n$ according to $B_mC \geq, =$ or $\leq DE_n$, That is, $m(\triangle ABC) \geq, =,$ or $\leq n(\triangle ADE)$ according to $m(BC) \geq, =,$ or $\leq n(DE)$. Hence, by the Eudoxian definition of proportion, $\triangle ABC:\triangle ADE = BC:DE$. So, Proposition 15.2 is established. Here, neither commensurable nor incommensurable quantities are mentioned, because the Eudoxian definition applies equally to both situations.

The Eudoxian revised theory of magnitudes and proportions has become one of the great mathematical masterpieces of history. Eudoxus' treatment of incommensurables coincides essentially with the modern exposition of irrational numbers that was first given by Richard Dedekind in 1872. It seems that this crisis in the foundations of mathematics is largely responsible for the subsequent formulation and adoption of the axiomatic method in mathematics.

The first 300 years of Greek mathematics, commencing with the initial efforts of demonstrative geometry by Thales, about 600 BC, and culminating with the remarkable elements by Euclid, about 300 BC, constitutes a period of extraordinary achievement. Looking back to this period, we notice three important and distinct lines of development. The first was the development of the materials that ultimately were organized into the elements by Euclid, began by the Pythagoreans and followed by Hippocrates, Eudoxus, Theodorus, Theaetelus, and others. The second line is the development of notions connected with infinitesimals and with limits and summation processes that did not attain final clarification until after the invention of calculus in modern times. The following Zeno paradoxes, the method of exhaustion of Antiphon and Eudoxus, and the atomistic theory associated with the name of Democritus belong to this line of development.

Zeno paradox 1 (the dichotomy): If a straight line segment is infinitely divisible, then motion is impossible, for to traverse the line segment, it is necessary first to reach the midpoint, and to do this one must first reach the one-quarter point, and to do this one must first reach the one-eighth point, and so on, ad infinitum. It follows that the motion can never even begin.

Zeno paradox 2 (the arrow): If time is made up of indivisible atomic instants, then a moving arrow is always at rest, for at any instant the arrow is in a fixed position. Because this is true of every instant, it follows that the arrow never moves.

These and some other paradoxes were devised by the Eleatic philosopher Zeno (ca. 450 BC) to address the question of whether a magnitude is infinitely divisible or is made up of a very large number of tiny indivisible atomic parts. There is evidence that since the time of Greek antiquity, schools of mathematical reasoning developed their thoughts using each of these two possibilities. The Zeno paradoxes above assert that motion is impossible whether we assume a magnitude to be infinitely divisible or made up of a large number of atomic parts. Because of these paradoxes, infinitesimals were excluded from Greek demonstrative geometry.

The third line of development is that of higher geometry, or the geometry of curves other than the circle and straight line, and of surfaces other than the sphere and plane, originated in the continued attempts to solve three construction problems: (1) Construct the edge of a cube that has twice the volume of a given cube. (2) Divide a given arbitrary angle into three equal parts. (3) Construct a square that has an area equal to that of a given circle.

To summarize, based on the Pythagorean theorem, it was found that for a right triangle with both legs of length 1, the hypotenuse $\sqrt{2}$ could not be represented by using whole numbers or ratios of whole numbers. The discovery fatally shook the foundation of the Pythagorean school's assumption and assaulted the common belief of Greek society. It made scholars panic and intranquil, and directly vacillated the foundations of mathematics of the time. The chaos created by this discovery and Zeno paradoxes has been called the first crisis in the history of mathematics. However, it was because of the occurrence of this crisis that the history of mathematics turned from intuition and experience-based arguments to deduction-based proofs. Axiomatic geometry and formal logic were born and developed, leading to the formation of the concept of irrational numbers and the expansion of the number field.

15.3 Second Crisis in the Foundations of Mathematics

The 17th century was very conspicuous in the history of mathematics. Early in the century, John Napier revealed his invention of logarithms, Thomas Harriot and William Oughtred contributed to the notion and codification of algebra, Galileo Galilei founded the science of dynamics, and Johann Kepler announced his laws of planetary motion. Later in the century, Gerard Desargues and Blaise Pascal opened a new field of pure geometry, Rene Descartes launched modern analytic geometry, Pierre de Fermat laid the foundations of modern number theory, and Christiaan Huygens made his distinguished contributions to the theory of probability and other fields. Then toward the end of the century, after many other 17th-century mathematicians had prepared the way, Isaac Newton and Gottfried Wilhelm Leibniz created calculus. It was during this century that many new and vast fields were opened up for mathematical investigation.

The first problems occurring in the history of calculus were concerned with the computation of areas, volumes, and lengths of arcs. In their treatment, one finds evidence of applying the two assumptions about divisibility of magnitudes. The famous Greek method of exhaustion is usually credited to Eudoxus (ca. 370 BC) and can be considered the Platonic school's explanation to Zeno paradoxes. The method assumes the infinite divisibility of magnitudes and has, as its foundation, the following proposition.

PROPOSITION 15.3
If from any magnitude there is subtracted a part not less than its half, from the remainder another part not less than its half, and so on, there will at length remain a magnitude less than any preassigned magnitude of the same kind.

Of the ancients, it was Archimedes who made the most elegant applications of the method of exhaustion and who came the nearest to actual integration. In his treatment of areas and volumes, Archimedes arrived at equivalents of a number of definite integrals found in our modern-day calculus textbooks. With time, such scholars as Kepler, Cavalieri, Torricelli, Fermat, Pascal, Saint-Vincent, Barrow, and others, employed the ideas of infinitesimals and indivisibles and produced results equivalent to the integrations of such expressions as x^n, $\sin\theta$, $\sin^2\theta$, and $\theta\sin\theta$.

On the other hand, differentiation has originated in the problem of drawing tangents to curves, and in finding maximum and minimum values of functions on the ideas set forth by Fermat in 1629. Isaac Barrow (1630–1677) was one main character who made important contributions to the theory of differentiation.

Calculus, aided by analytic geometry, was the greatest mathematical tool discovered in the 17th century. It proved to be remarkably powerful and capable of attacking problems quite unassailable in earlier days. It was the wide range and astonishing applicability of the theory that attracted the bulk of mathematics researchers of the day, with the result that papers were published in great profusion with little concern for the unsatisfactory foundations of the theory. The processes employed were justified largely on the ground that they worked. It was not until the 18th century had almost elapsed, after a number of absurdities and contradictions had crept into mathematics, that mathematicians felt the need to examine and establish rigorously the basis of their work. The effort to place calculus on a logically sound foundation proved to be difficult. Various ramifications occupied the better part of the next 100 years. A result of this careful work in the foundations of calculus was that it led to equally careful work in foundations of all branches of mathematics, and to the refinement of many important concepts. Chronologically, calculus was created in the latter half of the 17th century. The 18th century was largely spent in exploiting new and powerful methods of calculus. And the 19th century was largely devoted to the effort of establishing calculus on a firm, logical foundation.

Because the early calculus was established on the ambiguous and vague concept of infinitesimals without a solid foundation, many operations and applications of the theory were reproached and attacked from various angles. Among all the reproaches, the most central and able about the faulty foundation of the early calculus was brought forward by Bishop George Berkeley (1685–1753). In 1734, he published a book entitled *The Analyst: A Discourse Addressed to an Infidel Mathematician*. The "infidel mathematician" is believed to have been Edmond Halley or Newton. This book was a direct attack on the foundation and principles of calculus, specifically on Newton's notion of fluxion and Leibniz's notion of infinitesimal change. The following passage has been most frequently quoted from *The Analyst*:

And what are these fluxions? The velocities of evanescent increments? And what are these same evanescent increments? They are neither finite quantities nor quantities infinitely small, nor yet nothing. May we not call them the Ghosts of departed Quantities?

To understand what Berkeley was talking about, let us look at an example from a paper by Newton on how to compute the area of a region with a curved boundary. In his work, Newton claimed that he avoided using infinitesimals by going through the following steps: Give an increment to x, expand $(x + 0)^n$, subtract x^n, divide by 0, and compute the ratio of the increment in x^n over that in x. Then by throwing away the 0 term, he obtained the fluxion of x^n. Berkeley said that the variable was first given an increment, then the increment let to be 0. This involved the fallacy of a shift in the hypothesis. As for the derivative seen as the ratio of the disappeared increments in y and x, that is, the ratio of dy and dx, Berkeley called these disappeared increments "neither finite quantities nor quantities infinitely small, not yet nothing. May we not call them the Ghosts of departed quantities?" (Kline, 1972).

It was exactly because the early calculus did not have a solid theoretical foundation that many criticisms seemed at the time reasonable, so that conflicts between achievements of calculus in applications and the ambiguity in the foundation became more and more incisive. With the passage of time, a serious crisis in the foundations of mathematics became evident. The first suggestion to replace the sandy foundation of calculus on a more solid basis came from Jean-le-Rond d'Alembert (1717–1783), who observed in 1754 that a theory of limits was needed. It was Joseph Louis Lagrange (1736–1813) who was the earliest to attempt a rigorization of calculus. With his work, the long and difficult task of banishing intuitionism and formalism from calculus began.

In the 19th century, the theory of analysis continued to grow on an ever-deepening foundation due to the 1812 work of Carl Friedrich Gauss, who set new standards of mathematical rigor. A great stride was made in 1821 when Augustin-Louis Cauchy (1789–1857) successfully executed d'Alembert's suggestion by developing an acceptable theory of limits and then defining continuity, differentiability, and definite integral in terms of the concept of limits.

Such examples as a continuous function that has no derivative at any point in its domain (by Karl Weierstrass, 1874) and a function that is continuous for all irrational points and discontinuous for all rational points in its domain (by Georg Bernhard Riemann) suggested that Cauchy had not resolved all the difficulties in the foundation of analysis. Accordingly, Weierstrass advocated and materialized a program to first rigorize the real number system, and then establish the basic concepts of analysis in particular, and mathematics in general, on this system. Today, it can be stated that essentially all of existing mathematics is consistent if the real number system is consistent. In the late 19th century, Richard Dedekind (1831–1916), Georg Cantor (1845–1918), and Giuseppe Peano (1858–1932) showed how the real number system and the great bulk of mathematics could be derived from a postulate set for the natural number system. Then, in the early 20th century, it was shown that natural numbers can be defined in terms of concepts of set theory, and thus that the great bulk of mathematics can be made to rest on a platform in set theory.

Specifically, on the basis of Cauchy's theory of limits, Dedekind proved the fundamental theorems in the theory of limits using the rigorized real number theory. With the combined efforts of many mathematicians, the methods of $\varepsilon - N$ and $\varepsilon - \delta$ became widely accepted, so that infinitesimals and infinities could be successfully avoided. The mathematical analysis, developed along this line, has been called the standard analysis.

It has been a common belief that due to the satisfactory establishment of the theory of limits, calculus has ever since been constructed on a solid, rigorous theoretical foundation. So, in the history of mathematics, the second crisis in the foundations of mathematics is considered successfully resolved. In the following we will see that this belief cannot be further from the truth. That is, the Berkeley paradox did not really go away. And more paradoxes are created in the theory of limits and in the foundations of mathematics.

15.4 Third Crisis in the Foundations of Mathematics

Georg Cantor (1845–1918) was interested in number theory, indeterminate equations, and trigonometric series. This last interest somehow inspired him to look at the foundation of mathematical analysis. In 1874, he commenced his revolutionary work on set theory and the theory of the infinite, where he developed a theory of transfinite numbers based on the actual infinite. Because so much of mathematics is permeated with set concepts, the superstructure of mathematics can actually be made to rest upon set theory as its foundation.

However, in 1897, Burali-Forti brought to light the first publicized paradox of set theory. Two years later, Cantor found a nontechnical description of a very similar paradox. In his theory of set, Cantor proved that for any given set X, there is always another set Y such that the cardinality of X is less than that of Y. What is shown is similar to the situation that there is no greatest natural number. There also is no greatest transfinite number. Now, consider the set that contains all sets as its elements. Surely no set can have more elements than this set of all sets. But if this is the case, how can there be another set whose cardinality is greater than the cardinality of this set of all sets?

In 1902, Bertrand Russell discovered another paradox, which involves nothing more than just the concept of sets itself. In particular, let the set of all sets that are members of themselves be M, and the set of all sets that are not members of themselves be N. Now, let us consider this question: Is N a member of itself N? There are only two possible answers to this question: yes or no. If N is a member of itself, then N belongs to M and not of N. So, N is not a member of itself. A contradiction. On the other hand, if N is not a member of itself, then N is a member of N and not of M. So, N is a member of itself. Another contradiction. That is, in either case, we are led to a contradiction.

Since the discovery of the above contradictions within Cantor's set theory, more paradoxes have been produced in abundance. These modern paradoxes of set theory are related to several ancient paradoxes of logic. For instance, Eubulides (fourth century BC) is credited for the remark: "This statement I am now making is false." If what Eubulides said is true, then the statement must be false. On the other hand, if what Eubulides said is false, then it follows that his statement must be true. That is, Eubulides' statement can be neither true nor false without entailing a contradiction. As another example, Epimenides (sixth century BC) is claimed to have made the following remark: "Cretans are always liars." A simple analysis of this remark easily reveals that it, too, is self-contradictory.

The existence of paradoxes in set theory, like those described above, surely suggests that there must be something wrong with that theory. However, what is more devastating is that now the foundation of mathematics faces another major crisis, called the third crisis of mathematics, because these paradoxes naturally cast doubt on the validity of the entire fundamental structure of mathematics.

A close examination of the paradoxes considered above reveals the fact that in each case, a set S and a member m of S are involved, where the definition of the member m depends on S. That is, in a sense, a circular definition is involved. To avoid the paradoxes, it seems that one has to impose a restriction on the concept of set. That is why Russell introduced his vicious circle principle: no set S is allowed to contain members m definable only in terms of S, or members m involving or presupposing S. However, this approach of disallowing circular definitions met with one serious objection: many fundamental concepts of mathematics are introduced using such definitions. In 1918, Hermann Weyl undertook the task to find out how much of mathematical analysis can be constructed without using circular definitions. After successfully obtaining a considerable part of the analysis, he was stuck by being unable to derive the important theorem that every nonempty set of real numbers having an upper bound has a least upper bound.

To resolve the third crisis in mathematics, there have appeared three main schools of thought: the so-called logistic, intuitionist, and formalist schools. Each of these schools has attracted a sizable group of followers and generated a large body of associated literature. Russell and Alfred North Whitehead are the chief expositors of the logistic school, L. E. J. Brouwer led the intuitionist school, and David Hilbert developed the formalist school.

For the logistic school, its thesis is that mathematics is a branch of logic, and all mathematical concepts are to be formulated in terms of logical concepts, and all theorems of mathematics are to be developed as theorems of logic. The notion of logic underlying all science dates back at least as far as Leibniz (1646–1716). It was Dedekind (1888) and Frege (1884–1903) who were the first to actually reduce mathematical concepts to those of logic. And it was Peano (1889–1908) who was the first to rephrase the statement of mathematical theorems by means of a logical symbolism.

The logistic thesis was composed to push back the foundations of mathematics to as deep a level as possible. Because historically the foundation of mathematics was established on the real number system, and then on the natural number system, and thence into set theory, and because the theory of classes is an essential part of logic, the idea of reducing mathematics to logic certainly arises naturally. The documental *Principia Mathematica* of Whitehead and Russell (1910–1913) purports to be a detailed reduction of mathematics to logic. It starts with "primitive ideas" and "primitive propositions," corresponding to the "undefined terms" and "postulates" of a formal abstract development. These primitive ideas and propositions are not to be subjected to interpretation but are restricted to intuitive concepts of logic. They are accepted as plausible descriptions and hypotheses of the real world, so that there is no need to prove the consistency of the primitive propositions. Then *Principia Mathematica* develops mathematical concepts and theorems from these primitive ideas and propositions.

To avoid the paradoxes of set theory, *Principia Mathematica* sets up a hierarchy of levels of elements. The primary elements constitute those of type 0; classes of elements of type 0 constitute those of type 1; classes of elements of type 1 constitute those of type 2, and so on. So, one has such a rule that all the elements of any class must be of the same type. Because this rule precludes circular definitions, this new approach of constructing mathematics avoids the paradoxes of set theory. In the work of the logistic school, the great difficulty is on how to obtain the circular definitions needed for certain areas of mathematics.

The intuitionist school's thesis is that mathematics is to be built solely by finite constructive methods on the intuitively given sequence of natural numbers. From this intuitive sequence of natural numbers, all other mathematical objects must be built in a purely constructive manner, employing a finite number of steps or operations. This school of thought was originated in about 1908 by Brouwer with some earlier ideas proposed by Kronecker (in the 1880s) and Poincare (during 1902–1906).

For the intuitionists, the existence of an entity must be shown to be constructible in a finite number of steps. It is not sufficient to show that the assumption of the entity's nonexistence leads to a contradiction. So, to the intuitionists, a set cannot be thought of as a ready-made collection, but must be considered a law by means of which the elements of the set can be constructed in a step-by-step fashion. Because of this, such contradictory sets as the "set of all sets" cannot exist.

Also, the intuitionists' finite constructability leads to the denial of the universal acceptance of the law of the excluded middle. For them, this law holds for finite sets, but should not be employed when dealing with infinite sets. Based on the intuitionist thesis, as of this writing, large parts of present-day mathematics have been successfully rebuilt with a great amount of work still needed. In short, intuitionist mathematics has turned out to be considerably less powerful than classical mathematics, and in many ways, it is much more complicated to develop. Based on what has been done, it is now a general conviction that the intuitionist methods do not lead to known contradictions.

The formalist school's thesis is that mathematics is about formal symbolic systems, in which terms are mere symbols and statements are formulas involving these symbols. The ultimate foundation of mathematics does not lie in logic, but only in a collection of nonlogical marks (symbols) and in a set of operations with these marks. From this point of view, the establishment of the consistency of the various branches of mathematics becomes an important and necessary part of the formalist program. David Hilbert, together with Bernays, Ackermann, von Neumann, and others, founded this school of thought after 1899, when he published his *Grundlagen der Geometrie*.

The fate of the formalist school hinges upon its solution of the consistency problems. To Hilbert, freedom from contradiction is guaranteed only by consistency proofs, in which one must prove by the rules of a system that no contradictory formula can appear within the system. However, in 1931, Kurl Godel showed that it is impossible for a sufficiently rich, formalized deductive system, such as Hilbert's system established for all classical mathematics, to prove consistency of the system by methods of the system. This result reveals an unforeseen limitation in the methods of formal mathematics. It shows "that the formal systems known to be adequate for the derivation of mathematics are unsafe in the sense that their consistency cannot be demonstrated by finitary methods formalized within the system, whereas any system known to be safe in this sense is inadequate" (De Sua, 1956).

To summarize, it can be seen that because of the establishment of the theory of limits, calculus becomes a sound collection of powerful methods developed on a rigorous theoretical foundation. As to this point in time, the first and second crises in the history of mathematics were successfully resolved. However, the rigorous theory of limits is established on the system of real numbers. And to develop the system of real numbers, the theory of sets has to be employed as another deeper level of the foundation. However, in the maturing process of the naïve set theory, a series of paradoxes appeared one after another, constituting another, even greater crisis in the system of modern mathematics.

Until the year 1900, most mathematicians believed that the publicized paradoxes in set theory were only some technical problems. As long as some necessary details were fine-tuned, the problem would be resolved. Two years later, though, the well-known Russell paradox was publicized, astonishing the entire Western communities of philosophers, logicians, and mathematicians. This was because a simple analysis of the Russell paradox suggests that as soon as the language of logic is used to replace that of set theory, the Russell paradox will directly touch on the entire theory of the foundation, so that the common belief that mathematics and logic are two of the most rigorous and exact scientific disciplines is severely challenged. Facing the challenge, mathematicians and logicians were pressured to seriously treat and investigate the paradoxes of set theory. The chaotic situation, caused by the appearance of these set theoretic paradoxes, is adequately called the third crisis in mathematics. What needs to be pointed out is that to a certain degree, this third crisis is in fact a deepening evolution of the previous two crises, because the problems touched on by these paradoxes are more profound and involved a much wider area of human thought.

In the attempt to resolve the third crisis, mathematicians once thought of the possibility of giving up the naïve set theory as the theoretical foundation of mathematics, because the theory contains irresolvable antinomies, and identifying another theory as the foundation of mathematics. However, after careful analysis, it was realized how difficult it would be to adopt this approach. So instead, these mathematicians focused on modifying the naïve set theory in an attempt to make it plausible. As of the present day, there have been two plans to remold the set theory. The first is to employ the theory of types, developed by Whitehead and Russell. The second is to continue Zermelo's axiomatic set theory. Walking along the idea of Russell's "theory of extensionality," Zermelo in 1908 established his system of axioms for his new set theory. After several rounds of modification, Fraenkel and Skolem (1912–1923) provided a rigorous interpretation and formed the present-day Zermelo-Fraenkel (ZF) system. Because the ZF system accepts the axiom of choice, it is often written as ZFC system. From the structures and contents of the nonlogical axioms of the Zermelo-Fraenkel axiom of choice(ZFC) system, it can be seen that the ultimate goal is still about establishing a rigorous foundation for mathematical analysis. The specific route of thinking and technical details can be expressed as follows: By introducing the axioms of the empty set and of infinity, the legality of the set of all natural numbers is warranted. The legality of the set of all real numbers is derived by using the axiom of power sets. Then, the legality of each subset of those elements, satisfying a given property P, of the real numbers is based on the axiom of subsets. Therefore, as long as the ZFC system is consistent or contains no contradiction, the theory of limits, as a rigorous theoretical foundation of calculus, can be satisfactorily constructed on the ZFC axiomatic system. However, the remaining problem is that even though in the ZFC system, various paradoxes of the two-value logic, which have appeared in history, can be interpreted, that is, these paradoxes will not reappear in the ZFC system, as of this writing, we still cannot show in theory that no paradox whatever kind will ever be constructed in the ZFC system. That is, the consistency of the ZFC system or other presently available axiomatic set theoretic systems still cannot be shown, so that no theoretical guarantee exists that in the future no new paradoxes could be ever found in the current versions of modern axiomatic set theory. Of course, what is worth celebrating is that for over a century, since the time when the ZFC system was initially suggested, there fortunately have been no new paradoxes found in the ZFC system and other modern axiomatic set theoretic systems. Although it has been so, we still feel the anxiety over what Poincare once said about a century ago:

> We set up a fence to protect our flock of sheep from potential attacks of wolves. However, at the time when we installed the fence, there might have already been a wolf in sheep's clothing being enclosed inside the

flock of our sheep. So, how can we guarantee that in the future there will not be any problems?

We should notice that what Poincare said is absolutely not a few sarcastic remarks. Instead, they represent this great master's intuitive judgments on the essence of the matter. Harsdorff's comment below can be seen as another opinion that echoes what Poincare said. It was after several paradoxes were publicized in the naïve set theory; Hausdorff felt deeply grateful and reminded the community of mathematicians:

> These paradoxes made people feel uneasy. It is not because of the appearance of the paradoxes. Instead, it is because we did not expect these contradictions would ever exist. The set of all cardinalities seems so empirically indubitable just as the set of all natural numbers is that naturally acceptable. So, the following uncertainty is created. That is, is it possible that other infinite sets, all of them, are this kind contradictory, specious non-sets? (Hausdorff, 1935)

However, what attracted the most attention and what was most surprising are what Robinson expressed in 1964 (Robinson, 1964):

> In terms of the foundations of mathematics, my position (point of view) is based on the following two main principles (or opinions): (1) No matter which semantics is applied, infinite sets do not exist (both in practice and in theory). More precisely, any description about infinite sets is simply meaningless. (2) However, we still need to conduct mathematical research as we have used to. That is, in our work, we should still treat infinite sets as if they realistically exist.

To me, based on the yoyo model, I believe without any doubt that Robinson's statement (1) is an extremely deep, intuitive assessment instead of any irresponsible nonsense, and (2) is due to some kind of inability on his part to alter anything existing at the time of his remark.

Now, with the relevant history of mathematics in place, we will show in the next chapter the existence of the systemic yoyo structure in human thoughts, so that the human way of thinking is proven to have the same structure as that of the material world.

Chapter 16

Hidden Contradictions in the Modern System of Mathematics

From Chapter 15, by sorting through the ideas appearing in and concepts studied in the history of mathematics, we discover that some ideas and concepts, such as those of infinitesimals, infinities, limits, etc., underlie each crisis in the foundations of mathematics. After introducing the concepts of actual and potential infinities, in this chapter, we are able to uncover some hidden contradictions in the system of modern mathematics and see the return of the Berkeley paradox in the foundations of mathematics. By formally mapping the history of mathematics onto our yoyo model, it can be shown using available statistics that the periodic transitions between uniform and symmetric flows and chaotic currents existing in rotating fluids, the dishpan experiment, imply that the fourth crisis in the foundations of mathematics has occurred.

After we show in this chapter that the fourth crisis in the foundations of mathematics has appeared, we provide a plan of resolution of this new crisis by addressing two areas: select an appropriate theoretical foundation for modern mathematics and computer science theory, and choose an interpretation so that the known achievements of mathematics and computer science can be kept in their entirety. Related to this attempt, we have the internal set theory (Nelson, 1977), the nonstandard set theory (Hrbacek, 1978, 1979) and the Kawai (1981) set theory. Then, V. Kanovei of Moscow State University (Russia) and M. Reeken of Bergische Universitat (Germany) (2000) studied which transitive $\in$-models of the Zermelo-Fraenkel axiom of choice (ZFC) can be extended to models of a chosen nonstandard set theory. The spirit of all these works is that nonstandard models of natural and real numbers can be used to interpret the basic notions of mathematical analysis of the 17th and 18th century, including infinitesimals and infinitely large quantities.

This chapter is organized as follows. In Section 16.1, we look at the concepts of actual and potential infinities. In Section 16.2, we learn whether actual and potential infinities are equal. In Section 16.3, a theoretical foundation is established to show that no infinite set can actually exist. In Section 16.4, an intuitive picture is constructed to support the conclusions obtained in Section 16.3. In Section 16.5, examples are constructed to show that the well-known Berkeley paradox, which caused the second crisis in the foundation of mathematics over 300 years ago, has returned. And in the final section of this chapter, we employ the systemic yoyo model to show that the fourth crisis in the foundation of mathematics has already appeared.

16.1 The Concepts of Actual and Potential Infinities

By actual infinity, we mean a nonfinite process that definitely reaches the very end. For example, when a variable x approaches the endpoint b from within the interval $[a, b]$, x actually reaches its limit b. In this case, this limit process reflects the spirit of an actual infinity. In history, Plato was the first scholar to clearly recognize the existence of actual infinities (IFHP, 1962).

By potential infinity, we mean a nonfinite, nonterminating process that will never reach the very end. For example, let us consider a variable x defined on an open interval (a, b), and let x approach the endpoint b from within this open interval. Then, the process for x to get indefinitely close to its limit b is a potential infinity, because even though x can get as close to b as any predetermined accuracy, the variable x will never reach its limit b. In history, Aristotle was the first scholar to acknowledge the concept of potential infinities—he never accepted the existence of actual infinities. Similarly, Plato did not believe in the existence of potential infinities (IFHP, 1962).

Although these two kinds of infinites are different, where each actual infinity stands for a perfect tense and a potential infinity a present progressive tense, in the modern mathematics system, they have been seen as the same or applied hand by hand. For example, in the theory of limits, to avoid the Berkeley paradox, the definitions of limits, both infinite or finite, using $\varepsilon - N$ and $\varepsilon - \delta$ methods, are completely based on the thinking logic of potential infinities. In the naïve set theory and the modern (ZFC) axiomatic set theories, from Cantor to Zermelo, the existence and construction of infinite sets have been established on the concept of actual infinities. In many areas of the mathematics system, the existence of specific subsequences of a given sequence is guaranteed on the basis of mathematical induction and the recognition that actual infinities are the same as potential infinities. For example, in calculus, the following result is well known.

PROPOSITION 16.1
Any bounded sequence of real numbers has a convergent subsequence.

Let us look at a version of the proof for this result. Assume that $\{a_i\}_{i=1}^{\infty}$ is a given, bounded sequence of real numbers. So, there are numbers m and M such that

$$m \leq a_i \leq M, \text{ for } i = 1, 2, \ldots$$

Now, let us cut the interval $[m, M]$ into two equal halves, $\left[m, \frac{m+M}{2}\right]$ and $\left[\frac{m+M}{2}, M\right]$. Then, one of these subintervals must contain infinite many terms of the sequence $\{a_i\}_{i=1}^{\infty}$. Assume that the subinterval $\left[m, \frac{m+M}{2}\right]$ is such that it contains infinite many terms of $\{a_i\}_{i=1}^{\infty}$. Let us choose the minimum subscript k_1 such that $a_{k_1} \in \left[m, \frac{m+M}{2}\right]$.

Assume that for natural number n, we have picked $a_{k_1}, a_{k_2}, \ldots, a_{k_n}$ out of $\{a_i\}_{i=1}^{\infty}$ such that for any $1 \leq j \leq n$, the subscript $k_j > k_1, k_2, \ldots, k_{j-1}$, is the minimum so that $a_{k_j} \in [c_j, d_j] \subset [m, M]$, where $d_j - c_j \leq (M-m)/2^j$ and $[c_j, d_j]$ contains an infinite number of terms of $\{a_i\}_{i=1}^{\infty}$.

Now, we cut the interval $[c_n, d_n]$ into two equal halves, $\left[c_n, \frac{c_n+d_n}{2}\right]$ and $\left[\frac{c_n+d_n}{2}, d_n\right]$. Then one of these two subintervals must contain an infinite number of terms of the sequence $\{a_i\}_{i=a_{k_n+1}}^{\infty}$. Assume that $\left[c_n, \frac{c_n+d_n}{2}\right]$ is such an interval and we pick the least subscript k_{n+1} such that $a_{k_{n+1}} \in \left[c_n, \frac{c_n+d_n}{2}\right]$. Then, we have successfully constructed $a_{k_1}, a_{k_2}, \ldots, a_{k_n}, a_{k_{n+1}} \in \{a_i\}_{i=1}^{\infty}$, satisfying $k_1 < k_2 < \ldots < k_n < k_{n+1}$.

By using mathematical induction, we have constructed a subsequence $\{a_{k_j}\}_{j=1}^{\infty}$ such that it is convergent. (For our purposes here, we omit the rest of the proof.)

Because we now know the descriptive definitions of actual and potential infinities (for more details on these definitions, consult Zhu et al., in 2008a), all proven above by using mathematical induction, for any natural number j, a term a_{k_j} of $\{a_i\}_{i=1}^{\infty}$ can be picked to satisfy a set of desirable conditions. So, the nonterminating process of getting one more term a_{k_j} out of $\{a_i\}_{i=1}^{\infty}$ can be carried out indefinitely. That is, we proved the existence of a potential infinity. Right after that, what is claimed is that the actual infinity, the subsequence $\{a_{k_j}\}_{j=1}^{\infty}$, has been obtained. That is, in mathematical analysis, we assume that actual infinities are the same as potential infinities.

At this junction, a natural question is: Other than some scholars in history, such as Aristotle and Plato, who did not recognize both kinds of infinities at the same time, is there any evidence that these two kinds of infinities can in fact lead to different consequences? The answer to this question is yes. To this end, let us consider the following vase puzzle, which was first published in Lin (1999).

The Vase Puzzle

Suppose that a vase and infinitely many pieces of paper are available. The pieces of paper are labeled by natural numbers 1, 2, 3, ..., so that each piece has at most one label on it. The following recursive procedure is performed:

Step 1: Put the pieces of paper, labeled from 1 to 10, into the vase; then remove the piece labeled 1.

Step 2: Put the pieces of paper, labeled from $10n - 9$ through $10n$ into the vase; then remove the piece labeled n, where n is an arbitrary natural number 1, 2, 3, ...

Question: After the recursive procedure is finished, how many pieces of paper are left in the vase?

Some comments are needed here to make the situation practically doable. First, the vase need not be infinitely large; actually, any size will be fine. Second, the total area of the infinite number of pieces of paper can also be any chosen size. For example, Figure 16.1 shows how an infinite number of pieces of paper can be obtained. Third, the number labeling can be done according to the steps in the puzzle. Finally, the recursive procedure can be finished within any chosen period of time. The details here are similar to those of any chosen size for the total area of the pieces of paper above.

An Elementary Modeling

To answer the question of the vase puzzle, let us define a function, based upon mathematical induction, by

$$f(n) = 9n$$

which tells how many pieces of paper are left in the vase right after step n, where $n = 1, 2, ...$ Therefore, if the recursive procedure can be finished, the number of pieces of paper left in the vase should be equal to the limit of $f(n)$ as n approaches ∞. The answer is that infinitely many pieces of paper are left in the vase.

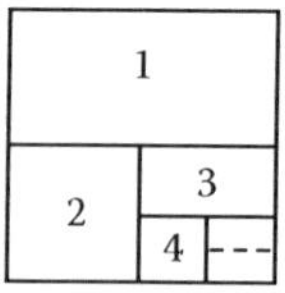

Figure 16.1 Obtain as many pieces of paper as needed out of a chosen area.

What we should note here is that this entire modeling process is based on the concept of potential infinities. The recursive procedure cannot really be finished. The step of taking the limit only represents the progressive growth in the number of pieces of paper in the vase and the (imaginary) limit state.

A Set Theoretic Modeling

Based on this modeling, the answer to the question of the vase puzzle is "no piece of paper is left in the vase." This, of course, contradicts the conclusion derived in the elementary modeling above.

For each natural number n, define the set M_n of the pieces of paper left in the vase right after the nth step of the recursive procedure:

$$M_n = \{x | x \text{ has a label between } n \text{ and } 10_n + 1 \text{ exclusively}\}$$

Then, after the recursive procedure is finished, the set of pieces of paper left in the vase equals the intersection

$$\cap_{n=1}^{\infty} M_n$$

That is, if x is a piece of paper left in the vase, x then has a label greater than all natural numbers. This contradicts the assumption that each piece of paper put into the vase has a natural number label. That is, the intersection $\cap_{n=1}^{\infty} M_n$ is empty.

What we should note here is that this set theoretic modeling is done on the basis of the concept of actual infinities. In particular, by using mathematical induction, we can show that for each natural number n, the set M_n can be well defined. After that, an actual infinity is assumed to exist. That is, the existence of the sequence $\{M_n\}_{n=1}^{\infty}$ is assumed. As soon as this assumption is made, the conclusion, completely opposite to that of the elementary modeling, is inevitably produced.

This vase puzzle vividly shows the difference between actual and potential infinities. Of course, our example is theoretical in nature. One might very well claim that in applications of mathematics, he will never run into such a problem. To this end, we should not be so sure, because in the study of quality control of the products off of an automated assembly line, for example, the concept of infinities is involved. In particular, when studying quality, we draw a random sample of the products and use the sample statistics to make inferences on the continually expanding population, which is theoretically the collection of all the products that have been and will be produced from the assembly line. To make the inferences more reliable, we often treat the ever-expanding population as an infinite population, either actually infinite or potentially infinite. For more impacts brought forward by the vase puzzle, consult Lin (1999), where a discussion about the connections between the vase puzzle and methodology, epistemology, and philosophy of science are given. And some comments on methodological indication of the fundamental structure of general systems, mathematical induction, and the knowability of the physical world can be found in Lin and Fan (1997).

16.2 Are Actual Infinities the Same as Potential Infinities?

First let us look at a situation where the assumption that actual infinities equal potential infinities is held in the system of modern mathematics. In particular, let us look at mathematical induction. As is known, the inductive step says that as long as a property P holds true for natural number n, one can show in theory that the property P holds true for $n + 1$. That is, what is implied by mathematical induction must be a present progressive tense—a potential infinity. However, when mathematical induction is used, the positive conclusion is always drawn that property P holds true for all natural numbers n—an actual infinity, which is a perfect tense. If, as we know, a present progressive tense is different from any perfect tense, then how can the proof on inductive reasoning, a potential infinity, lead to a conclusion of an actual infinity? Such a jump from a potential infinity to an actual infinity can only be materialized if mathematicians have implicitly admitted that potential and actual infinities are the same. Because such applications of mathematical induction appear all over the entire system of mathematics, we can conclude that in modern mathematics, the convention that potential and actual infinities are the same has been implicitly assumed.

Second, let us look at a situation where the convention that actual infinities are not the same as potential infinities is implicitly assumed in the system of modern mathematics. In particular, in formal logic, the concepts of contrary opposites and contradictory opposites have been distinguished since the time of Aristotle. Assume that P is a predicate. The contrary opposite of P is written $\daleth P$, and the contradictory opposite, $\neg P$. Let us now limit our attention to the system of modern mathematics, where the two-value logical calculus is the tool of deduction. The law of proof by contradiction can be expressed as $\Gamma, \neg A \vdash B, \neg B \Rightarrow \Gamma \vdash A$. This law is the logical foundation of the often used method of deduction, called proof by contradiction. Within the system of two-value logical calculus, from the law of proof by contradiction, one can prove the principle of excluded middle, $\vdash A \vee \neg A$, and the principle of no contradiction, $\vdash \neg(A \wedge \neg A)$. Now, in the two-value logical system, both contrary opposites (P, $\daleth P$) and contradictory opposites (P, $\neg P$) coincide. So, $\daleth P = \neg P$. Therefore, the potential infinities (*poi*) and actual infinities (*aci*), as the opposites in a pair of contrary opposites (P, $\daleth P$), must now be a pair of contradictory opposites (P, $\neg P$). That is, *poi* and *aci* must represent the affirmative and negative aspects, respectively, of a concept, and have to satisfy the principles of the excluded middle and no contradiction:

$$\vdash poi \vee aci \text{ and } \vdash \neg(poi \wedge aci)$$

That is, we can show *poi* ≠ *aci*. What is meant here is that in the system of modern mathematics, the thought convention that potential infinities are different from actual infinities has been implemented implicitly.

In short, the answer to the title question of this section is both yes and no in the system of modern mathematics. What a pair of hidden contradictions. For a more detailed discussion along this line, consult Zhu et al. (in 2008b).

16.3 Do Infinite Sets Exist?

As discussed in Section 15.4, such great minds as Hausdorff and Robinson of our modern time felt unease about the concept of infinite sets. So, the natural question is: Can we provide a definitive argument and an answer to address such a uncertainty? Our current section is about this end. In fact, we have the following result, assuming that actual infinities are different from potential infinities.

THEOREM 16.1
Each infinite set, as studied in the modern axiomatic set theory ZFC, is either nonexistent or a self-contradictory nonset.

Here, we only outline the proof of this theorem. For the relevant details, consult Zhu et al. (in 2008c, d).

Our outline of the argument consists of three parts: (1) We show that the set N of all natural numbers is a self-contradictory nonset. (2) We see why each countable infinite set in the ZFC system is a self-contradictory nonset. (3) We imagine the fact that each uncountable set in the ZFC system is either nonexistent or a self-contradictory nonset.

In part 1, assume that the set N is ordered naturally as follows:

$$\lambda: \{1, 2, 3, \ldots, n, \ldots\}$$

with the order type ω. Let $N^<$ be the following set of inequalities:

$$N^< = \{1 < \omega, 2 < \omega, 3 < \omega, \ldots, n < \omega, \ldots\}$$

So, we have

$$\forall n \ (n \text{ is in the } n\text{th inequality of } N^<) \quad (16.1)$$

and

$$\forall n \ \forall k \ (n \text{ is in the } k\text{th inequality of } N^< \rightarrow n = k) \quad (16.2)$$

Let k' be the counting variable that goes through N and approaches ω indefinitely, and $\lambda \mid k\omega$ stand for "N contains ω pairwisely different elements." Then, we have

$$\lambda \mid k'\omega \text{ iff } k' \uparrow \omega \wedge k' \top \omega \quad (16.3)$$

where ↑ means "approaches" and ⊤ "reaches." And we know that $\lambda \mid k'\omega$ holds true. Now, let k be the counting variable that passes through the inequalities in $N^<$ and increases indefinitely. So, similarly to Equation (16.3), we also have the true formula:

$$k \uparrow \omega \wedge k \top \omega \tag{16.4}$$

which stands for an actual infinity. Therefore, we have

$$\forall n\, \forall k\, (n \text{ is in the } k\text{th inequality in } N^< \to (k \uparrow \omega \wedge k \top \omega))$$

From Equation (16.2), we replace k with n and obtain

$$\forall n\, \forall n\, (n \text{ is in the } n\text{th inequality in } N^< \to (n \uparrow \omega \wedge n \top \omega))$$

This leads to

$$\forall n\, (n \text{ in the } n\text{th inequality in } N^<) \to \forall n\, (n \uparrow \omega \wedge n \top \omega) \tag{16.5}$$

Now, let n be the natural number contained in the inequalities in $N^<$. Then we have $(n \uparrow \omega) \wedge (n \bar{\top} \omega)$, a potential infinity, where $\bar{\top}$ means "do not reach." That is, we have

$$\forall n\, (n \text{ in the } n\text{th inequality in } N^< \to (n \uparrow \omega \wedge n \top \omega))$$

which implies

$$\forall n\, (n \text{ in the } n\text{th inequality in } N^<) \to \forall n\, (n \uparrow \omega \wedge n \bar{\top} \omega) \tag{16.6}$$

Combining Equations (16.1) and (16.5) and applying the law of separation leads to

$$\forall n\, ((n \uparrow \omega) \wedge (n \top \omega)) \tag{16.7}$$

Similarly, Equations (16.1) and (16.6) imply

$$\forall n\, ((n \uparrow \omega) \wedge (n \bar{\top} \omega)) \tag{16.8}$$

which contradicts that in Equation (16.7). That is, the set $N^<$ is not consistent. So, λ and N are self-contradictory nonsets.

In part 2, let G be a countable infinite set in the ZFC system. Then, the elements of G can be indexed by natural numbers in N. Now, the result in part 1 can be applied to show G is a self-contradictory nonset.

In part 3, From the axioms in the ZFC system, it follows that for any ZFC set $A \neq \emptyset$, each element of A is a ZFC set, and each subset and the power set of A are ZFC sets. And only if A is a ZFC set, can the power set P(A) be constructed using the power set axiom.

Now, if at the time when the ZFC framework was initially constructed, it were found that any countable infinite set A is a nonset, then it would not be possible for us to employ the power set axiom to construct the power set P(A). So, in this case, each uncountable set in the ZFC system would not exist at all. However, what is just discussed is not the historical reality. Instead, the truth of the matter is that the ZFC system was developed before we found that any countable infinite set is a self-contradictory nonset. To this end, let A be an arbitrary ZFC uncountable set. Then, by our discussion on mathematical induction above and the axiom of choice, we can obtain a countable infinite subset $A_C \subset A$. Because A_C is a self-contradictory nonset as implied in part 2, the original set A has to be a self-contradictory nonset. Otherwise, A_C would have to be a ZFC set. This ends the argument of Theorem 16.1.

To summarize, we have provided our definite answer to the title question of this section. No, there does not exist any infinite ZFC set!

16.4 The Cauchy Theater Phenomena

Here, we outline an imaginary theater, called Cauchy theater, to provide a second version of the argument for Theorem 16.1.

Let N = {x|x is a natural number} be called the rigid set of natural numbers, which represents an actual infinity, and let N = {x|x is a natural number) be a symbol representing the process of endlessly constructing more and more natural numbers. Here, the ending brace, }, in N is replaced by a parenthesis,), to indicate the ongoing process of including more natural numbers. That is, N stands for a potential infinity. In the following, the symbol for N, as originally used by Cantor, means that the totality of all natural numbers has formed a closed field, of which nothing goes in or out. The potentially infinite set N is called the spring set of natural numbers. The symbol

$$\text{C-N:} \{1, 2, 3, \ldots, \vec{n})$$

represents the potentially infinite process of the finite set {1, 2, 3, ..., n}, with n increasing indefinitely. This potential infinite process C-N is called the special Cauchy theater, where each natural number represents an individual seat in the theater. One important characteristic of the Cauchy theater C-N is that it can never have ω (an actual infinity) seats.

Now, consider the situation of holding a movie reception for our friends—all the natural numbers—in the Cauchy theater (Figure 16.2). The theater has a rest area and two entrance doors watched by A and B, respectively. Their assigned duties are as follows: A checks on each guest's identification, and B makes sure that the theater is not overloaded, so that each guest has his or her own seat when watching the movie. Before the scheduled time to show the movie, each of the natural

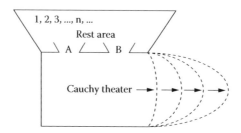

Figure 16.2 The dynamics of the special Cauchy theater.

numbers holds a ticket, relaxing in the rest area and waiting to enter the theater. When A confirms the identifications of all the guests, he okays them to enter the theater. Now, B checks on the number of the guests waiting in the rest area and realizes that these ω guests cannot enter the theater at the same time, because the theater does not have so many seats. Right after he refused the guests to enter the theater, he remembers the fact that each natural number is a finite ordinal number. So, he reasons, for each chosen natural number m, the C-N theater has at least $m + 1$ seats. So, every guest should have his or her own seat in the theater. That is, these ω many guests can enter the theater at the same time to watch the movie.

Facing the contradictory, opposite conclusions, B is really confused. Because he does not want to be responsible for any unexpected outcomes and chaos, he cancels the movie reception, complaining the event planner did not doing his job right, while the guests complain too about why the reception has to be cancelled because no one would not have his or her own seat in the theater. What is more troublesome than all the complaints is that the well-accepted fact in set theory that "there are ω many natural numbers" does not hold true. In other words, the set N of all natural numbers is a self-contradictory nonset.

To introduce the general Cauchy theater, let:

$$x(\text{In})A =_{df} \text{"the object } x \text{ is included into set A"}$$

$$x(\neg \text{In})A =_{df} \text{"the object } x \text{ is not included into set A"}$$

Ex P(x) $\uparrow$ x(In)A $=_{df}$ "each object x, satisfying predicate P, can be included into set A without any restriction"

$\forall x$ P(x)T x(In)A $=_{df}$ "all objects x, satisfying predicate P, can be affirmatively included into set A"

$$\forall x\, P(x) \mathbf{\overline{T}}\, x(\text{In})A =_{df} \neg(\forall x\, P(x) T\, x(\text{In})A)$$

where E stands for the listing quantifier, read as "for each" or "for any," and $\forall$ is the conventional universal quantifier with the restricted interpretation "for all."

Hidden Contradictions in the Modern System of Mathematics ■ 285

For any given predicate P and a set A out of an infinite background world, if the following condition holds true,

$$(Ex\ P(x) \uparrow x(In)\ A) \wedge (\forall x\ P(x) \mp x(In)\ A) \wedge (Ex\ \neg P(x) T x(\neg In)\ A)$$

then the set A is called the spring set of the predicate P, written $A = \{x|P(x)\}$. If the predicate P and the set A out of an infinite background world satisfy

$$(Ex\ P(x) \uparrow x(In)A) \wedge (\forall x\ P(x) T x(In)A) \wedge (Ex\ \neg P(x) T x(\neg In)A)$$

then A is called the rigid set of the predicate P, written $A = \{x|P(x)\}$.

Now, for any uncountable transfinite ordinal number Ω, the general Cauchy theater can be defined as follows, similar to the case of the special Cauchy theater:

$$\text{C-N-I: } \{1, 2, 3, \ldots, n, \ldots, \omega, \omega+1, \ldots, \omega \times 2, \ldots, \omega^2, \ldots, \vec{\eta})$$

Similar to the discussion on the special Cauchy theater, we can show the fact that any uncountable ZFC set A is a self-contradictory nonset. The details are omitted for the reader to work out. As a clue, the reader will need the following results from the ZFC axiomatic set theory.

THEOREM 16.2
For each well-ordered set A, it cannot be isomorphic to any segment A_a, and $\overline{A_a} < \overline{A}$, for any $a \in A$.

THEOREM 16.3
The ordinality of the well-ordered set w_α, consisting of all ordinal numbers less than α, is the ordinal number α. That is, $\overline{w_\alpha} = \alpha$.

For relevant technical details related to our discussion in this section, consult Zhu et al. (in 2008e, f).

16.5 The Return of the Berkeley Paradox

As seen in Section 16.2, it is an inherent part of the modern system of mathematics to allow both of the following contradictory conventions: actual infinities ≠ potential infinities, and actual infinities = potential infinities. However, our previous discussion has clearly shown that these two kinds of infinities are two different concepts. They might lead to completely opposite outcomes. So, in this section, let us treat these infinities as different concepts.

If we have $a\hat{\uparrow}b \wedge a T b$, then we say that the variable a approaches the limit b in the fashion of actual infinities. If we have $a\hat{\uparrow}b \wedge a \mp b$, then we say that the variable a approaches the limit b in the fashion of potential infinities. It can be seen that for any variable a, it approaches its limit b, if it exists, in the fashion of either actual infinities or potential infinities, but not both. For the conventional expressions of limits,

$$\lim_{x \to x_0} f(x) = A, \lim_{x \to 0} f(x) = A, \text{ or } \lim_{x \to \infty} f(x) = A$$

there are two sets of variables in each: $x \to x_0$ (respectively, $x \to 0$ or ∞) and $f(x) \to A$. If the variable x approaching its limit x_0 (respectively, $x \to 0$ or ∞), in the fashion of actual infinities, causes trouble or leads to contradictions, that is, $x \uparrow x_0 \wedge x T x_0 \vdash B, \neg B$, then the variable x has to approach its limit x_0 in the fashion of potential infinities. That is, it must become that $x \uparrow x_0 \wedge x \not\vdash x_0$.

Let us now use the example of computing the instantaneous speed of a free-falling object at the time moment t_0 to illustrate the Berkeley paradox of past and present time.

The well-known distance formula for a free-falling object is given by

$$S = \frac{1}{2} g t^2$$

where g is the gravitational constant. When $t = t_0$, the distance the object has fallen through is $S_0 = \frac{1}{2} g t_0^2$. When $t = t_0 + h$, the distance of falling is $S_0 + L = \frac{1}{2} g (t_0 + h)^2$. This end implies that during the h seconds, the object has fallen through the distance L given by

$$L = \frac{1}{2} g (t_0 + h)^2 - S_0 = \frac{1}{2} g (t_0 + h)^2 - \frac{1}{2} g t_0^2$$

$$= \frac{1}{2} g (2 t_0 + h) h$$

So, within the h seconds, the average speed of the falling object is

$$V = \frac{L}{h} = g t_0 + \frac{1}{2} g h$$

Evidently, the smaller the time interval h is, the closer the average speed is to the instantaneous speed at $t = t_0$. However, no matter how small h is, as long as $h \neq 0$, the average speed is not the same as the instantaneous speed at $t = t_0$. When $h = 0$, there is no change in the falling distance. So, $V = \frac{0}{0}$ becomes meaningless. So, it is impossible to compute the instantaneous speed of the free-falling object at $t = t_0$.

To this end, both Newton and Leibniz provided several explanations to get rid of this difficulty. One of the explanations is as follows: Assume that h is an infinitesimal. So, $h \neq 0$ and the ratio

$$\frac{L}{h} = \frac{\frac{1}{2} g (2 t_0 + h) h}{h}$$

is meaningful. This ratio can be simplified to $\frac{L}{h} = gt_0 + \frac{1}{2}gh$. Because the product of the infinitesimal h and a positive, bounded value $\frac{1}{2}g$ can be ignored, the term $\frac{1}{2}gh$ can be erased, producing gt_0. That is, the instantaneous speed of the falling object at $t = t_0$ is $V|_{t=t_0} = gt_0$.

As we can see, this explanation, like all others provided by Newton and Leibniz, cannot really resolve the following contradictions:

(A) To make $\frac{L}{h}$ meaningful, one must have $h \neq 0$. (16.9)

(B) To obtain gt_0 as the outcome, one must assume $h = 0$.

This is the so-called Berkeley paradox in the history of mathematics. After the theory of limits is established, this problem of computing the instantaneous speed of the free-falling object is solved as follows:

$$V|_{t=t_0} = \lim_{\Delta t \to 0} \frac{\Delta S}{\Delta t}$$

$$= \lim_{\Delta t \to 0} \left(gt_0 + \frac{1}{2}g\Delta t \right)$$

$$= gt_0 + \frac{1}{2}g \lim_{\Delta t \to 0} \Delta t$$

$$= gt_0 + \frac{1}{2}g \times 0$$

$$= gt_0 + 0 = gt_0 \qquad (16.10)$$

Within the current theory of limits, the Berkeley paradox is successfully resolved. However, for now, we have shown the need to separate the concept of a variable approaching its limit in the fashion of potential infinities from that in the fashion of actual infinities; we can expect a return of the Berkeley paradox to the theory of limits.

Here is a detailed explanation. The limit expression in Equation (16.10) consists of two important limits:

(1) $\lim_{\Delta t \to 0} \frac{\Delta S}{\Delta t} = gt_0$

(2) $\lim_{\Delta t \to 0} \Delta t = 0$

The first variable $\Delta t \to 0$ is the same in both of expressions. However, the limit expression in (1) is not realizable, meaning that $\Delta t \uparrow 0 \wedge \Delta t \uparrow 0$. That is, the variable

Δt must approach its limit 0 in the fashion of potential infinities. Now, the limit expression in (2) is realizable, meaning that $\Delta t \hat{\uparrow} 0 \wedge \Delta t \top 0$. That is, the variable Δt must approach its limit 0 in the fashion of actual infinities. So, a natural question is: For the same process of solving the same problem and for the same variable $\Delta t \rightarrow 0$, how can both $\Delta t \top 0$ and $\Delta t \top 0$ be allowed at the same time?

What is suggested in Equation (16.10) is that to make $\lim_{\Delta t \rightarrow 0} \frac{\Delta S}{\Delta t} = gt_0$ meaningful, we take the assumption $\Delta t \hat{\uparrow} 0 \wedge \Delta t \top 0$. On the other hand, to obtain $\frac{1}{2} g \times \lim_{\Delta t \rightarrow 0} \Delta t = \frac{1}{2} g \times 0$, we change our mind and allow $\Delta t \hat{\uparrow} 0 \wedge \Delta t \top 0$. This kind of deduction is difficult for people to feel comfortable with. To resolve this dissatisfaction, can we suggest permitting Δt to approach its limit 0 in the fashion of actual infinities? It is impossible, because it would make $\lim_{\Delta t \rightarrow 0} \frac{\Delta S}{\Delta t} = \frac{0}{0}$ meaningless. Because this is so, can we unify the situation by allowing $\Delta t \rightarrow 0$ in the fashion of potential infinities? This end is also impossible, because under the assumption that $\Delta t \hat{\uparrow} 0 \wedge \Delta t \top 0$, it must cause the limit expression $\lim_{\Delta t \rightarrow 0} \Delta t = 0$ to be not realizable. (The expression $\lim_{x \rightarrow x_0} f(x) = A$ is said to be realizable if (i) $x \hat{\uparrow} x_0 \wedge x \top x_0$, $f(x) \hat{\uparrow} A \wedge f(x) \top A$ and (ii) $f(x_0) = A$; otherwise, the limit expression $\lim_{x \rightarrow x_0} f(x) = A$ is not realizable.) But it can be checked that the limit $\lim_{\Delta t \rightarrow 0} \Delta t = 0$ is realizable. From the definition for a limit expression to be realizable, it follows that each limit expression has to be either realizable or not realizable, but not both. So, the expression $\lim_{\Delta t \rightarrow 0} \Delta t = 0$ should not be an exception.

To summarize, what we have obtained previously is that as soon as we come to the realization that potential infinities and actual infinities are different, various explanations for the variable Δt to approach its limit 0, employed in the process of calculating the instantaneous speed of a free-falling object at $t = t_0$, can no longer be plausible. With that said, what is implied is that as soon as we introduce and distinguish the concepts in the theory of limits of approaching limits in the fashion of either actual or potential infinities, the Berkeley paradox comes back.

For more detailed discussion along this line, consult Zhu et al. (in 2008g).

16.6 The Fourth Crisis in the Foundations of Mathematics?

In the recorded history of mathematics, there have appeared three crises in the foundations of mathematics. The first crisis arose in the fifth century BC out of the discovery of the Pythagorean school. In particular, the Pythagoreans discovered the Pythagorean theorem regarding right triangles, which led to the proof of the existence of irrational numbers. The existence of these new numbers was a mortal blow to the Pythagorean philosophy that all in the world depend on whole numbers. The final resolution of this crisis was successfully achieved in about 370 BC by Eudoxus by redefining proportions. What good comes out of this crisis is that the axiomatic method is formulated and adopted in the system of mathematics.

During this time period, notions connected with infinitesimals, limits, and summation processes were considered. Zeno paradoxes were devised (ca. 450 BC) to address the question of whether a magnitude is infinitely divisible or made up of a very large number of tiny indivisible atomic parts. (Lin and Ma [1987] show that under ZFC, each system must be finitely divisible). There is evidence that since the Greek time, schools of mathematical reasoning developed their thoughts using both of these possibilities.

The second crisis in the foundation of mathematics appeared around the end of the 18th century, when a number of absurdities and contradictions had crept into mathematics. The early calculus was established on the ambiguous and vague concept of infinitesimals, attracting reproaches from various angles, the most representative of which is the Berkeley paradox published in 1734. In 1754, Jean-le Rond d'Alembert observed that a theory of limits was needed to resolve this serious crisis. In 1821, a great stride was made when Cauchy successfully executed d'Alembert's suggestion by developing an acceptable theory of limits. From our discussions in Section 16.1, it can be seen that the Berkeley paradox is avoided by using the modern theory of limits to assume that actual infinities equal potential infinities. This assumption beautifully hides the "fallacy of a shift in hypothesis," as said Berkeley, so that the increment of a variable can be assumed to be zero or nonzero at will (see discussion in Section 16.5 for more details).

The third crisis in the foundations of mathematics appeared at the end of the 19th century out of the continued effort of developing the foundation of mathematics onto the level of the naïve set theory, where a set is defined as a collection of objects. The Russell paradox (1902) proved that this definition of sets is the source of a lot of troubles. Soon after that, it was found that throughout history, similar paradoxes in logic have appeared at various times, and the root to all these paradoxes is the application of circular definitions.

To resolve this crisis, several schools of thought emerged. However, each of these schools met with difficulties in its attempt to reconstruct the edifice of the system of mathematics. As a joint effort, several axiomatic set theoretic systems were established, including the ZFC system. To avoid the known paradoxes, for instance, in the ZFC system, the concept of sets is left undefined. Similar treatment was repeatedly employed throughout the history of mathematics when introducing elementary concepts, such as that of points in geometry. Even though such a procedure has been criticized for merely avoiding the paradoxes without explaining any of them, and carries no guarantee that other kinds of paradoxes will not crop up in the future, the community of mathematicians has been feeling pretty fortunate that it has been a century without any new paradox publicized.

Because the ultimate goal of reconstructing the naïve set theory is to lay the foundation of the modern system of mathematics on it, the new theory of sets has to be rich enough and powerful enough to handle infinities. Like we discussed earlier, this is exactly where new contradictions and paradoxes are found, and where the Berkeley paradox made its way back.

Now, if we model the history of mathematics mentally as an abstract systemic yoyo, then we can see that the black hole side sucks in empirical data and puzzles we collect and experience in daily lives. Through our human capability available to us naturally, we process the data with rational thinking (the narrow neck of the yoyo), then produce (out of the Big Bang side) all kinds of concepts, theories, and products that we can physically use in our lives. Now, some concepts, theories, and products become obsolete and disregarded over time, while the rest are recycled back into the black hole side. These recycled concepts, theories, and products, combined with new observations and newly collected empirical data, through the tunnel of rational thinking again, lead to new or renewed concepts, theories, and products, which help to bring the quality of daily lives to another level. That is, we have the yoyo model shown in Figure 16.3 for the general structure of the system of mathematics.

Now, if we stand at a distance away from this spinning yoyo structure and look at the spinning field from either the black hole side or the Big Bang side, what we see is like a huge whirlpool of data, observations, puzzles, concepts, theories, products, etc. What the dishpan experiment suggests is that this huge whirlpool goes through pattern changes periodically, alternating between uniform and symmetric flows to chaotic currents with local eddies formed.

Naturally, we can identify the chaotic currents with local eddies with one of the crises in the history of mathematics, while equating the uniform and symmetric flow pattern with one of the quiet periods in the history between two consecutive crises. Then, what we can expect is that after the relatively quiet 20th century, we should see another major crisis of mathematics in the making. Like the previous crises, this forthcoming event will also be as devastating, as frustrating as any of the known crises. Its impact will be felt throughout most major corners of the modern system of mathematics. Here, we notice that some of the major concepts and puzzles, such

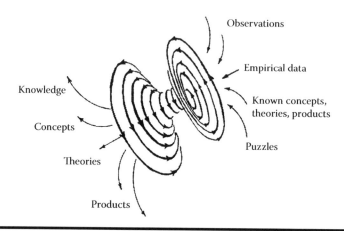

Figure 16.3 The yoyo structure of the system of mathematics.

as those of infinitesimals, limits, and summations of numbers, etc., played a role in each of the past crises. So, they can be seen as some of the knowledge, concepts, theories, puzzles, etc., that are sucked into the black hole of our yoyo structure of the system of mathematics. After being processed by human thinking and reasoning, they are spit out of the Big Bang side in their renewed forms.

The discussions in Sections 16.2 to 16.5 indicate that the impact of the hidden contradictions existing in modern mathematics, the problem with the self-contradictory nonsets existing in the foundation of modern mathematics, and the return of the Berkeley paradox should be seen and felt in all major corners of the system of modern mathematics in the years and decades to come. To be clear, what we try to say is that as predicted by our yoyo model, established mentally for the structure of the system of modern mathematics, the fourth crisis in the foundations of mathematics has appeared. Here, how do we explain the periodic pattern change in the dishpan experiment with our model of mathematics? The reason why we need to address this question is that in the history of mathematics, the time span between the appearance of the first crisis and that of the second crisis was about 23 centuries. The time span between the appearances of the second and the third crisis was about two centuries. And the time span between the third crisis and the fourth, which we just declared, should be about one century. With such a drastic decrease in time spans, how can we understand the appearance of the crises as periodic pattern changes in the dishpan experiment?

We look at the word *periodic* from the angle of mathematical research produced. In particular, we have the following statistics (Eves, 1992). Moritz Cantor's history of mathematics stops at the end of the 18th century. This work consists of four volumes of about 1,000 pages each. It is conservatively estimated that if the history of mathematics of the 19th century were to be written in the same detail as Cantor's work, it would require at least 14 more such volumes. No one has tried to estimate the number of volumes needed for a similar treatment of the history of mathematics of the 20th century, which is by far the most active era of all.

Here is another piece of statistics (Eves, 1992). Prior to 1700, there were only 17 periodicals containing articles of mathematical contents. In the 18th century, there were 210 such periodicals, and in the 19th century, 950 of them. The number has increased enormously during the 20th century—by one count done before 1990, to some 2,600.

Here is still another piece of statistics (Eves, 1992). It has been estimated that more than 50 percent of all known mathematics was created during the past 50 years, and that 50 percent of all mathematicians who have ever lived are alive today.

It is surely our hope that these statistics well define the word *periodic* for any of our doubtful readers.

As in the past, with the previous crises in the foundations of mathematics, one possible resolution for this current crisis is to select an appropriate theoretical foundation for modern mathematics and to choose a right interpretation so that modern mathematics can be kept in its entirety. In particular, we construct the needed mathematical system of potential infinities, which is different from that constructed on the basis of intuitionism, and the mathematical system of actual infinities. When putting these two systems together, we cover the entirety of the current mathematics plus whatever else might be gained. For more details about this end, consult Zhu et al. (in 2008h, i, j, k, l, m).

ROLLING CURRENTS AND PREDICTION OF DISASTROUS WEATHER

5

Chapter 17

V-3θ Graphs: A Structural Prediction Method

In the previous chapters we have learned how the systemic yoyo model and its methodology can be beautifully applied to:

- Study classical physics
- Answer such urgent problems as what time is, whether it occupies any material dimension, etc.
- Shed new lights on extremely difficult mathematics problems, such as the three-body problem and hidden inconsistencies in the system of mathematics
- Deepen the study of household economics and corporate governance

In this final part, we will learn how the systemic yoyo model and its methodology can be successfully applied to the prediction of disastrous weather conditions. The forecasting of most of these zero-probability disastrous events has been a world-class hard problem in modern science.

17.1 The Fundamentals

Structural prediction methods, such as **V**-3θ graphs, were initially established by S. C. OuYang and his colleagues (OuYang, 1998; Wu and Lin, 2002), aimed at the forecasting of weather conditions, especially disastrous ones. The first hint of such methods came from OuYang's discovery of the phenomenon of the so-called ultra-low temperatures existing at the upper layer of the troposphere in the 1960s. To distinguish heavy fog weather from torrential rains, he established the concept of rolling currents in the vertical direction of the atmosphere. Through 30-plus years of practical experiments and testing with nearly 2,000 case studies and practical real-life forecasting exercises, several structural prediction methods are formalized and shown to be effective in terms of the prediction of major disastrous weather conditions, such as torrential rains, strong convections, windstorms and sandstorms, heavy fog with low visibility, abnormally high temperatures, etc.

Currently, many meteorological stations across China employ these methods for their practical forecasts of weather. Because these methods have helped to greatly improve forecasting accuracy, compared to the conventional methods of forecasting, definite social and economic benefits have been materialized. However, due to the constraints of the conventional beliefs and concepts, these methods still have not been widely accepted by front-line meteorologists and academic professionals, because they are designed on the concepts and principles of blown-up theory and whole evolutions of systems. The system of these methods is brand new, instead of some clever-but-minor changes to the system of conventional methods. In particular, difference among this system and the conventional system include the following:

1. The concept of predicting disastrous weather. The new system emphasizes confrontational movements instead of the first push system initially established by Isaac Newton. Speaking in ordinary language, investigations on evolution are about how movements of materials are "fighting" against each other instead of being stepped over passively under the will and plan of God.
2. Converting quantitative forces back to the original uneven structures of materials with a focus on materials' irregular or peculiar structures. The reasons for doing so are that (1) well-established computational schemes in the Newtonian system cannot handle irregularities, and (2) materials' structures are prior to the existence of quantitative forces, so that by using structures of materials or events, we can gain valuable lead time in forecasts. What is important is that these prediction methods include the structure of ultra-low temperatures, because without an ultra-low temperature, such disastrous weather as heavy rain gushes, strong convections, etc., would not appear.
3. To represent confrontational movements, we have to find out how such movements move. That end is exactly what has been missing from Newton's laws of physics. Based on the uneven structurality of materials, confrontational movements have to produce rotations. So, according to the conservation law of stirring energy, there must exist differences in rotational directions, leading to the appearance of confrontations. So, discontinuities in movements have to appear. Confrontations are not only the key for materials to evolve, but also make the concept of continuity in Newton's system not appropriate in the analysis of evolution and prediction of disasters. In Newton's system

of the first push, transportations of materials by stirring motions are totally missed.

4. The problem of order structure. Traditionally, or in Newton's system, forces have been seen as the reason for materials to move. In form, this convention has no problem. However, when dealing with practical problems, it involves how to recognize forces and apply them. For example, even if we treat the atmospheric movement as an acting force, then we still have to deal with the rotational force of the Earth and the stirring effect of the uneven structure of the atmosphere. In other words, the acting force that causes the atmospheric movement involves the interaction (or confrontation) of at least two major acting forces. And in the confrontation, there exist at least two kinds of airstreams, one cold and the other warm, which fight against each other, instead of a single pushing force. What is interesting is that whatever moves on Earth changes its direction of motion. So, directions are shown prior to quantities. Also, in terms of directions, rotations either inwardly or outwardly produce different effects so that problems of directional structures appear. Besides, quantities cannot deal with irregular structures, while realistic rotational structures are exactly irregular. In particular, when two materials rotate in the same direction, the quantities representing their combined forces are relatively large (the enclosed area in Figure 17.1(a)). When the materials rotate in different directions, the quantities representing their combined forces will be relatively small (the enclosed area in Figure 17.1(b)). In the traditional system, these small quantities might very well be ignored. However, in terms of structures and confrontations, it is the latter case with small quantities, that the stirring energy is greater due to the acting moments of forces, so that evolution is caused.

Nowadays, the conventional weather forecast is based on the pushes of the gradients of barometric pressures. The corresponding order of information is pressure, temperature, humidity, and wind, with the barometric pressure system as the core. However, in practical applications, the numerical values for pressures (or altitude fields) are statically modified. Plus, in meteorology, atmospheric densities are not observed. So, the usable information of pressures becomes postevent, lagging behind

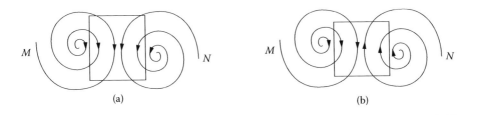

Figure 17.1 Interactions of spinning yoyos.

what is appearing. Hence, we can see that actual important weather often occurs before the observed pressure systems. For example, after precipitation occurs, one observes the relevant pressure system, causing frequent and erroneous forecasts of transitional weather. So, the traditional order of information in structural prediction methods is reversed, using wind (its direction and speed), humidity, temperature, and pressure, with emphasis placed on wind direction. So, an inverse order structure is applied. The reasons for such an inversion of order are:

1. Wind directions in reality not only are a product of the push of pressures, but also appear at almost the same time as the direction of movements in uneven structures of materials due to the existing subeddies produced jointly by the rotational effects of Earth and other factors. Besides, errors in the observation of wind directions are relatively minor, making it a piece of useful information collectable ahead of the appearance of the event of the forecasting interest. Next, in the wind factor, because of wind speed, there is a quantity. Wind directions together with different wind speeds provide structures. So, winds provide two pieces of useful, prior-to-event information in structural prediction methods.
2. The reason why humidity information is moved to an earlier position than temperature information is because of the need of predicting torrential rains. To this end, one has to know the difference between general cloudy weather and torrential weather.

What needs to be noted is that for predicting disastrous weather related to torrential rains, one has to pay particular attention to the sources of water vapor. For 24-hour short-term forecasts, one only needs to trace the sources to about 10–15 longitudinal and latitudinal distances.

To summarize, the key points used in structural prediction methods are: Natural disasters are originated in confrontational movements of materials; quantitative analysis consists of postevent reportability as well as the impossibility of dealing with irregularities. This new system of methods does not employ the regularized quantitative computational schemes of Newton's first push. Instead, it uses an irregular structural analysis method of the second stir with rotational directions. If we see Newton's system as dissolving shapes into quantities, then what is embodied in the new system of methods is an informational structure analysis, briefly called structural analysis, in which quantities are converted back to shapes. The central idea of the methods is to seek out irregular information existing ahead of the appearance of the events of interest.

In terms of epistemology, these methods are established on the basis that rotational constraints are the fundamental principle of movements in the universe. Quantities in comparison are merely tools for formal analysis. They are neither the origin of all things nor the laws of all matter. And in terms of predictions, they do not possess any leadability in time.

17.2 Roles of Rolling Currents and Ultra-Low Temperature in Weather Evolution

In the designs of the traditional weather forecasting methods and in the study of meteorology, the functions and impacts of eddy motions and vortices are already known. However, due to historical reasons and constraints of conventional concepts, the kinds of vortices that are considered for practical applications are only limited to horizontal ones, without clearly noticing the roles of transformations of vortices in evolution. In particular, vertical vortices (or horizontal vorticities or eddies) have not even been seriously looked at in both theory and applications. Although in dynamics, the study of vorticity equations has mentioned the concept of horizontal vorticities, the functions of these vorticities and how to practically apply them are not clearly pointed out. Especially in the current dynamics, the forms of motion under the effects of baroclinic solenoids are named baroclinic waves, making eddy motions in flow fields be seen as wave motions of barometric pressures or in the altitude field. (The pascal [Pa] is the SI-derived unit of pressure or stress. It is a measure of perpendicular force per unit area, i.e., equivalent to one newton per square meter or one joule per cubic meter. In everyday life, the pascal is perhaps best known from meteorological air pressure reports, where it occurs in the form of hectopascal [1 hPa = 100 Pa]. One hectopascal corresponds to about 0.1 percent, and one kilopascal to about 1 percent of atmospheric pressure [near sea level].) The weather maps used in the profession of weather forecasting are produced after smoothing treatments with irregular information greatly damaged, where a piece of information is irregular, provided that the chance for such information to appear is very small (small-probability information). (For most meteorological stations, only four maps, for the altitudes of 500, 700, and 850 hPa and the ground surface, are used. So, for areas of high elevation, there are only three available maps.) In other words, in the current profession of weather forecasting, not only are vertical vortices not practically used, but also they did not even get noticed, wasting all the available irregular information beneath 500 hPa and all available meteorological data above 500 hPa. Speaking more specifically, the present practice of weather forecasting does not even sufficiently utilize the currently available resources of meteorological data.

What is most important is that at the upper layer, 300–100 hPa of the troposphere, which is the layer of the atmosphere about 9,000–12,000 m above sea level, there might exist an ultra-low temperature that is closely related to disastrous weather. (The Earth is blanketed by an atmosphere consisting of 99 percent oxygen and nitrogen. The atmosphere has five layers: troposphere, stratosphere, mesosphere, thermosphere, and exosphere. Seventy-five percent of the atmosphere's gasses are in the bottom-most layer, the troposphere [Adams and Lambert, 2006].) This phenomenon is an important weather factor ignored and missed by traditional meteorological science. It involves a drop in temperature of troposphere and touches on the problem of under what physical and photochemical mechanisms the temperature slides suddenly. Evidently, only with a temperature drop at the upper layer of the troposphere do the rising water vapors condensate. And a severe drop in the temperature naturally leads to increased condensation of water vapors, causing windstorms or torrential rain. Hence, for such weather conditions as strong convections, torrential rains, etc., the existance of such a severe ultra-low temperature becomes a necessary requirement and condition. That is why, in employing the **V-3θ** graphs, a structural prediction method, we make use of irregular information collected from the ground level to 100 hPa. More details will be given in the next section.

What needs to be pointed out is that the discoveries of ultra-low temperatures and functions of rolling currents were the initial motivations for us to recognize that mutual reactions are not a problem of the first push, that forces are originated from materials' structures, and that the forces under Newton's quantitative manipulations cannot lead to the actual forces existing in uneven structures of materials, leading to the eventual establishment of the systemic yoyo model (Wu and Lin, 2002). Also, the quantitative forces, studied in Newton's system of the first push, come into being after the existence of materials' structures. What can be known are only the forces of the present moment, without involving any knowledge about potential changes in the forces and their future. That is, these quantitative forces do not have the leadability of time, which is badly needed in prediction practices. In particular, mutual reactions or confrontations operate in the form of stirring motion of the structures of moments of forces. So, these operations are no longer the simple pushes of particles (without any size and volume) of the inertial system.

Because realistic movements of materials are almost entirely rotational under confrontational stirrings, the corresponding disasters only represent the most intensive and hottest spots of confrontation. So, to comprehend disastrous changes or unrepeatable changes, we must investigate confrontational movements. That is where we have to study evolutions and the concept of rolling currents.

The meaning of the word *evolution* in English is "gradual change," while the Latin root of the word originally contained the meaning of "rolling forward." The phrase "rolling current," which we have been using, is a term from the hydraulics of rivers. In rivers, not only can rolling currents cause silts to rise from the river bottom, but also collisions of rolling currents can cause boat-capsizing accidents. So, the reason why OuYang chose this hydraulics phrase is to reflect the fact that vertical vortices (or horizontal vorticities) were mentioned in meteorology without either catching much attention or being practically applied. As of this writing, no method on this concept is established in conventional meteorology.

Practical and empirical experiments and tests show that atmospheric rolling currents play extremely important roles in transitional changes in weather evolution. It can even be said that by knowing the existing rolling currents, one can basically foretell transitional changes in weather. What is practically significant is that changes in the direction of rolling currents occur ahead of weather changes. That is, directions are the epistemology and the methodology. When combined with the distribution and transportation of water vapor, one can basically foretell torrential rains, severe convections, and other disastrous weather, and transitional changes in weather systems.

Because rolling forward has at most quasi-repetition instead of the pure repetition of a wave motion, and because of the changes in states and properties caused by confrontations of the rolling current, rolling forward is also shown with nonrepetitious and nonperiodic changes. So, rolling currents are closely related to evolution with practical significance of applications. Changes in atmospheric rolling currents occur ahead of changes in horizontal vortices. So, directional changes in atmospheric rolling currents can be applied to forecast transitional changes in weather systems (including subtropical high pressures).

To analyze atmospheric rolling currents, we can no longer use only the four traditional weather maps, because each rolling current reflects the entire structure of the troposphere. Considering the current available pool of meteorological information, because we have to as much as possible consider the sounding data of each station, including that on the troposphere and the possibly existing ultra-low temperature near the top of the troposphere, the **V**-3θ graphs were designed using the significant sphere. For our purpose, we focus on the relationship between rolling currents and weather evolution:

1. Rolling currents can help to form the overall self-contained tropospheric circulation systems made up of many different, scattered, and multitudinous eddy currents. That is exactly the necessary condition for convective systems to appear and develop. In other words, it is exactly because rolling currents have self-contained circulation systems that they can constitute relatively closed systems independent of weather scale systems. General convective systems that last at least six hours and affect human lives can all be described by using overall rolling currents within the troposphere.

    What needs to be pointed out is that even though hailstones, heavy precipitation, and torrential rains, within the category of convective weather, can all be shown by using clockwise rolling currents, their structures in **V**-3θ graphs are different. And what is truly great is that these differences can be easily distinguished before the appearance of the event of concern.

2. In regional west wind systems (regional east wind systems should be treated inversely), the discontinuous areas before the troughs and after the ridges are places in the flow field for rolling currents to appear and develop. Or speaking more specifically, rolling currents appear before the discontinuity that appears before troughs and after the ridges. Thus, after a rolling current appears, a discontinuity in the horizontal flow field appears. So, rolling currents have the desired prediction leadability of time. Due to the difference between the rising and sinking airflows in the front and back of a rolling current, the disposition and distribution of horizontal weather scale systems are changed. And under the influence of strong rolling currents, breaking-up movements of weather scale systems result. Because of

the lag of barometric pressure systems, rolling currents frequently initiate transitional changes in weather, leading to transitional changes in weather systems.

The corresponding principle for counterclockwise rolling currents is similar, with opposite effects. The strength of rolling currents is determined by the unevenness of the atmospheric structure. So, using structural analysis, we can be not only ahead of the appearance of the event of concern, but also ahead of "forces," so that the direction and strength of the forthcoming movement can be foretold.

What needs to be pointed out is that in the profession of weather forecasting, what is used currently is extrapolation forecasting in the first push system, where fluid movements are constrained to wave motions of regularized barometric pressure and altitude fields. Not only has the information of the altitude or pressure field below 500 hPa already been damaged, but also, the profession of weather forecasting has been restricted to the use of four horizontal isobaric surface maps below 500 hPa. Almost all the available information collected for above 500 hPa is not utilized, so that whatever effects rolling currents may produce are ignored. So, the involved conceptual changes are that disastrous weather represents transformational movements in rotational flows instead of a "project of clearing a traffic jam" of the first push.

3. Other than the unevenness in the horizontal field of masses, the effects of rolling currents also contain the unevenness of the vertical structure. In particular, the uneven structure, made up of the existing ultra-low temperature near the top of the troposphere, combined with a rising ground-level temperature, is an important reason for a rolling current to appear and develop. At the mid- to low-latitudinal area in the middle of summer, the difference in temperature between the top of the troposphere and the ground level can reach 100–120°C.

4. No matter what positioning system is used, rolling currents or vortices can only do one of two things: spin inwardly or rotate outwardly. This simplicity no doubt provides convenience to cognitive understanding and practical applications, and can be traced back over several thousand years to the concepts of Yin and Yang, first contained in the *Book of Changes*. This end reveals that directions of rotation are not only a theory helpful for us to understand nature, but also a tool for applications. They are more straightforward and effective than quantitative analysis, without the need to worry about the possibility of trapping into quantitative irregularities and complexities. This end touches on materials and quantities, and such epistemological problems as whether materials determine quantities or quantities determine materials. Because the quantities of the first push do not push back, there should not exist quantitative irregularities or complexities. Evidently, with cut much traffic in the first push system, how can it be possible for us to see

"traffic jams" and disasters, and nature should not experience any change in directions.Nonuniformity in directions has to cause confrontations in materials' movements and the existence of constraints and transformations. No matter how complicated pure quantitative computations can be, there are only three forms of complications: changes in quantities are either increasing, decreasing, or staying constant.

Obviously, when changes in quantities are increasing, we will have to deal with the problem of the conceptual infinity, which is the traditional instability problem unavoidable by Newton's quantities. The situation where changes in quantities are decreasing does not have any fundamental difference from that of changes in quantities staying constant. They represent the results of Newton's analysis system under the constraint of stability. So, complexity is not about quantities themselves. Instead, it is about the reason for the appearance of the complexity. Quantities, as measurements of materials, are evidently determined by the materials and properties of their movements, and originate in confrontational irregularities of materials. What is interesting is that no matter how complicated an irregularity is, the direction of its underlying rotation is very simple. To avoid possible confusion about rotational directions in applications, let us use the following right-hand rules and make the following conventions:

1. In terms of the northern hemisphere (for the southern hemisphere, inverse the order), if within the troposphere, the airstream directions at the top and bottom layers are different, point the four fingers of your right hand to the direction of the southern or eastern wind direction, while taking the origin of the coordinate system at the left-hand side (latitudinal direction) or the right-hand side (longitudinal direction) of the observation station. The rolling current flowing along our right-hand finger spiral is called in the clockwise direction, or a clockwise rolling current (Figure 17.2). (The symbols for wind vectors, including both direction and speed, are defined as in Figure 17.3, where the symbol on the left stands for a wind vector that exists in the atmosphere such that the bottom, horizontal bar represents the wind

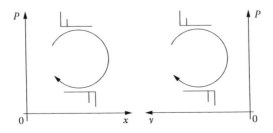

Figure 17.2 Clockwise rolling currents of winds of different directions.

Figure 17.3 Symbols for wind direction and speed.

direction, blowing from the end with vertical bars to the other end, called head; each taller vertical bar stands for 4 m/s, and a shorter bar for 2 m/s; the triangle in the right-hand symbol represents 10 m/s; and each of these wind vectors stands for horizontal winds at their individual altitudes, indicated by their wind head positions.) Otherwise, the rolling current is counterclockwise (Figure 17.4).

A clockwise rolling current indicates such transitional changes in weather that transform from clear, sunny (good) conditions to more gloomy conditions, such as traveling clouds, planted rains, strong winds, and others. A counterclockwise rolling current stands for the opposite: gloomy weather will change into better conditions. Here, the wind vectors are first decomposed into components, then in the overall analysis the greater components are used to determine the direction of the rolling current, if any. For the northern hemisphere, upper-level northwest winds combined with lower-level southeast winds are the clearest indicator for good weather to turn into gloomy weather. And that is an often seen transitional change. In applications, to determine the severity of the forthcoming gloomy weather, we need to pay attention to the entire internal tropospheric structure, including temperature differences, water vapor compositions, and, of course, the rolling current. That is why we have produced V-3θ graphs.

2. When the wind directions are the same, due to friction and decreasing air density with the altitude in the atmosphere, the wind speed in general is smaller at lower levels and greater at upper levels. (Exceptions to this general rule can happen due to rotations or cloud layers involved.) Because shear effects can also produce rolling currents, the air in the troposphere is mixed; in other words, the well-mixed air in the troposphere is produced by stirrings. In this case, let the four fingers of your right hand point to the direction of the strongest wind direction, while the origin of the coordinate system is placed to the left of the observation station. This resultant right-hand spiral is used for the corresponding direction of the rolling current (Figures 17.5 and 17.6).

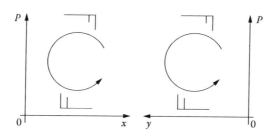

Figure 17.4 Counterclockwise rolling currents of winds of different directions.

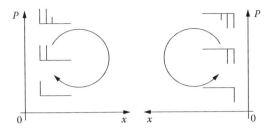

Figure 17.5 Rolling currents of same north or west directional winds.

Here, please pay attention to the cases where the uniform directional winds are in either the southeast or northwest direction. Because east, north winds produce counterclockwise rolling currents, while south, west winds present clockwise rolling currents, the effects of the winds will consequently cancel each other. So, we will have to consider the strengths and origins of either the east, west winds or the south, north winds to decide which directional winds to keep and which to ignore. This is because the effects of a directional wind that originates in a dry inland or an ocean are totally different. So, the application of wind directions is not entirely a problem of quantities. Remember not to mix uniform and nonuniform winds in applications, and also pay attention to the possibility that some wind speed decreases with altitude. When the speed of either north winds or east winds decreases with altitude, we also have a clockwise rolling current. When the speed of either west winds or south winds decreases with altitude, we have a counterclockwise rolling current. In short, rolling currents can change the structure of the atmosphere. So, even as a force, they can also change due to rotations. The tendency for the atmospheric structure to become more even indicates that the forces (or energies) are decreasing. If the atmospheric structure becomes more uneven, the forces (or energies) are increasing. So, the rotational direction of a rolling current not only is related to transitional changes, but also shows the timely leadability of its observational information. This is why the directions of rolling currents can be used to foretell the event before it happens.

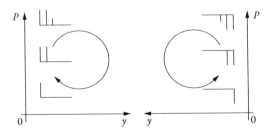

Figure 17.6 Rolling currents of same south or north directional winds.

17.3 The Design of V-3θ Graphs

Although the method of **V**-3θ graphs is designed for forecasting disastrous weather, as a prediction tool, it can also be employed to analyze transitions existing in general weather conditions or nontransitional changes. The first reason for such a convenience and power is that these graphs can be easily applied to make comparisons, and the second reason is that for general weather conditions, there is no need to combine this tool with other methods.

Currently, there are three structural methods of prediction in use, including the **V**-3θ graphs, twist-up structures (and blown-up charts), and phase space representations. In this section and book, we will only look at the construction and applications of **V**-3θ graphs.

Based on structural irregularity, **V**-3θ graphs use as much realistic information available in the data resources as possible. In other words, what is different from the traditional methods is that these graphs are tools of analysis designed on the principle of making use of all the irregular information, which is ignored by traditional methods. In terms of meteorological problems, the available information of the significant sphere is mainly considered and used. Irregularities in the vertical structure of this information are employed to reveal the characteristics of rolling currents and the transformation of vortices of the evolutionary tendency. That is, the graphs help to analyze:

1. The effects of the existing rolling current in the flow field
2. The potential effects of the rolling current

Item 1 stands for the momentarily existing rolling current in the atmosphere. All of the maintenance and development periods of cloudy, rainy weather have clockwise rolling currents, while the development and maintenance of clear skies are backed up by counterclockwise rolling currents. Item 2 implies that during good (or gloomy) weather, the appearance of a clockwise (or counterclockwise) rolling current in 24–36 hours will make the good (or gloomy) weather turn to gloomy (or good) weather. If an uneven structure in the θ (3θ) field is combined with a counterclockwise rolling current, or an even structure in the θ (3θ) field is combined with a clockwise rolling current, then the existing counterclockwise rolling current will change to clockwise, or the clockwise rolling current will change to counterclockwise; when the θ (3θ) field becomes uneven, the existing clockwise rolling current is further strengthened. Here, we need to pay attention to the difference in wind vectors and changes in wind speeds. In particular, differences in wind vectors should be quite clear. According to the traditional methods, both items 1 and 2 stand for the baroclinic effects of the atmospheric structure. In our situation, we carefully utilize these effects and the available baroclinic information in both vertical and horizontal directions. The conceptual difference reflected between our method and traditional methods is that in theory, the atmospheric baroclinity represents eddy effects instead of wave motions. In terms of our method, the principle is to use all available information, with emphasis placed on showing irregularity and informational discontinuity. That is very different from smoothing and continuity of the wave motion system.

Figure 17.7 is an example of a **V-3θ** graph for Shanghai station. The symbol **V** stands for the direct observation data of wind direction and speed available in the sounding data without being artificially processed, including initialization. These wind vectors are placed on the θ* curve. The symbol 3θ stands for the three θ, θ_{se}, and θ* curves, respectively, in the vertical direction plotted on the P-T coordinate system for each observation station, where P stands for hPa and T is the absolute temperature in K.

In our computation, the parameter θ_{se} is calculated using the dew point at each layer. So, its value, in general, is greater than that computed by using the condensation temperature. This difference is introduced from considering the states of distribution of water vapor over either a subtropical high pressure or a drought region, and the lagness of the humidometer on the sounding device. This modified value of θ_{se} is denoted θ_{sed}.

The symbol θ represents the potential temperature, and θ* is the imagined potential temperature computed under the assumed saturated state, designed for the purpose of comparing the state of water vapor in the atmospheric structure.

Because each **V-3θ** graph is shown on the P-T coordinate system instead of the lnP-T system, with information of the significant sphere included, it contains three irregular curves with wind vectors at different layers; that is why it is named the **V-3θ** graph. What deserves our attention is that each **V-3θ** graph includes the

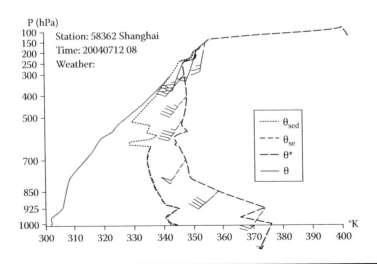

Figure 17.7 A typical V-3θ graph.

significant and required spheres' data from the ground level all the way to ~100 hPa, plus the information of possible ultra-low temperatures.

The main function of **V-3θ** graphs is established on the theoretical foundation that uneven structures can cause rotational movements in the atmosphere, and the problem of how vertical rolling currents in the troposphere appear, maintain, and transform, directly involving the theory and recognition that eddy motions cause divided flows. The design of **V-3θ** graphs is not purely geometrical. Instead, it is done mainly by using an understanding on topological equivalences, the meaning of the structure of each physical quantity, and their effects.

1. If the three θ curves lean to the right along with P in the form of a linear growth, then the structure within the troposphere is even, representing a non-realistic state. During 1963–1970 and 1981–2000, S. C. OuYang checked over 1,600 real-life cases and did not see even one case in which the three θ curves increase with P linearly. Instead, what was found is the tendency for the three θ curves to bend over to the right nonlinearly.
2. If the three θ curves decrease with P toward the left, or are invariant or change slightly with P, then the vertical structure of the atmosphere is extremely uneven. Here, what is often seen is in some atmospheric layers, such a situation appears. For example, before cloudy, rainy weather appears, the $θ_{sed}$, $θ^*$ curves in the mid to low altitudes almost always behave like this. Once this situation is combined with the θ curve leaning left or forming a right angle, or an angle between 70 and 80° with the T-axis, and there is an ultra-low temperature in the upper layer and a clockwise rolling current, then there will definitely appear a severe weather condition.

Because the potential temperature is defined by

$$\theta = T\left(\frac{P_0}{P}\right)^{R/C_p}$$

where T is the air absolute temperature (in K), P_0 the air pressure at sea level (hPa), P the pressure at any altitude, R the gas constant of air, and C_p the specific heat capacity at a constant pressure, the potential temperature θ increases as P decreases. So, if there appears one of the following situations—θ suddenly decreases with P, or stays constant, or suddenly slows down its increase with P so that a sharp turn angle is formed with the average value—then there is the phenomenon of an ultra-low temperature at the top level of the troposphere. Figure 17.7 is an example of a situation where θ stays constant and suddenly slows down its increase with P at the level 250–200 hPa, indicating the existence of an ultra-low temperature at the top of the troposphere. In studies done since the 1960s, the lowest ultra-low temperature reached −83°C. Evidently, it is difficult to plausibly explain sudden appearances of such low temperatures near the top of the troposphere by employing either long-wave radiation of the Earth or physical molecular processes. After consulting with relevant experts, it has been concluded that the appearance of such a phenomenon might have something to do with the ionizing effects of photochemistry. That is, under the direct effects of ultraviolet lights and Rontgen rays, water molecules are ionized, constituting plasma states. The corresponding reaction equation of the ionization is

$$4H_2O \xrightarrow{\text{ultraviolet light}} 4OH^-\uparrow + 4H^+\uparrow \qquad (17.1)$$

The reaction process, as shown in Equation (17.1), can be carried out entirely in a laboratory environment. Because the air density is low at high altitude, the four decomposed hydrogen atomic nuclei cannot as easily acquire four electrons from their surrounding environment as in a laboratory. So, the decomposed hydrogen atomic nuclei must quickly float upward to the plasma sphere with their enlarged volumes. The corresponding four electrons, carried by the negative oxyhydrogen ions, also due to the low density of air, cannot be absorbed by the surrounding materials as easily as in a laboratory. So, they have to sink to the bottom of the stratosphere.

Evidently, the reaction, as described in Equation (17.1), decomposes each four-volume molecule into four volumes of negative oxyhydrogen ions and four volumes of hydrogen atomic nuclei, a total of eight volumes. That is, Equation (17.1) is a reaction that absorbs heat. In general, such a reaction can make the temperature drop as much as 15–25°C. That end is one explanation for the appearance of ultra-low temperatures near the top of the troposphere. Even though, as of this writing, we sill cannot determine if this explanation agrees with the reality, it indeed provides a way of thinking for further investigation of the phenomenon of ultra-low temperatures. If the physical mechanism underneath the phenomenon of ultra-low temperatures is well understood, then it is not only a cognitive breakthrough but also a discovery with wide-ranging practical applications. The phenomenon of ultra-low temperatures makes the vertical structure of the atmosphere extremely uneven, so that it causes a rolling current in the troposphere to appear and develop, which in turn leads to the appearance and development of torrential rains and strong convective cloud systems. This understanding can be used as a method to artificially control and operate the weather. As for weather forecasts, it also provides an explanation for the necessary conditions under which disastrous weather can occur. So, what possesses the significance of direct applications is that the phenomenon of ultra-low temperatures not only makes strong convective weather predictable, but also can be employed to distinguish such strong convective weather conditions as hailstones, strong winds, tornadoes, and sandstorms, from ordinary torrential rains, and distinguish ordinary torrential rains from extraordinarily heavy rain gushes. So, such weather as torrential rains, heavy rain gushes, and extraordinarily heavy rain gushes become predictable.

At this juncture, we point out that those four negative oxyhydrogen ions from the decomposition reaction can also be returned to water molecules with heat released. That is, we can have

$$4OH^- \uparrow = 2H_2O \downarrow + O_2 + 4e \qquad (17.2)$$

This end indicates that if there are negative oxyhydrogen ions in the atmosphere, there will be heat-releasing effects that are slower than the opposite decomposition reactions. This can be used to explain why there exists the phenomenon of temperature increase at the lower layer of the stratosphere, and why after thundershowers, the existing ultra-low temperatures are weakened or disappear. This observational information can surely be employed to predict when bad weather is turning clear. Also, we have already employed the information of nonweakening or strengthening ultra-low temperatures after thundershowers to predict such disastrous weather as heavy or extraordinarily heavy rain gushes with very satisfactory results. This fact can at least provide the clue that the effects of solar radiation might very well magnify atmospheric photochemical processes. At the same time, when the sun increases the temperature near the ground level, it also helps to decrease the temperature near the top of the troposphere, making the heat or temperature structure of the troposphere extremely uneven. This uneven structure causes the moving atmosphere to drastically readjust itself, producing disastrous weather in the form of rolling currents. In this sense, disastrous weather processes also help to weaken or readjust uneven atmospheric structures, in which heat transformations are accomplished through subeddy currents. As for weather forecasts, in terms of the 1,600-plus cases we went through, it could be said that without an ultra-low temperature, no severe weather phenomenon will appear, and that ultra-low temperatures appear ahead of the consequent disastrous weather.

What we also found is that if we substitute the actual data of an ultra-low temperature into currently applied models in meteorology, the computations of the models will have to stop due to quantitative instability. If we limit or weaken the observed ultra-low temperature, the computational outcomes no longer agree with the reality.

1. Speaking generally, at mid to low altitudes, the three θ curves can slightly lean to the left; in particular, the θ_{sed} and θ^* curves can more clearly lean to the left, or form a right angle with the T-axis in the form of straight lines. In general, the θ curve leans to the right or slightly bends to the left, the θ_{sed} curve stays in the middle, and the θ^* curve can lean to the left and form an obtuse angle with the T-axis. At mid to high altitudes, the three θ curves are mostly leaning to the right, and near the top of the troposphere, they almost coincide with each other. This end has something to do with the slightly higher temperature and slightly greater air pressure at the mid to low altitudes and the relative concentration of water vapor at low altitudes.

2. In the northern hemisphere, partially southern winds can make the three θ curves lean to the left, while partially northern winds make the curves lean to the right. This combination is called reasonable and is a positive combination. If the situation is reversed, that is, with partially northern winds, the three θ curves, especially the θ curve, lean to the left, while with partially southern winds, they lean to the right; this is called a reversed combination. In this case, one has to check the attributes and origin of the winds. For reversed combinations, in general the data contains misguided cases.
3. In general, the three θ curves lean left, with either θ agreeing with θ*, or θ leaning right while both $θ_{sed}$ and θ* lean left; these are all reasonable combinations. If the θ curve leans left, while both $θ_{sed}$ and θ* lean to the right, then this is an unreasonable combination. One needs to double-check for possible misguided cases by using the θ curve as the standard to check on $θ_{sed}$.
4. Left-leaning θ curves, especially at the mid to low altitude, mean uneven structures and potential clockwise rolling currents. In this case, even if the existing rolling current is counterclockwise, it will change to clockwise within 12–24 hours. If the three θ curves lean to the right and show an even structure, a potential counterclockwise rolling current or no rolling current appears. In this case, even if the existing rolling current is clockwise, it will soon change to counterclockwise. This indicates that structures are deterministic, while directions are only a representation of the tendency of the movements.
5. If the three θ curves, especially the θ curve, suddenly lean to the left at around 300–100 hPa or quasi-parallel to the P-axis, or in the right-leaning trend there exist turning points for slight left leanings, then at the top of the troposphere there exists the phenomenon of ultra-low temperature. In this case, check on other conditions and be warned about possible soon-to-appear disastrous weather.
6. From **V-θ** graphs, one can more conveniently spot cloud clusters and strati and their altitudes than from cloud charts. When the $θ_{sed}$ and θ* curves are close to each other (≤3K), and the three θ curves show a multilayered irregular structure of zigzagged turns, the implication is that there exist strati at the exact altitude where the zigzagged turns appear. The corresponding wind speed has the tendency to decrease with altitude. If the three θ curves are nearly over each other, and turn to the right almost parallel to the T-axis, then there is the phenomenon of a strong inversed temperature and a stratus. The altitude of the stratus is where the strong inversion is. In particular, above the stratus, there might appear multilayered zigzagged turns in the three θ curves, and if the $θ_{sed}$ and θ* curves are close to each other, then above the stratus there are indigested clouds. In this case, be prepared for increased precipitation, which will at least double what has appeared.

With the relevant theories in place, we now are ready to examine case studies of successful practical applications.

Chapter 18

Case Studies Using V-3θ Graphs

In this final chapter, we will learn how to make real-life predictions using the idea of the systemic yoyo model and its methodology for:

- Suddenly appearing severe convective weather
- Small, regional, short-lived fog and thunderstorms
- Windstorms and sandstorms
- Abnormally high temperature

For a complete parade of case studies on predicting various kinds of disastrous weather conditions using this new method and its complementary methods, consult Lin (1998) and OuYang et al. (2002a).

18.1 Suddenly Appearing Severe Convective Weather

Considering the fact that suddenly appearing disastrous weather is a hard problem, which has not been solved since the time modern science was initiated over 300 years ago, and that peculiar events cannot be dealt with by using quantities, in this section, we look at how to employ digital structure transformations to analyze the special structural characteristics of the suddenly appearing strong, convective precipitation and windstorms in Beijing on July 10, 2004, and Shanghai on July 12, 2004, and the structural characteristics of the regional, especially heavy torrential rains for the northeast area of Sichuan. Even with the current network of radiosonde stations, the method shown here can be effectively applied to forecast local severe convective precipitation, windstorms, and regional heavy torrential rains. And as effective precautions for metropolitans to take against these natural calamities, such emergency measures as drainage works, storage facilities, wind prevention, etc., should be considered. This section is mainly based on OuYang et al. (in press).

18.1.1 Background Information

Severe convective weather is traditionally known as suddenly appearing disastrous weather. In essence, it means the phenomenon of drastic changes in local weather, including hailstones, tornadoes, strong winds, black winds, sandstorms, and heavy precipitation. The precipitation of severe convective rains changes quickly, contains a large degree of unevenness, and can reach the strengths of heavy or extraordinarily heavy rain gushes within several or fewer than 20 hours. Due to their regionality and strength in precipitation, these severe convective rains possess damaging effects on metropolitan networks of transportation and small- to medium-sized reservoirs. What is most important is that by continuing to employ the method of extrapolation, the scientific community has met with the difficulty of being unable to predict suddenly appearing events.

Since the 2004 flood season in China, one by one there appeared local severe precipitation and windstorms in Beijing, Shanghai, and other locations, and suddenly appearing regional extraordinary heavy torrential rains in the northeastern areas of Sichuan. Because of the unexpectedness of these events and the non-promptness of the relevant measures, people started to worry about potential losses in human lives and properties, because to stand 100 to 200 mm torrential rains or strong winds of about magnitude 10, major metropolitans should be equipped with relevant and adequate emergency measures. In fact, such disastrous severe rainy weather conditions have already occurred frequently in history. For example, in July 1998, Wuhan City experienced a suddenly appearing torrential rain. Its course precipitation reached over 250 mm (the measurement of the hydrologic station reached 500 mm), setting the new regional record of severe precipitation. Because the widely accepted method had missed this event, and our method, based on the structural characteristics of the event, was applied in time to modify the forecast so that an effective and accurate prediction was produced, practical social and economic benefits resulted. Our work was consequently highly appraised by meteorologists, residents, and the city officers of Wuhan. To this end, in this section, we illustrate how to analyze severe convective weather and regional torrential rains using the structural characteristics of the irregular information existing before the occurrence of past disastrous weather conditions. And accordingly, we propose several recommendations regarding the corresponding emergency measures.

18.1.2 The Blown-Ups Principle

We established the theory of blown-ups for the purpose of investigating transitionalities existing in such processes of changes that reflect "injuries, aging, and damages" in materials. It belongs to the study of evolution science (OuYang, 1998).

The transitional changes existing in the processes of materials' destruction is the blown-ups we introduced. Weather conditions belong to those that change quickly among easily destructible materials and represent problems that belong to noninertial systems. The corresponding durable goods are problems of estimation on the design and manufacture of the inertial system (OuYang et al., 2002b). So, we established the corresponding theory and method to deal with the easily destructible products. The materials' evolution process from aging to death (the transitionality) should first be shown, according to the blown-up principle, in the changes of the materials' structural mechanisms. Then and only after then, quantitative changes appear. Because quantities cannot deal with peculiar events, evolutional blown-up analysis is rooted in the structures of materials and events. In particular, structural transformations of irregular information are employed to describe evolutionary transitional changes. That is, we employ the method of event analysis to substitute the currently fashionable quantitative analysis system. The direct reason for us to do so is that quantities cannot deal with peculiar information, whereas the essential reason is that events are not the same as quantities (OuYang and Lin, in press b). No doubt, the establishment of such a new structure-based methodology represents a major reform in our conceptual understanding 300 years after modern science was initiated.

18.1.3 Structural Analysis Method and the V-3θ Graphs

The **V-3θ** graphs are designed on the basis of the blown-up theory for the purpose of discovering the characteristics and functions of irregular information in the form of structures (OuYang, 1998; OuYang et al., 2002a). This structural prediction method is developed particularly for the study of evolutionary transitional processes of weather systems by employing irregular information, where the structural states of the θ- (geo-potential) curves are used to represent the slopes in the corresponding P–T phase space. Because quantities cannot be used to replace the specifics of events, the structural characteristics of events do not follow the rules of the formal logical calculus of quantities, quantities can increase unstably and lead to the quantitative infinity ∞, the quantitative ∞ can be perpendicular to the horizontal T-axis in the form of slopes in a structural phase space, and the quantitative ∞ in Euclidean spaces is exactly a transitional change in a curvature space (OuYang et al., 2002a), **V-3θ** graphs can betoken forthcoming transitions in the form of quasi- perpendicularity between the θ curve and the T-axis. They ingeniously resolve the extremely difficult problem of transitional changes of modern science. That is why the blown-up principle also reveals the law of philosophy that at extremes, matters will evolve in the opposite direction. The essence of sudden appearance is about acute transitionality of events. Because quantities cannot represent the directionality of events, while structures can reveal changes in the direction of rolling currents, structures can forecast sudden appearances of weather systems that experience transitional changes.

At the height of concepts, materials' structures and related changes are the source of forces and exist prior to any change in the forces. The structural characteristics of irregular information are also the representation of materials' structural features. When combined with the inverse sequential structure of the information series on wind directions, we can gain the needed valuable prediction time compared to the current method that is mainly based on pressure information series. For the purpose of practical applications, observational errors in wind directions are the smallest among all the meteorological factors observed. Because ultra-low temperature is a piece of irregular information, when combined with the direction of rolling currents, etc., it forms the physical mechanism of convective weather systems. This end explains why several stations, such as Chengdu station, which employs our blown-up software, as long as the information provided does not contain misguided cases, can successfully forecast disastrous weather conditions, such as heavy and extremely heavy torrential rains (Chen et al., 2003).

The symbol **V** in **V**-3θ graphs stands for the wind vector available in the sounding data, containing the wind directions and speed, and is positioned on the θ* curve. The three θ's in **V**-3θ graphs are, respectively, θ, $θ_{sed}$, and θ*, which constitute three curves in the vertical direction in the P-T coordinate system of the phase space. That is, they are the vertical section graph for each individual observational station. Here, θ stands for the potential temperature, and $θ_{sed}$ and θ* represent the state of distribution of water vapor in the atmosphere—the structural characteristics. These two later variables can be employed to determine the severity or the magnitude of a disastrous weather condition (OuYang, 1998; OuYang et al., 2002a). All details are omitted here.

18.1.4 Structural Characteristics of Suddenly Appearing Convective Weather

Forecasting small-scale suddenly appearing weather conditions has been listed in modern science as a world-class difficulty problem. Some experts even believe that this problem will not be resolved for at least 20 to 50 years. Their reasoning is that the scales are too small, the unexpectationality is too high, and the current observation systems have low resolutions. In particular, observations based on the current monitoring radar systems can only provide several hours, or even a few minutes, of advanced time. In fact, the most fundamental reason is that unexpectationality is not part of the content of modern science; modern science is only about tracking what has already happened. Based on the blown-up principle, each material's sudden change must first go through a process of accumulation, which is revealed by the appearance of some irregular information. As long as the irregular information constitutes the physical mechanism, equivalent to a medium- or small-scale weather system with about 6 hours of life span, we can make use of the **V**-3θ graphs to produce our accurate prediction about 6 to 24 hours ahead of the event. It is because from the **V**-3θ graphs, one can clearly see the structural characteristics and related changes before the appearance of the weather phenomenon. This end is something unachievable by formal quantitative analysis, which is shown to be postevent.

In terms of the structural characteristics of severe convective weather, one should at least pay attention to the following before he makes the forecasting:

1. Other than the uneven structure of the 3θ curves, what is noticeable is the obtuse angle formed by the T-axis and both θ_{sed} and θ^*. This formation can distinguish the forthcoming severe convective weather from a torrential rain. However, the condition of clockwise rolling currents holds for both scenarios.
2. Ultra-low temperature is not a single station or individual regional, local phenomenon. So, in the prediction analysis, other than noticing the phenomenon of ultra-low temperatures, one should also pay attention to the combination of the 3θ curves at low attitudes in the atmosphere. That is, ultra-low temperatures can be seen as a necessary condition for severe convective weather to occur, but it is not a sufficient condition. In terms of convective weather, their corresponding ultra-low temperatures, compared to other weather conditions, possess a deeper, thicker, and more leftward structure with even lower temperatures. All these features are most particularly obvious before the appearance of the black windstorm or a sandstorm.

18.1.5 Suddenly Appearing Weather over Major Metropolitans

18.1.5.1 The Severe Precipitation Process over Beijing, July 10, 2004

In the afternoon of July 10, 2004, Beijing City experienced a severe convective precipitation process. The precipitation had a strong regionality and an uneven distribution. It mainly focused on the center and the southern area of the city. For example, the precipitation at Tiananmen was 104 mm; at Mentouguo, 84 mm; and at Shijingshan, 74 mm. However, northern areas like Yaqing, Minyun, and Pinggu basically did not receive any precipitation. This regional torrential rain flooded the bottom roads of many elevated highway bridges and paralyzed the transportation system of the entire city for four hours.

This local severe precipitation process was not shown on the weather map for July 10. However, from the V-3θ graphic structures of the 20th hour on July 9 and the 8th hour on July 10, one could have already seen structural abnormalities and changes. At the 20th hour on July 9, there had appeared an overall unstable clockwise rolling current in the troposphere, except that the water vapor distribution still did not provide the necessary condition for precipitation (see Figure 18.1). However, the V-3θ graph of the eighth hour of July 10 had already shown the structural characteristics of a severe precipitation (Figure 18.2). In the V-3θ graph of the eighth hour of July 10, one can also see the structure of the wind vector. Other than the effects of slight northeastern winds near the surface level due to the night-day change, in

318 ■ *Systemic Yoyos: Some Impacts of the Second Dimension*

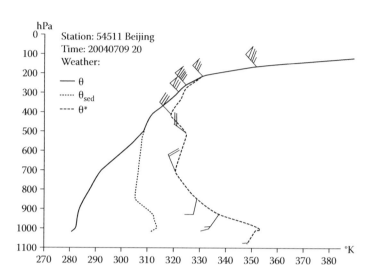

Figure 18.1   V-3θ graph for the 20th hour of July 9.

the low-altitude atmosphere there existed southwestern winds. At the level above 700 hPa there were northwestern winds, making the entire atmosphere a clockwise rolling current. The ultra-low temperature was of the thin-layer structural characteristic of precipitation. The water-vapor distribution provided the condition for convective precipitation, where the humidity at the low altitude was increased, while the higher altitude was dry (there existed almost no water vapor at above 500 hPa). The lower

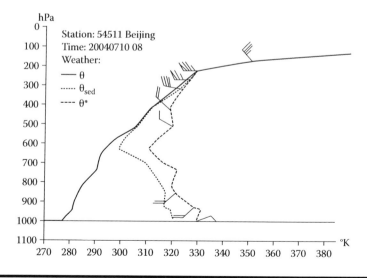

Figure 18.2   V-3θ graph for the 8th hour of July 10.

end of the θ curve formed a bow facing the left and a 75° acute angle with the T-axis and contained several folding turns. This formation, according to meteorology, is customarily seen as an unstable atmospheric structure full of energy. The corresponding θ_sed and θ* curves form obtuse angles with the T-axis at the mid to low altitudes with many folding turns (which are exactly the characteristics of development of concentrated massive amounts of nimbus). As for the source of water vapor, from Xingtai to Taiyuan to the Beijing area, there existed a uniform southwestern airstream. So, a passageway of water vapor was already opened. That is, the precipitation in the Beijing region had a reserved source of water vapor. So, based on what was discussed above, we predicted that in the timeframe from 6 to 24 hours, the Beijing region should experience a distinct, local, severe precipitation process. From the **V-3θ** graph of the 20th hour on July 10 (Figure 18.3), it could be seen that even though there still existed some structural instabilities, the wind at the high altitude had already turned to a southwest direction, with the wind at the lower layers turned to the northwest. The water vapor in the lower levels of the atmosphere was reduced and in the higher levels increased. At the altitude from 700 to 600 hPa, there appeared the structure of strati, including cold strati (notice that the appearance of cold strati is also a characteristic of convective weather). At this junction, let us compare the three **V-3θ** graphs created for the 20th hour of July 9, the 8th hour of July 10, and the 20th hour of July 10. It is not hard to see that the atmospheric structure varied throughout the entire precipitation process, with clear contracting structural differences between before and after the occurrence of the convection. Therefore, meteorological problems are evolutionary noninertial system problems. Weather phenomena change along with the atmospheric structure

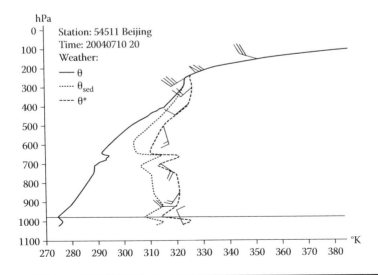

Figure 18.3 V-3θ graph for the 20th hour on July 10.

and cannot be fully understood by simply employing materials' invariant extrapolations. As a structural analysis, from Figure 18.3, it can also be seen that the θ_{sed} and θ^* curves in the mid to low altitude are quasi-parallel, form a right angle with the T-axis, and contain folding turns, indicating that there were still convective cloud clusters and the structure of strati, and that the ultra-low temperatures were not distinctively weakened. However, the right-leaning θ curve from 400 to 300 hPa (rising temperature) constrained the convective activities in the mid- to low-altitude atmosphere. To analyze the **V**-3θ graphs successfully, one must pay attention to the structural details and first recognize direction changes in the airflow field.

To help the reader understand how to employ our **V**-3θ graph analysis to predict suddenly appearing local severe convective weather, we also attached the **V**-3θ structure graph for Wuhan on July 20, 1998. Evidently, the Beijing **V**-3θ graph of the eighth hour on July 10, 2004, is very similar to this Wuhan graph. The similarity in the structural mechanism indicates that the underlying principles should be the same. The only difference is that the strength of the rolling current and the water vapor distribution of Beijing were weaker than those of Wuhan.

From Figure 18.4, it can be clearly seen that Wuhan City's convergent layer of water vapor was very high, reaching above 500 hPa, providing a more sufficient water vapor condition than the July 10 severe precipitation in Beijing in 2004. So, the daytime precipitation on July 20, 1998, in Wuhan, already reached 154 mm, whereas the Beijing precipitation at the most concentrated area only reached 104 mm on July 10. When analyzing the **V**-3θ graphs, one has to not only pay attention to the rolling currents and the phenomenon of ultra-low temperatures, but also analyze carefully cold air, the passageways of water vapors, and other like conditions. By doing so, he or she can not only foretell the types of the forthcoming disastrous weather, but

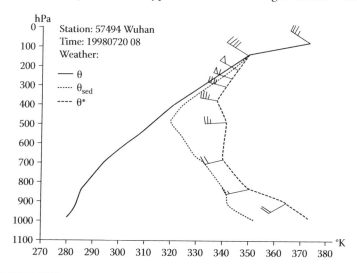

Figure 18.4 V-3θ graph of Wuhan, July 20, 1998.

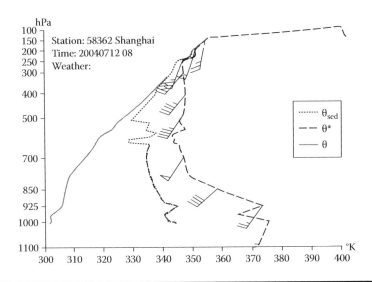

Figure 18.5 V-3θ graph, Shanghai, 8th hour, July 12.

also accurately forecast the intensity and severity of events. The prediction omens for the Beijing severe precipitation had appeared at least 20 hours ahead of the event. As a prediction technique, considering the **V-3θ** graph of the 20th hour on July 9, where the southwest winds' θ* curve at the low altitudes formed an obtuse angle and the northwest wind layer was lowered from above, the meteorologist could contact the local observation field to obtain the sounding data at the same time when these data were submitted. By doing so, he or she could obtain needed information ahead of the available time of the sounding data so that his or her prediction could be announced about 6 to 12 hours before the time he or she would otherwise be making the announcement.

18.1.5.2 The Shanghai Process of Windstorm and Thundershower, July 12, 2004

At the 16th hour 20 minutes on July 12, hard-blowing winds and heavy sheets of falling rain suddenly hit Shanghai. In less than 1 hour, the highest-level precipitation reached 30 mm and the strongest winds reached the magnitude 11. This suddenly arriving windstorm and thundershower made such areas as Jiading, Putuo, Pudong, etc., suffer disastrous aftermath at various degrees. A total of 7 human deaths and over 40 injuries was reported for Shanghai City.

From Figure 18.5, it can be seen that this process represented such a convective weather that the entire troposphere was a clockwise rolling current of strong southwest winds. And the structural mechanic information for a severe windstorm had already appeared at the 20th hour on July 11 (Figure 18.6). That is, the whole clockwise rolling current of southwest winds had already been formed together with the existing violently unstable atmosphere at low and near ground levels.

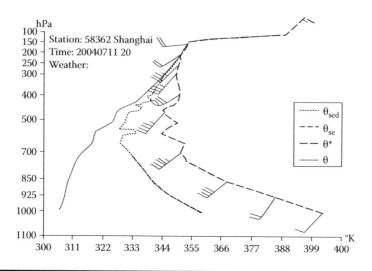

Figure 18.6 V-3θ graph, Shanghai, 20th hour, July 11.

Here, the θ curve at mid and low altitudes formed a rough 80° angle with the T-axis. The angle between the θ* curve and the T-axis could reach 140°. And the ultra-low temperature showed a structure of a deep angle, indicating an extreme severity of the low temperatures. However, because the water vapor condition was poor, the forthcoming convection process should be mainly made up of strong winds, and accompanying the water vapor concentration between 700 and 500 hPa (Figure 18.5), a thundershower should also appear.

For the purpose of practical predictions, in specific analyses, one should look at the wind source and analyze the wind structures and water vapor distributions of the nearby observation stations. Also, it should be easier to forecast the Shanghai thunder rain and windstorm than the Beijing severe precipitation, because the structural characteristics and the clockwise rolling currents of strong southwest winds were very clearly shown. The prediction could have been made over 20 hours in advance.

18.1.6 Regional Suddenly Appearing Extraordinarily Heavy Rain Gushes

For extraordinarily heavy rain gushes or regional suddenly appearing extraordinarily heavy rain gushes, even though their area coverage and time duration are larger and longer than those of an individual city's suddenly appearing convection weather, their ranges of effects in general only reach about 200 to 300 km^2, and their timescales can last about 24 to 48 hours. This class of weather conditions has also been a difficult problem for the current meteorological predictions. In particular, they are an urgent problem for local stationary predictions, such as district or county stations, to resolve.

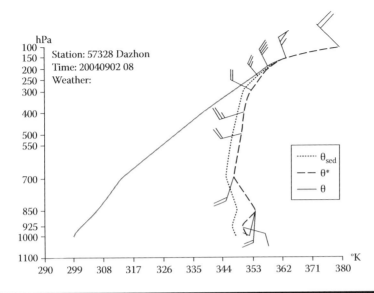

Figure 18.7 V-3θ graph of Dazhou station at the 8th hour, September 2, 2004.

In early September 2004, at Kai county area, which is located in the northeastern portion of Sichuan Province between Dazhou and north of Chongqing, there appeared an extraordinarily heavy rain gush whose daily precipitation reached 200 to 340 mm, creating mud-rock flows and heavy economic and human life losses.

What is shown in Figure 18.7 is the **V-3θ** graph of Dazhou station at the eighth hour on September 2, 2004. The meteorologist with our structural prediction software could definitely predict this specific process of extraordinarily heavy rain gush. Figure 18.7 represents a typical heavy to extraordinarily heavy rain gush's structure. The fully developed, powerful clockwise rolling current, combined with the convergent water vapor layer, reaching as high as above 400 hPa, indicates that at least a heavy rain gush was imminent.

Even if we employed the cloud chart, we would also clearly see this process (Figure 18.8), where the convection cloud cluster exactly covered the northeast Sichuan Province and the northern area of Chongqing. It is generally about 24 hours after the **V-3θ** structural information becomes available that the cloud cluster starts to form on the cloud map for suddenly appearing weather. Also, based on cloud charts, no one seems to be able to foretell the severity of the precipitation. On the other hand, using the **V-3θ** graph information, based on the visible water vapor source and the convergent layer's thickness, we can predict the magnitude of the forthcoming disastrous torrential rain. It can also be said that under the condition that there is no misguided case, on the basis of the current sounding station networks, we can forecast the precipitation of rains above the magnitude of heavy rain gushes within the area range of 200 to 300 km² by applying the irregular

Figure 18.8 Cloud chart for the 12th hour on September 3, 2004.

information contained in the **V-3θ** graphs. Additionally, as long as the available information does not include misplaced data, we can accurately predict the said precipitation with absolute confidence.

18.1.7 Discussion

Prediction stands for a branch of evolution science. So, the current main problem is that on the basis of the theoretical system of modern science, monitoring is identified with prediction. That is why the prediction of suddenly appearing events becomes a difficult problem. It can be said that without first resolving this conceptual confusion, there would not be any breakthrough in the study of prediction problems.

The first step in any specific technique designed to resolve the prediction of suddenly appearing events is still the epistemological problem of how to understand irregular information. The core of this problem lies is the fact that events are not quantities, and that irregular information is about changes. In other words, to address the problem of prediction meaningfully, we must gain new understanding about events and make use of irregular information without treating events in the manner of numbers. The method of structural analysis is established beyond the system of modern science and represents a brand new tool useful for problems unsolvable by using quantitative analysis. The second step is about the observation of the so-called small-scale weather phenomena in meteorology. The current observation system needs to be modified so that the radiosonde stations are arranged at least one in each 100 km (notice that the information on significant spheres must be kept), and that the number of observations changed to four times in each 24 hours. With these modifications in place, the goal of accurately predicting convection weather with at least six hours of life history can be accomplished immediately.

For the prevention and reduction of disasters for the benefit of economic constructions, there is no doubt that accurate weather prediction is necessary. However, to effectively accomplish the long-term effect of preventing disasters and reducing the aftermath of disasters, emergency protection measures are needed. Particularly, for such major metropolitan areas as those with a population over 100 million, these measures should be seen as a necessity. In terms of suddenly appearing severe precipitation, other than the indispensable means of flood drainage systems, there should also be specially designed water storage facilities, such as the planning and construction of municipal forests, wetlands, and alternate water storage lakes (Chen et al., 2005). To meet special needs, such as local black windstorms or sandstorms, forest fires, etc., we should consider those methods and facilities that can be employed to locally reduce the level of severity of the ultra-low temperatures located on top of the troposphere to weaken the strength of the atmospheric convection (OuYang, 1998), so that the severity of the disaster can be reduced. That is, on the basis of knowing the mechanism of atmospheric convections, apply effective long-term methods to prevent disasters or reduce the disastrous consequences. Besides, concentrated, concrete tall constructions in cities can lead to the petrifaction of the irregular bottom layer. They not only signal the destruction and imbalance of ecological environments in cities, but also become a factor triggering sudden appearances of local severe convective weather.

So, for world-class metropolitan areas, their city planning and construction should aim at building an ecologically balanced self-sustained system with relevant laws and regulations in place.

18.2 Small, Regional, Short-Lived Fog and Thunderstorms

Aiming at the special needs of aviation of flights—the atmospheric visibility—in this section, we look at the problem of predicting fog. On the basis of the blown-up principle of materials' evolutions, we analyze the physical mechanism and the problem of prediction using real-life cases of fog's appearance, development, and disappearance on three scales: large area, regional, and local at the observation station. The results of this section indicate that the **V-3θ** graphs' structural characteristics can be employed to forecast the regional distributions and scales of fog and to clearly clarify fog into categories. Through practical testing, it is found that this method can substantially improve the accuracy of visibility forecasts for airports. This section is mainly based on Bao et al. (in press).

18.2.1 The Background Information

As is well known, thick fog of low visibility directly affects the normal operations of air traffic and highway transportations, constituting such weather conditions that are disastrous to human society. As economies are more advanced, the degree of disastrous consequences of low visibility can be more than those of wars or pestilences. The ability to predict low visibility and strong convective thunderstorms directly affects the number of calamities in space and related major economic losses. However, because such predictions for the purpose of aviation are about small-scale atmospheric events with specified location and time, they are under more stringent requirements than predicting meteorological systems. Such small-scale atmospheric events' prediction has always been among the most difficult, world-class unsettled problems in meteorological science and technology. Especially for the prediction of concrete magnitude of low visibility or the particular location of a thunderstorm, no reliable method has been established. Discussions in the previous chapters indicate that meteorological disastrous weather originates from transformations of eddy currents in the atmospheric evolution and cannot be understood as the transitional propagation of wave motions. In particular, as specific events, atmospheric phenomena cannot be identified with quantities.

Evidently, in terms of the prediction of low visibility or strong convective thunderstorms for the purpose of aviation, if anyone can provide useful and clear insights on the location, intensity, and category, it will be much needed in the relevant applications. In this section, we focus on the analysis and prediction, by using V-3θ graphs, of low-visibility fog that directly jeopardizes aviation flights. The results prove the effectiveness of this method.

What needs to be pointed out specifically is that in terms of fog as studied in the blown-up system of evolutions, the actual divergent fog and advective fog existing in the atmosphere are mostly intertwined together, unlike what is described in textbooks, where these fogs are formed statically without any dynamic structural evolution. Like other disastrous weather conditions, fog is such a phenomenon that appears in transitional changes of the atmosphere. These transitional changes are not extrapolations of initial value systems. So, in terms of predictions, even though there is the concept of advective fog, forecasts with desirable accuracy still cannot be produced by simply applying the method of extrapolation. As a natural phenomenon, fog is exactly such an event that appears at the moment of transitional changes in the atmosphere. It is a problem of evolution and also involves the intensity or magnitude of the evolution. Hence, predicting fog itself is a typical case of resolving atmospheric transitional changes.

The blown-up principle was initially established for the investigation of transitional changes existing in materials' evolutions and is exactly a principle needed for the prediction of disastrous weather. This is because all disastrous weather systems occur in transitional changes (OuYang, 1998; OuYang et al., 2002a,b; Lin, 1998). To not make this section too long, we will only discuss the analysis and prediction of regional fog connected to radiosonde stations.

18.2.2 Case Studies on Fog

18.2.2.1 A Description of an Actual Fog Situation

In this section, we employ the actual case of a large-scale thick fog, which occurred at the eighth hour on December 12, 2002, in China, as our example to study the atmospheric structural characteristics of fog, and the problem of predicting the start and end of a major fog phenomenon. For this specific fog process, it covered a large area, including the northern plain of China, the region between Yangtze River and Huaihe River, mid and lower reaches of Yangtze River, the Yun-Gui plateau, and the eastern part of Sichuan basin. And because of the uneven distribution in the fog's density, the ability of our method to predict can be more easily shown. In particular, for this large-area fog coverage, the radiosonde stations involved included:

In heavy fog area: Xi'an, Changsa, Nanning, Quzhou, Baise, Exi, Fuyang, Ankang, Da County, Chongqing

In light fog area: Nanchang, Jiangpu, Kunming, Xuzhou, Guiyang, Yibing, Hongzhou, Zhenzhou, Sheyang, Anqing

18.2.2.2 Counterclockwise Rolling Current Analysis of the V-3θ Graphs

Evidently, by using the traditional weather maps, we could only see two troughs and one ridge on the 700 and 850 hPa situation maps and the cold front pushed southward by the northwestern airstream in front of the high-pressure ridge located in Siberia. However, cold air is only one of the factors that produce fog. So, the weather maps in essence are only a way of monitoring fog after it has already appeared. To truly forecast the specific coverage area and the magnitude (the heavy and light fog areas) of such a large-scale indigested fog, and to effectively serve air and ground transportation, the traditional method of weather maps is not enough. One coauthor of this piece of research (Yuzhang Bao) has been a front-line forecaster for many years with rich practical experience. Even though by applying weather maps, one can outline the rough distribution, as of this writing there does not exist any reliable method to specify particular airlines and locations. To this end, under the guidance of OuYang, we organized our students to test and analyze over 100 real-life cases. We find that the method of analysis using the structural characteristics of irregular information is indeed a much better and more definite alternative than that of weather maps. In the following, we will use our example to illustrate how this method works.

From the **V-3θ** graphs, it can be seen that from 700, 850, and 925 hPa to the ground level, there is a counterclockwise rolling current area consisting of sinking airstreams of the northwestern winds. This rolling current area controlled from

Sichuan, Chongqing, and the southern part of Yunnan and Guangxi, to Guizhou, Hunan, and the Hubei area. By combining the characteristics of the water vapor layer near the ground, we can list the areas the fog affected and distinguish the fog's magnitudes (heavy or light). What is more significant is that by employing the **V-3θ** graphs, we can not only reveal the fog's territorial distribution by looking at where the large-scale counterclockwise rolling current controls and how the water vapor distributes, but also predict that the region from south of Yangtze River to the west of the southwestern area will also be controlled by the northwestern airstream, even though this region is controlled by a southwestern wind. Because the details involve the graphs of over 200 situations, we will not list them one by one.

18.2.2.3   V-3θ Structures of Heavy Fog: A Case Study

Fog, a phenomenon of weather, in fact is such a concrete event that does not follow the logical calculus of quantities (OuYang et al., 2002a; OuYang and Lin, in press a). Due to physical, microscopic static effects, traditional meteorology specifically emphasizes the difference between divergent and advective fogs. In fact, there is not a clear-cut separation between divergent and advective fogs in nature. As a weather phenomenon, fog is an evolutionary process event.

To predict, one must analyze the characteristics of the atmospheric structure and their tendencies of change before a fog appears, during the process of the fog, and after the fog has disappeared. So, by applying **V-3θ** graphs in terms of a specific region or area (if a radiosonde station is available, one can be as specific as that station), if one can detail the characteristics for the appearance and disappearance of a fog, his or her work will possess the practical significance of prediction. For accuracy and timely management of air and ground transportation, this is a very important problem. That is, practically effective prediction consists of the forecasting of a fog's appearance and disappearance. To limit the length of this section, we will only use several stations' situations to illustrate how the **V-3θ** graphs work.

18.2.2.3.1   The Problem of Predicting a Heavy Fog

Xi'an was the observation station that foretold the fact that this specific process could have a heavy fog. The detailed fog information is given below.

At 02:00 on the July 12, there was a light fog, which changed to a heavy fog at the eighth hour on the same day. At the 14th hour, the fog turned to light. The structural characteristics of the **V-3θ** graph drawn on the information of the day prior to the appearance of the fog are given in Figure 18.9. That is, at the Xi'an station, there was no ultra-low temperature. The θ curve below 700 hPa leaned to the right, indicating a stable lower layer. The entire troposphere consisted of a counterclockwise rolling current with northwestern winds. Below 850 hPa, it

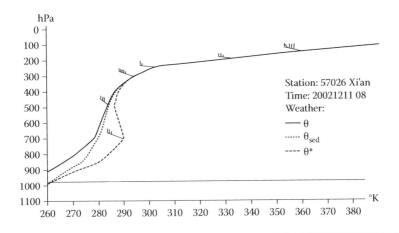

Figure 18.9 The V-3θ graph at the eighth hour on December 11, 2002, Xi'an.

was the layer of concentrated water vapor, and the northeastern winds at the bottom of the layer provided a source of water vapor (graph omitted here). Also, the weak southwestern winds at 850 hPa were also one source of water vapor. Near the ground level, both θ_{sed} and θ^* almost coincided and structurally leaned to the right. That means that at or near the ground level, the atmospheric structure was stable and full of sufficient water vapor. Combined with the overall counterclockwise rolling current, the weather condition would turn to clear with the bottom cushion divergence strengthened, which would inevitably lead to heavy fog for the next day.

Based on this and many other case studies, the structural characteristics for predicting heavy fog are summarized as follows: There exist no ultra-low temperature or only a weak ultra-low temperature and a counterclockwise rolling current of roughly northern winds. The mid to low portion of the θ curve tilts to the right with stable atmospheric structure. The θ_{sed} and θ^* curves are close to each other or coincide with each other near the ground level, indicating that the lower-level atmosphere contains a relatively thick layer of water vapor (for about 1 km).

18.2.2.3.2 The Problem of Predicting a Light Fog

Kunming was also one of the observation stations of this fog weather process. Its fog situation was as follows: on July 12, from 05:00 to 11:00, it was a light fog, clear weather. The V-3θ graph's structural characteristics are the following, where the V-3θ graph at the eighth hour on December 11 is given in Figure 18.10: The θ curve tilts rightward, indicating a stable atmospheric structure. At levels below 700 hPa, the θ_{sed} and θ^* curves are quasi-parallel. However, θ_{sed} is located in between θ^* and θ slightly toward the right, indicating the existence of water vapor

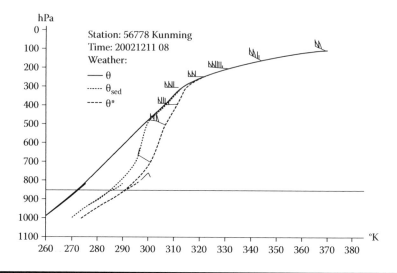

Figure 18.10 The V-3θ graph at the eighth hour on December 11, 2002, Kunming.

but not in abundance (Figure 18.10). As for Guiyang and Menzi stations, which are on the route of water vapor source of Kunming and located in the east and south, respectively, to Kunming, there were ground-level east winds carrying abundant water vapor with thinner water vapor layers than that of heavy fog weather. Also, there was such a counterclockwise rolling current that consisted of northwest-west winds at high altitudes and northeast winds at low levels (Figures 18.11 and 18.12).

The V-3θ graphic characteristics for light fog are: There exists no ultra-low temperature or weak ultra-low temperature. The entire atmospheric structure is stable. Near ground level at the observation station there is abundant water vapor (or when the water vapor is not abundant at the station, the surrounding stations can provide water vapor and their water vapor layers are thin), but thinner than that of heavy fog (about 500 m). And there mostly exists a counterclockwise sustained rolling current with partially west or southwest winds on the top and partially north or northwest winds on the bottom. Or, it is the situation that a clockwise rolling current is transformed into a counterclockwise rolling current.

18.2.2.4 The Problem of Predicting an After-Rain Fog

The reason why we picked after-rain fog as a separate problem to discuss is because such foggy days involve mostly heavy fog (or thick fog with extremely low visibility). For our purpose, we will use the Nanning station to illustrate our method with the chosen weather process as outlined above.

Case Studies Using V-3θ Graphs ■ 331

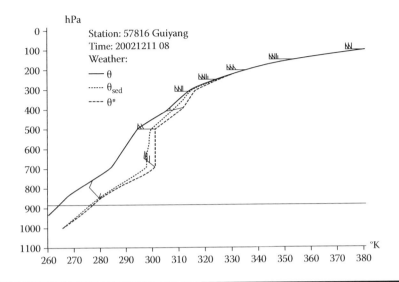

Figure 18.11 The V-3θ graph at the eighth hour on December 11, 2002, Guiyang.

The fog situation at Nanning station was: During 08:00 to 11:00 on July 11, there was a light rain with 12 mm of precipitation. Instantly, the area was covered by a heavy and thick fog during 00:00 to 11:00 on the 12th. It gradually dissipated after 11 hours.

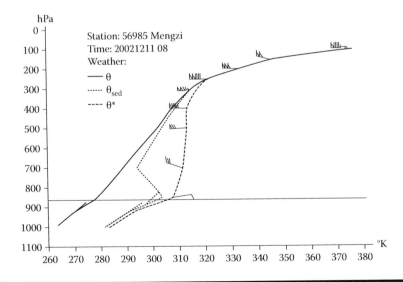

Figure 18.12 The V-3θ graph at the eighth hour on December 11, 2002, Mengzi.

332 ■ *Systemic Yoyos: Some Impacts of the Second Dimension*

As for the **V-3θ** graphic structure (Figure 18.13), even though a counterclockwise rolling current was formed and the ultra-low temperature was weaker than that before the precipitation, we need to pay more attention to that, because at below 900 hPa there was still some unstable energy to be released, and its thick layer of strati still covered the sky over Nanning, indicating that the process of precipitation was not over. However, because of this, in the mid and low atmospheric layers above Nanning, there was abundant concentrated water vapor. So, from the **V-3θ** graph of the Nanning station, it can be seen that the $θ_{sed}$ and $θ^*$ curves almost entirely coincided at the mid and low levels, indicating that there was abundant water vapor at the bottom layer. (In the analysis of Figure 18.13, we noticed that the $θ_{sed}$ and $θ^*$ curves were the same from 900 hPa to 700 hPa, indicating the existence of strati. This is exactly the advantage of using the **V-3θ** graphs to analyze and predict weather conditions. As is shown, these graphs can not only quite accurately tell the altitudes of the clouds, but also spell out the properties of the clouds.) What needs to be specifically noticed is that the existing counterclockwise rolling current betokened that the air of the troposphere would be changed to a sinking movement, causing the cloud layer to dissipate. Combined with the diverging effect of the ground surface, heavy fog weather would be greatly stimulated. This end is both the formation mechanism of heavy fog and the method of predicting such thick fog conditions, showing the advantage of the prediction method based on structural analysis.

Coordinating with surrounding observation stations (Figures 18.14 to 18.17), we saw that at the Baise station, there were mid- to low-level northwest winds; at

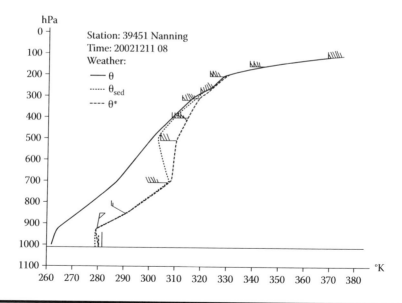

Figure 18.13 The V-3θ graph at the eighth hour on December 11, 2002, Nanning.

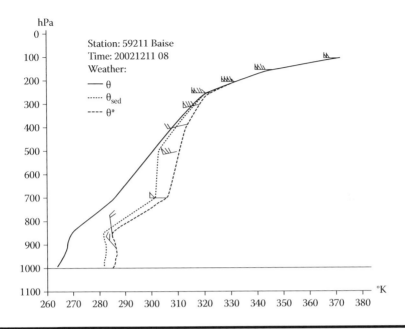

Figure 18.14 The V-3θ graph at the eighth hour on December 11, 2002, Baise.

the Guilin and Wuzhou stations, low-altitude quasi-north winds; and at the Beihai station, northwest winds. So, we were very positive that the structure of the counterclockwise rolling current at the Nanning station would be maintained, that the northwest winds of 800 hPa at Nanning were not an isolated situation, and that when the strati dissipated at Nanning, a heavy after-rain fog would be formed. So, it can also be said that the problem of predicting heavy fog involves the forecasting of the end of a precipitation. In other words, the problem of weather forecast is not limited to the prediction of a certain weather phenomenon before it happens, but also involves the prediction of the end of a weather process.

Here are the V-3θ graph's structural characteristics for after-rain heavy fog: The ultra-low temperature is weaker than that before the rainfall, or the temperature increases (called an inversed ultra-low temperature). The θ curve at the mid and low levels tilts to the right and is stable, or tilts to the left with unstable energy, which weakens and dissipates after the rainfall. The $θ_{sed}$ and $θ^*$ curves are close to each other or coincide at mid and low levels or for the entire layer, indicating abundant water vapor. After the rainfall, the rolling current of the mid and low altitude or the entire layer becomes a counterclockwise rolling current structure. For most cases, there is a layer of strati, which are weakened and dissipated due to the effect of cold airstreams at low altitudes before a wide-ranging fog appears, or at the same time, the counterclockwise rolling current strengthens. It can be said that fog, as a phenomenon of the atmosphere, is an event of transitional changes

334 ■ *Systemic Yoyos: Some Impacts of the Second Dimension*

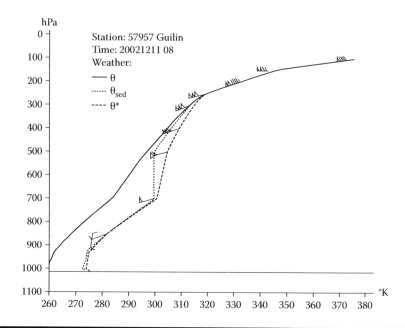

Figure 18.15 The V-3θ graph at the eighth hour on December 11, 2002, Guilin.

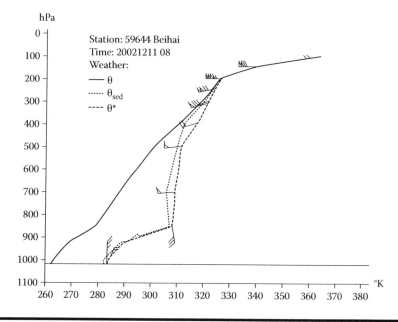

Figure 18.16 The V-3θ graph at the eighth hour on December 11, 2002, Beihai.

Case Studies Using V-3θ Graphs ■ 335

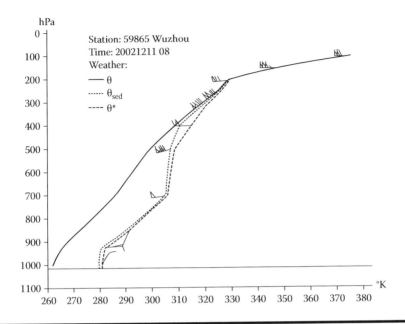

Figure 18.17 The V-3θ graph at the eighth hour on December 11, 2002, Wuzhou.

in the atmosphere. Hence, as long as we can grasp transitional changes appearing in the atmosphere, we will be able to resolve the problem of predicting fog. At the end, weather is a problem of evolution, where the fog phenomenon is both a content of prediction and a problem of information closely related to the prediction of atmospheric evolution.

18.2.2.5 Prediction of No-Fog Weather or Dissipation of Fog

During this large-area indigested fog weather process, Xichang, Yibing, Kunming, Lijiang, and other stations suffered from no fog or dissipating light fog. So, fog also stands for a problem of unevenly distributed evolution, where under the thick fog coverage, there were still spots that were missed. For the **V-3θ** graphs of Xichang and Yibing (Figures 18.18 and 18.19), the structural characteristics are that the bottom layer did not contain sufficient water vapor; however, because θ_{sed} was located between θ and θ^* and more toward the right, there was an adequate amount of water vapor. So, a light fog should have appeared the next day. There were south winds in the bottom atmospheric layer, making the entire atmospheric layer a weak clockwise rolling current with relatively strong northwest winds without any ultra-low temperature. The three θ curves tilted to the right, showing a stable atmospheric structure. The Kunming and Lijiang stations

336 ◼ *Systemic Yoyos: Some Impacts of the Second Dimension*

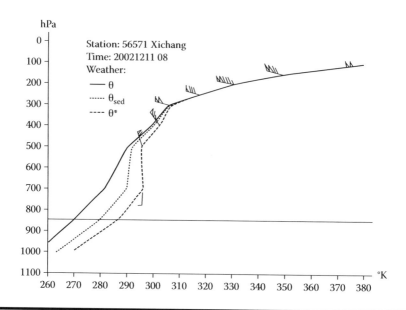

Figure 18.18 The V-3θ graph at the eighth hour on December 11, 2002, Xichang.

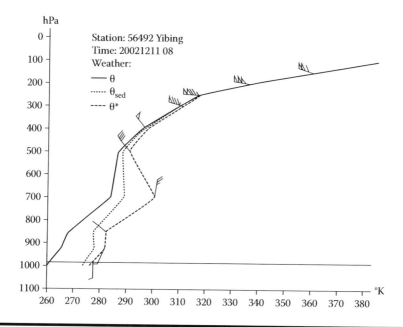

Figure 18.19 The V-3θ graph at the eighth hour on December 11, 2002, Yibing.

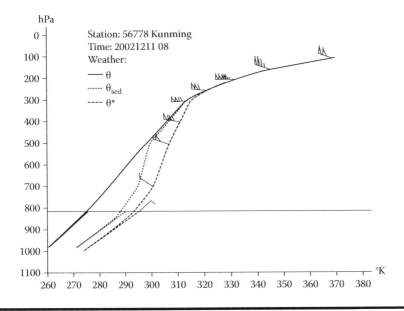

Figure 18.20 The V-3θ graph at the eighth hour on December 11, 2002, Kunming.

(Figures 18.20 and 18.21) both had a counterclockwise rolling current made up of partially north winds. The three θ curves also leaned to the right, indicating a stable atmosphere with little water vapor (so Kunming should have experienced a light fog) or no water vapor at all (because for Lijiang, θ_{sed} was located to the left between θ and θ*, no fog would appear).

The **V-3θ** graphs' structural characteristics for no-fog weather or dissipating fog are the following: There is a counterclockwise rolling current with a stable structure and insufficient water vapor. And under the conditions of stable structure and insufficient water vapor, even the structure of a clockwise rolling current will not create fog, and even when a counterclockwise rolling current is changed to the structure of a clockwise current, no fog will be produced.

As for the prediction of magnitudes or categories of heavy fog at the locations of radiosonde stations, one should pay attention to the following in terms of the structure of **V-3θ** graphs:

1. The strength and thickness of the atmospheric counterclockwise rolling current.
2. The thickness of the bottom atmospheric layer with abundant water vapor, and how abundant the water vapor is. For heavy fog weather, the thickness can range from 500 m to 1 km or more.
3. From the mid to low levels of the atmosphere, the structure is stable.

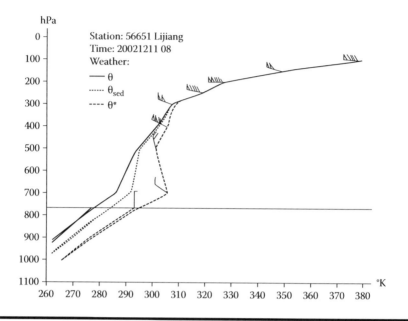

Figure 18.21 The V-3θ graph at the eighth hour on December 11, 2002, Lijiang.

Here, light fog might satisfy one or two of these conditions, with the thickness of the water vapor layer being about 500 m beneath 1 km. What are common for both light and heavy fogs is that there exists no ultra-low temperature or only a weak ultra-low temperature, and that the wind speed near the ground level is less than 4 m/s.

18.2.3 Discussion

As a weather phenomenon, fog stands for an evolution problem of eddy currents and appears in transitional changes of weather. Even for a same weather process, fog is unevenly distributed. By only relying on the static, microscopic physics concepts or the method of extrapolation, one cannot fathom the formation mechanism and produce desirable predictions of fog and other disastrous weather conditions.

The reason for us to introduce the point of view of structures is that there are definite causes for the formation of fog, and the information involved with fog is deterministic. That is why in designing the operational method, we have to keep the originality of information as much as possible. Also, in terms of the direction of rotation, it is, of course, a problem of structure, and it can appear before the underlying event so that valuable time is gained for making true predictions about the forthcoming appearance of the event. In the method of structural prediction, an emphasis is placed on the variable: wind direction. In particular, the introduction and application of the direction of rolling currents is new and not mentioned in the traditional methods (OuYang, 1998; OuYang et al., 2002a; Lin 1998). It turns out that this new concept is very effective in applications. That exactly verifies the fact that changes in weather represent a problem of transformation of eddy currents, while the traditional propagation of wave motions stands for only the problem of extrapolation with invariant materials' properties and states.

In terms of forecasting the magnitudes and properties of fog for specific observation stations, the structures of the **V-3θ** graphs provide a much finer method of analysis than weather maps and the significant weather factors used in numerical forecasting. These structures employ detailed information on the significant spheres instead of the interpolated values of grid-mesh, with the original information invalidated. That explains why this new method has greatly improved the timeliness and accuracy of weather forecasting. There is no doubt that how to comprehend irregular or peculiar information involves a change in basic cognitive concepts. However, even so, there is no need to continue an endless debate, because what is important is which comprehension and method can solve urgent problems in life and stand the test of history and time. To this end, what is presented in this section has been tested with over 100 real-life cases.

What needs to be pointed out is that the location prediction, as discussed in this section, is only limited to radiosonde station points. Because the layout of radiosonde stations is done according to the traditional concept of large scales, most commercial and military airports are not located at the points of the stations. So, to provide the needed service to these airports, we will have to employ OuYang's third-defense plan, which is the method of how to deal with irregular information by combining sounding data with that automatically recorded (OuYang et al., 2005). All the details are omitted here.

As an epistemological problem, it can be said that to truly understand the theory of blown-ups and the method of structural analysis, one must recognize that "from quantitative changes to a qualitative change" is not materialistic. That is why quantities are formal postevent measurements, events are not quantities, and events do not follow the formal logical calculus of quantities. These understandings should constitute major breakthroughs of our modern time.

18.3 Windstorms and Sandstorms

On the basis of the prediction method of the **V-3θ** graphic structures of the blown-up theory and the discussions in previous chapters, in this section, which is mainly from Du et al. (in press), we sort through our predictions of the four sandstorm weather processes of 2002 and 2003 and obtain the mechanic characteristics of the structures prior to the occurrence of strong windstorms and sandstorms. The presentation below shows not only that the **V-3θ** graphs can well predict the locations and severities of forthcoming major sandstorms ahead of time with a high degree of accuracy, but also that weather problems are about structural evolutions.

18.3.1 Background Information

Science stands for a process of exploration. And in this process, it often meets with challenges of various problems. In terms of weather problems, science is faced with the challenge of how to resolve transitional changes, where disastrous weather mainly appears in the process of transition. Disastrous weather is an irregular, small-probability event, which cannot be understood by using the quantitative method of regularization or large probablization. So, transitional changes have always been a difficult problem in the area of weather forecasting. The traditional quantitative analysis is only employed to describe generality. However, the generality of events is not the same as the quantitative averages. Specifics are also those of events and do not follow the formal logical calculus of quantities. But they can be reflected in the characteristics of structures. Because over 86 percent of our information comes from structural images and figures, one has to deal with the problem of how to comprehend and handle the information of peculiarity. By using quantities to substitute for events, modern science established the formal logical calculus of quantities. However, OuYang believed that events are not quantities and are not equal to quantities (OuYang, in press c; OuYang et al., 2002a), leading to a system of concepts and methods completely different from those currently employed in the profession of meteorology. That is how he introduced the method of structural analysis. Because in the past 300 years, people have used quantitative analysis, the originally very intuitive and plain concepts, are now messed up. In their originality, the structural characteristics of events provide people with an impression and information about the events. And events do not follow the formal logical calculus of quantities, which appear after the events. However, since the start of modern science, people just prefer to use quantities to replace events, pushing the actual events to the world of nihilation. In fact, it should be common knowledge that quantities are postevent formal measurements.

No matter which side of the debate one is on, at least most people admit that no matter whether it is dynamic equations or statistical methods that are employed, irregular, small-probability events have not been handled successfully. Evidently, due to the helplessness in solving practical problems, there is no harm for us to try the available nonquantitative methods to deal with irregular or small-probability events. The V-3θ graphs are one of such available methods, where V stands for the direction of the wind vector (here direction is not a quantity). With different wind directions, the corresponding weather situations will be different. Because wind direction appears before the corresponding weather phenomenon, the structural characteristics of wind can be employed to foretell the degree of severity of forthcoming disasters. That is how the analysis method of the inversed order structure of wind direction, wind speed, humidity, temperature, and barometric pressure is established, where order structure is also a concept of organization. What is important in the structure of any V-3θ graph is the specific information of the significant spheres. And in the construction of a V-3θ graph, more than three times the amount of information is used than in the construction of traditional weather maps. The more irregular information there is, the better effect the consequent **V**-3θ graph produces. Second, structural characteristics can handle quantitative instability or quantitative ∞ (OuYang et al., 2002a,b). This end is exactly one of the problems needing urgent solutions for the purpose of predicting disastrous weather systems.

By the way, quantitative ∞ is a problem that mathematics cannot resolve satisfactorily. However, by employing the method of structures, one can directly spot whether the slope in the phase space is vertical or not, to easily avoid the trouble of quantitative ∞. So, structural irregularities in **V**-3θ graphs should not only be understood as the degree of stability of the quantification in traditional meteorological science, but instead, these irregularities show that the structural characteristics of informational symbols are not quantities. In the calculation of the amount of water vapor, what is applied is the dew point temperature instead of the temperature at the condensation altitude. That is why $θ_{sed}$ instead of $θ_{sc}$ is used. The purpose for this is to reveal the structural characteristics of the water vapor in the atmosphere before its condensation, constituting a piece of information available before the appearance of the event one wants to predict. Hence, prediction and durable product design are different, where prediction has to be done by relying on information available ahead of the event to be predicted, instead of that which becomes available after the event. The **V**-3θ graphs are ominous information. They are prediction figures of evolution, characterized with wind direction and informational order structure, where no quantity can be used to modify the available information and no postevent information is employed. So, it is a misunderstanding if the structural characteristics of irregular information are seen as a degree of the traditional quantitative stability. The traditional degree of stability is a problem of quantitative stability in the regularized quantity system, where no direction is introduced and no transitionality can be given using quantities. For detailed discussions, consult OuYang et al. (2002b), OuYang (1998), and Lin (1998).

Modern science also recognizes that quantities cannot resolve the problem of transitions. It is exactly because the **V**-3θ graphs can handle the transitionalities in evolutional information that, in this section, the prediction method of the **V**-3θ structural characteristics is employed to analyze the prediction of large wind- and sandstorm weather. What is similar to other disastrous weather is that sandstorm systems also appear in transitional changes of the atmosphere. So, predicting sandstorm systems is also one of the difficult problems faced by the method of extrapolation. In current times, due to large-scale damages of vegetation coverage, huge windstorms and sandstorms have come to be seen as relatively disastrous weather in northern China. Their impacts have gone beyond the Chinese border.

In this section, we use the historical data of nearly 10 years and the method of structures to systematically analyze windstorms and sandstorms. In particular, we use the real-life sandstorm processes of 2002 and 2003 to specifically illustrate how to practically predict such processes. What is shown clearly is that other than having the characteristics of strong convective weather, major wind- and sandstorms also possess some characteristics that are different from those structures of strong convective rainfall. What is important is that in the spring season, the atmospheric structure has to be extremely strongly irregular under the influence of ultra-low temperatures. The structural characteristics of irregularity must be far beyond those of convective weather in the summer season. This end is a new discovery Du et al. (in press) made from using the **V**-3θ graphic structures beyond what has been known from weather maps (OuYang, in press c). By employing the **V**-3θ graphs, one can predict the location and severity of a windstorm/sandstorm of any chosen area with the needed lead time, a high degree of accuracy, and practically beneficial consequences.

18.3.2 Practical Applications

18.3.2.1 On Practical Applications of V-3θ Graphs

In applications, irregular information exists in the observed data of the significant spheres. This information follows the determinacy of events, cannot be quantitatively modified, and satisfies the condition of discontinuity of wind direction. Therefore, this information can objectively represent transformations of eddy currents under eddy effects. At the same time, one can predict the degree of disasters based on the irregular structural characteristics of structure information and its accuracy. Like structural irregularity, the structures of the curves in **V**-3θ graphs are analogous to the slopes in the two-dimensionoal phase structures and cannot be represented quantitatively, because when a slope is quasi-perpendicular to the T-axis, it will be a quantitative quasi-infinity ∞, an impossible problem to deal with by using quantities. However, such a quantitative inconvenience brings forward a structural advantage. When it is combined with the direction of an existing rolling current, one will be able to predict transitional changes. By comparing weather maps with the **V**-3θ graphs of the corresponding time moments, it is found that the **V**-3θ graphs can show transitionality ahead of the weather maps. This gained lead time is also ahead of the actual appearance of the corresponding weather phenomenon. The weather systems shown on traditional weather maps tend to be after the actual events; by using the **V**-3θ graphs, one can avoid this problem. So, **V**-3θ graphs cannot be understood as the degree of traditional stability, because in the creation of these graphs, ominous information on transitions, which are not recognized and employed by traditional meteorological science, are introduced. For more detailed explanations, consult OuYang (1998), OuYang et al. (2002), and Lin (1998).

18.3.2.2 Case Studies

To not make this section too long, in the following, we will only provide the real-life situation of the sandstorm weather process that occurred during March 18 to 21, 2002, to illustrate how to employ the **V**-3θ graphs to predict large windstorms and sandstorms.

18.3.2.2.1 The Path and Affected Areas of the Sandstorm

This weather process is the most severe sandstorm recorded in the past 10 years in China. It swept through a landmass of over 1,400,000 km², covering eight provinces, autonomous districts, and municipalities that are directly under the national government, in northern China. This event affected the lives of over 0.13 billion people. From the map of the regions affected by this sandstorm (Figure 18.22), it can be seen that starting on March 18, 2002, the sandstorm in our example began its attack on the area of Beijing and Tianjing along two paths—from north and from west. In the west, the sandstorm was originated in Aheqi, Bachu, and other locations in the southern basin of Xinjiang. And in the north, the sandstorm started in Zhurihe, Zhangwu, and the Duolun areas in Inner Mongolia. In China, after a sandstorm is formed, it blows to the Beijing and Tianjing areas along roughly three routes. From the north, the sandstorm starts in the areas of Erlianhaote, west of Hunshandeke sandy land, Zhurihe, and others, travels through Siwangziqi, Huade, Zhangjiako, and Xuanhua, and reaches Beijing and Tianjing. From the northwest, the sandstorm starts at the border area, Alashan between China and Mongolia, in Inner Mongolia, Wutela, Hexi Corridor, or others, travels through Wusu sandy land in Helanshan Mountain or Wulanbuhe desert, Huhhot, Datong, and Zhangjiako, and reaches Beijing and Tianjing. From the west, the sandstorm starts in Hami or

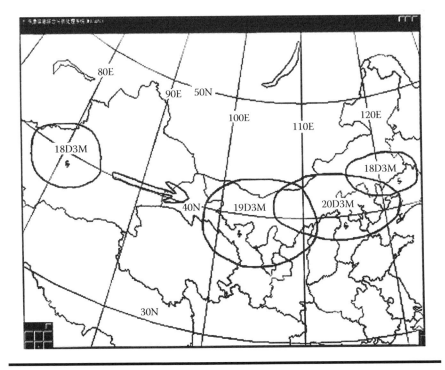

Figure 18.22 Regions affected by the sandstorm during March 18–20, 2002.

Mangya, travels through Hexi Corridor, Yinchuan or Xi'an, Zhangye, Mingqin, Taiyuan, and other areas, and reaches Beijing and Tianjing.

On March 19, a strong cold airstream started from Xinjiang and affected the west of China. Regions in Gansu, such as Wudi City on the west of the Yellow River, Zhangye, Lanzhou, etc., suffered from a sandstorm, which turned out to be the largest such event since 2002. From the 8th to the 17th hour on March 19, in Zhangye, Mangya, Lanzhou, and other areas, sand and dust weather appeared, with some regions hit by a sandstorm. Regions such as Jinta, Jinchang, and Wuwei experienced an extraordinarily severe sandstorm. The visibility in Jinta was once 0 m, while the visibility in Lanzhou was only 400 m. Starting at the 14th hour on the March 19, in the west of Hexi Corridor, Gansu, a severe snowfall accompanied with a sudden drop in temperature appeared. The drop in temperature at Dunhuang, Jinquan, Yumen, Zhangye, Shandan, and other areas reached 12°C to 19°C within 24 hours. On the same day, in the north of Ningxia, there also appeared a severe sandstorm covering a large area, including Yinchuan and Shizuishan City. Gradually and one by one, Shaanxi, Shanxi, and north of Hebei saw a large-scale severe sandstorm.

Starting the night of March 19, there appeared flying dust and sandstorm weather in Ejinaqi and Bayinmaodao of west Inner Mongolia, and Linhe and most part of Dongsheng of central Inner Mongolia. From the west to the east, the sand dust weather went through the autonomous district Alashanmeng in Inner Mongolia, north of Hetao plain, Erduosi City, Baotou City, north of Wulanchabumeng, and west and south of Xilingelemeng. Eventually, Huhhot City also saw the sand dust weather.

On March 20, the two sandstorms from the north and the west met in the mideast part of Inner Mongolia, which affected eastern and central Inner Mongolia, Shanxi, north Shaanxi, and most of Hebei. Then, the joint sandstorm approached the region of Beijing and Tianjing. At the ninth hour on March 20, sand dust weather appeared in Beijing. At the 10th hour, the visibility dropped down to less than 1 km and the weather turned to that of a sandstorm. At the 11th hour, the visibility dropped to less than 500 m and the weather became a strong sandstorm.

On March 21, the sandstorm that had sustained for two days started to gradually weaken after a further brief continuation at several regions, such as east Inner Mongolia, areas near Beijing and Tianjing, and south Xinjiang basin. The entire wind- and sandstorm weather began to end.

This process of sandstorm was quite strong. Its effect was also felt in Korea and Japan.

18.3.2.2.2 Characteristic Analysis of the V-3θ Graphic Structures for Wind- and Sandstorms

From the V-3θ graph information of all the observation stations located in the weather process during March 17 to 20, it can be seen that the characteristics of all these V-3θ graphic structures were similar, so that they pointed out the same underlying mechanism. In the following, we will use Ruoqiang, Jinzhou, Yinchuan, Pingliang,

Huhhot, Linhe, Xi'an, and Beijing as examples to demonstrate the structural characteristics of the **V**-3θ graphs prior to the appearance of a wind- and sandstorm.

18.3.2.2.2.1 Ruoqiang Sandstorm and Jinzhou Flying Dust Weather—Figure 18.23 is the **V**-3θ graph of Ruoqiang at the eighth hour on March 17, 2002. Its structural characteristics were:

1. The θ* and θ$_{sed}$ curves almost coincided and were very steep, almost vertical, from 870 to 300 hPa. The structure of the water vapor layer was not stable and was far beyond the instability of such convective weather as hailstones, tornadoes, etc. So, such strong winds as those that can lift sand and move gravel would have to be created. Therefore, the mechanism of the strong winds accompanying each sandstorm is revealed. The θ* curve above 870 hPa formed an obtuse angle with the T-axis, representing an unstable structure. And when combined with θ$_{sed}$ − θ ≈ 0, it showed the reality that the vegetation coverage of the region was nearly nonexistent. Only at the level near the ground surface was there a relatively stable layer (that is, the θ, θ$_{sed}$, and θ* curves formed acute angles with the T-axis). However, near-ground-level stable layer would disappear after the sun rose and the ground temperature increased.
2. There existed an ultra-low temperature at the upper layer at 300 hPa of the troposphere, where a sharp turn was shown.

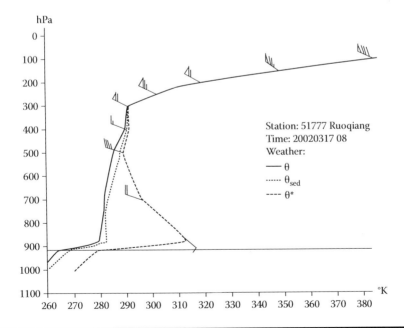

Figure 18.23 Ruoqiang at the eighth hour of March 17, 2002.

3. From 700 hPa up, there was a northwest wind. Combined with the southeast wind at 850 hPa, an overall clockwise rolling current was formed.
4. The θ and θ_{sed} curves were very close to each other, indicating that at this moment, in Ruoqiang, there was very little amount of water vapor in the atmosphere. So, the unstable energy could not be released through the form of precipitation, and had to be released through the form of strong winds.

With such an overall structure in place, because Ruoqiang is located on the southeast edge of Taklimakan desert, a sufficient source of sand dust was provided for Ruoqiang. Also, because there existed unstable structures at the surrounding stations, such as Mangya, Lenghu, a supportive environment was formed for a sandstorm in Ruoqiang to appear. So, 24 hours later, sandstorm weather broke out in Ruoqiang.

Figure 18.24 is the **V**-3θ graph in Jinzhou at the eighth hour on March 17. It structural characteristics were:

1. Near the ground level, all three θ curves leaned to the right, representing a relatively stable layer. However, after the ground temperature increased, this stable layer was broken. Between 980 and 700 hPa, the θ curve was almost perpendicular to the T-axis, forming an angle greater than 80°, indicting that there existed unstable energy.

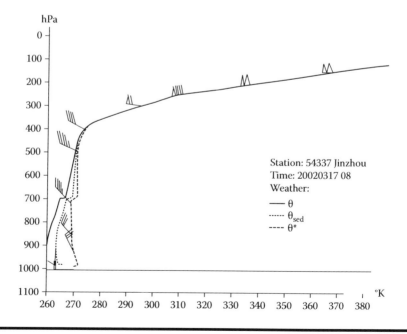

Figure 18.24   Jinzhou at the eighth hour of March 17, 2002.

2. The θ_{sed} curve was positioned in the middle, with slight leaning toward the left, meaning that there existed water vapor in the atmosphere, but in a small amount.
3. Both the θ and θ_{sed} curves formed obtuse angles with the T-axis, showing that the structure of the atmospheric layer below 700 hPa in Jinzhou was very unstable. So, there could be a potential clockwise rolling current structure.

Because the existing ultra-low temperature above Jinzhou was weak and there existed a layer of strati at 700 hPa, even though there existed some water vapor in the atmosphere, it was relatively dry, and in the surrounding area, there did not exist similar structures to support the wind field in Jinzhou; 24 hours later in Jinzhou area, there appeared flying dust weather caused by strong winds, without any sandstorm.

Jinzhou and Ruoqiang are located in one of the two origins of this large-scale sandstorm, respectively. On March 18, other than two stations, in Mangya, Chifeng, and other areas, there also appeared similar structures. And 24 hours later, similar weather phenomena appeared in these places, too.

18.3.2.2.2.2 The Sandstorm Weather in Huhhot and Linhe

Figure 18.25 is the **V-3θ** graph of Huhhot at the 20th hour on March 18, 2002. Its structural characteristics were:

1. For below 700 hPa, the θ curve and the T-axis formed nearly a right angle, indicating the characteristic of being ready for a transitional change.
2. The θ and θ_{sed} curves were very close to each other, showing an extremely low amount of water vapor at the region.
3. At the 300 hPa level, the upper layer of the troposphere, there existed a shape turn, meaning the existence of an ultra-low temperature.
4. The southwest winds near the ground level, together with the northwest winds above 700 hPa, constituted an overall clockwise rolling current structure.

Other than the extremely unstable structure at Huhhot, there were also similar structures at the upper stream stations, such as Zhurihe, Erlianhote, etc. When the strong winds passed the sandy land at Hunshandake, they brought with them a large amount of sand dust. Twenty-four hours later, this area started to experience sandstorm caused by the strong winds. Figure 18.26 is the **V-3θ** graph of Linhe at the 20th hour on March 18, 2002, where there existed two layers of ultra-low temperature. At 300 hPa, there was a turning ultra-low temperature; at about 280 hPa, there was another one. These two layers of ultra-low temperature strengthened the irregularity of the overall structure of the atmosphere, which helped strong convective weather to develop. With this structure in place, 18 hours later, a sandstorm, caused by the strong winds, appeared in this region.

348 ■ *Systemic Yoyos: Some Impacts of the Second Dimension*

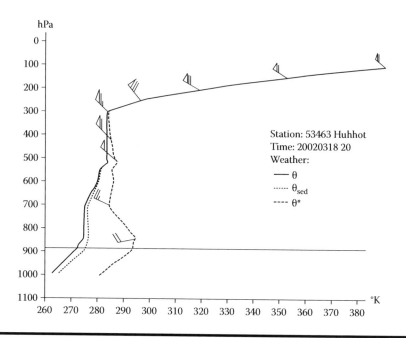

Figure 18.25 Huhhot at the 20th hour of March 18, 2002.

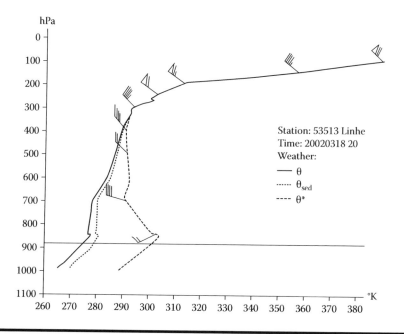

Figure 18.26 Linhe at the 20th hour of March 18, 2002.

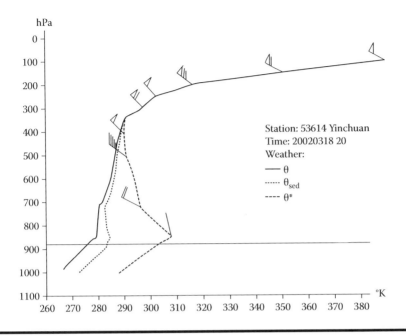

Figure 18.27 Yinchuan at the 20th hour of March 18, 2002.

18.3.2.2.2.3 The Sandstorm Weather in Yinchuan and Pingliang

On the 19th day, other than Huhhot and Linhe, in other locations, including Dongsheng, Zhurihe, Zhangye, Erlianhote, Lanzhou, etc., there also appeared similar unstable structures on their respective **V-3θ** graphs. With the desert sitting right behind these places as their source of sand dust, sandstorm weather with disastrous consequences developed. Figures 18.27 and 18.28 are the **V-3θ** graphs of Yinchuan and Pingliang, respectively, at the 20th hour on March 18, 2002. In comparison, their structures had a lot of similarities. For example:

1. At 700 hPa, Yinchuan had a weak ultra-low temperature. At 600 hPa, Pingliang had a layer of strati. With the inversed temperature or the layer of strati as boundaries, respectively, at these two places, at the lower altitudes, the θ curve and the T-axis formed an acute angle of >85°, indicating a piece of very obvious irregular information.
2. At 300 hPa, the upper layer of the troposphere, both places had ultra-low temperatures of the bending form.
3. At Pingliang, the low-level southwest winds and the high-altitude northwest winds formed a clockwise rolling current, while at Yinchuan, all three θ curves leaned left, indicating a potentially existing clockwise rolling current structure.

350 ■ *Systemic Yoyos: Some Impacts of the Second Dimension*

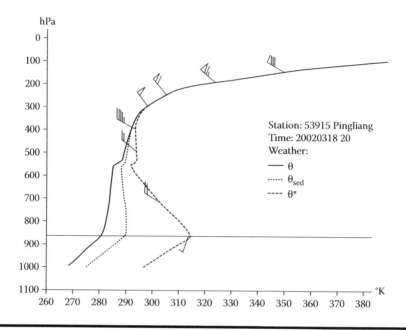

Figure 18.28 Pingliang at the 20th hour of March 18, 2002.

4. For both of these places, the θ_{sed} curve was located slightly to the left of the middle between the θ and θ^* curves, showing that there was a small amount of water vapor in the atmosphere, which was relatively dry.

The actual happening indicated that 24 hours later, Yinchuan experienced a sandstorm caused by strong winds, while Pingliang saw flying sand, also caused by strong winds. The reason why these places had different kinds of weather was because Yinchuan is located at the southeast side of Tenggeli desert, and Bayinmaodao and Zhizuishan are located on the upper stream; one after another experienced a strong wind- and sandstorm that provided the source of sand dust for Yinchuan, located on the lower stream. This explains why there appeared a sandstorm in Yinchuan. Because its atmosphere contained a small amount of water vapor, the sandstorm only lasted a relatively short time. For Pingliang, not only were its surrounding observation stations not able to provide it with a sufficient source of sand dust, but also there was a small amount of water vapor in the atmosphere. So, the degree of the sand dust disaster was relatively higher.

18.3.2.2.2.4 Beijing Sandstorm and Xi'an Flying Dust Weather

Figure 18.29 is the **V**-3θ graph of Beijing at the 20th hour on March 19, 2002. Its structural characteristics were:

1. Beneath 900 hPa, the three θ curves are obviously leaning to the left, indicating that the bottom layer is extremely unstable. However, above 900 hPa, the angle between the θ curve and the T-axis is slightly less than but near 70°. So, the amount of unstable energy at higher altitude is slightly weakened. What is shown is that the sandstorm in Beijing would not be deep and thick.
2. For the entire layer, both of the θ and θ_{sed} curves were very close to each other, so that the atmosphere contained very little water vapor.
3. At 200 hPa, there was an ultra-low temperature of the bending form.
4. The southeast winds at the ground level and the southwest winds at 850 hPa, together with the northwest winds at higher altitudes, formed an overall clockwise rolling current structure.

At the same time, Huhhot area was experiencing a sandstorm, Zhangjiako had flying dust caused by strong winds, and all other observation stations northwest of Beijing suffered from sandstorm weather. Because at the bottom layer of all these stations there were northwest winds of extremely high speeds, sufficient sand dust was supplied for Beijing, so that starting at the ninth hour on March 20, Beijing was also affected by sand dust. At 10 o'clock in the morning, the visibility dropped to below 1 km; a sandstorm was formed. At 11 o'clock in the morning, the visibility fell further, to below 500 m. The weather condition became a severe sandstorm. However, this condition did not last long.

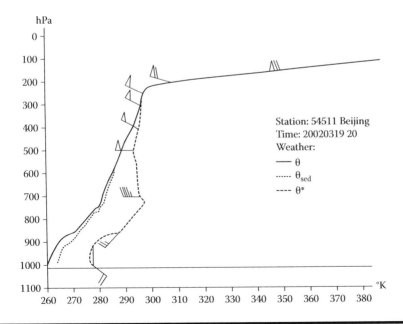

Figure 18.29 Beijing at the 20th hour of March 19, 2002.

352 ■ *Systemic Yoyos: Some Impacts of the Second Dimension*

Figure 18.30 is the **V-3θ** graph of Xi'an at the 20th hour on March 19, 2002. Its structural characteristics were:

1. Above 850 hPa, the θ curve and the T-axis formed an > 80° angle and the θ$_{sed}$ and θ* curves formed obtuse angles with the T-axis, showing that there was a relatively strong irregular structure. At the same time, beneath 850 hPa, all three θ curves leaned to the right, showing a stable state.
2. The southeast winds at the ground level and the southwest winds at 850 and 700 hPa transported some water vapor. So, the θ$_{sed}$ curve was located slightly to the left of the middle, between the θ and θ* curves, showing that there was some water vapor in the atmosphere. But the overall condition was relatively dry.
3. There was an ultra-low temperature of the strong bending form at 200 hPa, the upper layer of the troposphere.
4. The northwest winds at 500 hPa and above, together with the southwest winds at 700 and 850 hPa, formed a clockwise rolling current. There existed a layer of counterclockwise rolling current below 850 hPa. However, the overall structure showed an irregularity.

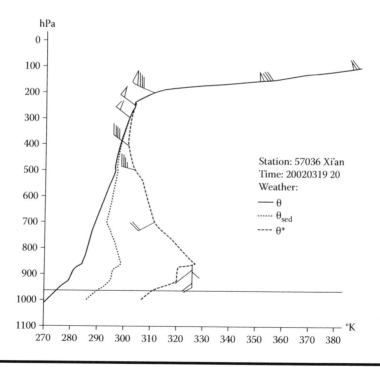

Figure 18.30 Xi'an at the 20th hour of March 19, 2002.

Even though the overall structure in Xi'an was irregular, a strong wind was created. However, because the condition of vegetation coverage in Xi'an was relatively good, and the bottom layer contained a certain amount of humidity with a relatively regular structure, during the daytime of March 20, Xi'an only experienced flying dust, caused by strong winds. This case study shows that the appearance of a sandstorm is not only a problem of the atmospheric structure, but also dependent on the vegetation coverage of the area and other conditions.

After March 20, the sandstorm paused temporarily in Beijing and Tianjing. Then, it went south and affected some areas of Henan, Shandong, and Hebei. After that, the storm quickly weakened. After settling in Qinghai, Gansu, Hunshandake desert, and the surrounding areas of south Xinjiang basin, the sandstorm also quickly weakened. After 8 a.m. on March 21, 2002, this sandstorm weather process began to round up.

18.3.3 Discussion

Before the occurrence of strong wind- and sandstorm weather, the structural characteristics of the **V-3θ** graphs are the following:

1. The θ and $θ_{sed}$ curves are very close to each other, indicating that the area's atmosphere contains very little water vapor.
2. The angle between the θ curve and the T-axis is greater than 75° and can reach almost 90° in some specific areas. That is, in terms of structures, strong windstorms and sandstorms are characterized by strong irregularities in atmospheric layers, including the quasi-vertical structures of transitional information. That is the structural mechanism of these weather conditions.
3. At the upper layer of the troposphere, there exists an ultra-low temperature in either the folding form or the bending form. In some incidences, there can exist two layers of ultra-low temperatures at the same time.
4. There is a clockwise rolling current structure made up of lower-level southwest winds and upper-level northwest winds. Or, due to the instability of the overall structure of the atmosphere, there is a potential clockwise rolling current structure.

When these four conditions are satisfied, there will appear strong wind weather. If at the area of the strong winds there exists a source of sand dust, strong wind- and sandstorm weather will occur. If one of the previous conditions is relatively weak, for example, the atmosphere contains a small amount of water vapor, or the ultra-low temperature is relatively weak, or the sand dust supply is not sufficient, or the vegetation coverage of the bottom cushion is good, etc., the severity of the resultant sandstorm weather will be weakened. Instead, flying dust weather caused by strong winds will develop, or the sandstorm will only be sustained for a relatively short period. If one of the conditions does not exist at all, for example, the atmosphere contains sufficient water vapor, or the layer structures are stable, or it is a counterclockwise structure, or there does not exist any ultra-low temperature, or there does not exist any source of sand dust, etc., there will not be any strong wind- and sandstorm weather or flying dust weather.

Based on our analysis above, it can be seen that by employing structural characteristics of events, the prediction of strong wind- and sandstorm weather becomes a pretty simple matter. And one can predict the severity of the future weather condition about 24 hours ahead of the traditional method of extrapolation.

18.4 Abnormally High Temperatures

High-temperature weather, as a problem of meteorological forecasting, did not attract nearly as much attention as the prediction of precipitation, due to its slow and gradual changes. In fact, not only does air temperature directly affect the condition of the objective environment in which we live our day-to-day lives, but also, peculiar or abnormally high-temperatures are disastrous and jeopardize the existence of humankind. Air temperature, especially abnormally high-temperature weather, also involves such important problems as how to control and allocate energy and how to make use of efficient facilities of systems engineering.

What needs to be clear is that in essence, changes in air temperature can also be sudden, although to a certain degree, the method of extrapolation can be employed. However, to truly forecast air temperature with efficiency is not an easy task. Through an analysis of over 30 years' worth of historical data, we find that changes in air temperature have something to do with the evolution of weather, and that all important high-and low-temperature weather is connected with transitional changes in weather. These transitional changes are a difficulty confronting the traditional methods of weather forecasting. Although by using the method of structural prediction, we can deal with transitional changes in weather relatively well (OuYang, 1998; Lin, 1998), our prediction accuracy is not the same as us being able to effectively or substantially prevent disastrous or abnormal weather, or knowing how to essentially make use of or control the related sources of energy. In particular, recent development and expansion of urbanization have brought forward new problems. That is, for the air temperature to rise, there exists a natural overall reason, but human effects should not be excluded. Also, at such a time that meteorological science and technology work to comprehensively serve the development of economy, they are no longer limited to providing forecasts on light, heat, and water services for crops or the agricultural sector. Instead, they are involved in how to create and manipulate environmental conditions. Corresponding to this end, air temperature, especially abnormally high-temperature weather, has become a problem of people's attention. In particular, to a certain degree, sustained high temperatures or abnormally high temperatures can be listed in the category of disastrous weather.

The Chinese Bureau of Meteorology regulated that if the daily high temperature is greater than or equal to 35°C for three or more consecutive days, the weather system is classified as high temperature (39°C or higher is classified as abnormally high temperature). High-temperature weather leads to increased electricity burdens for metropolitan areas, has severely affected the development of economy, and has also caused inconveniences to all people involved. In July 2003, there appeared such a tragedy in Nanjing, where some human lives were lost due to two consecutive hot days.

During July and August 2003, China experienced rarely seen, large-area, long-lasting, high-temperature weather. Fuzhou, Hongzhou, Shanghai, Nanchang, Chongqing, and other cities had seen abnormally high temperatures ranging from 40°C to 42°C. Some local highs reached 43°C. The middle and lower reaches of the Yangtze River and most of the South China area, which were under the control of a subtropical high pressure, continuously reached their new regional 30-year, 60-year, and even 100-year highs. At the same time, when we forecasted abnormally high-temperature weather by employing the method of structural prediction, we discovered the peculiar structures and characteristics of change in the atmospheric layers over major metropolitans located in inland China. So, our work directly touches on the problem of urban planning and construction. That is, our work also deals with the relationship between the layout of urban structures and high-temperature weather. Currently, such a problem has also spread to coastal cities in eastern China. This section is mainly based on OuYang et al. (in press).

18.4.1 Background Information

The theory of blown-ups is established for the study of the materials' evolution process of injuries, ages, and breakups. It belongs to evolution science, where breakups or transitional processes are the collapse—blown-up—of the initial value stability, and are also called transitional changes. According to the principle of evolutional blown-ups, weather problems are about variant fragile products instead of the eternality of durable products (OuYang et al., 2002a, b). To this end, there should be corresponding theories and methods for dealing with the fragile products. The materials' process of aging and death (transitionality) is prominently shown by changes in structures. Abnormally high-temperature weather among disastrous weather conditions is also embodied in changes of structures of the atmosphere. Through our analysis of historical data, we discover the near-ground low-altitude arid atmospheric layer, which was not mentioned in traditional theories. Initially, we only employed this arid layer in the low altitude to forecast abnormally high-temperature weather, combined with counterclockwise rolling currents in the atmospheric structure. However, through comparative studies of the historical data, we found that this arid layer near ground level is in fact a structural characteristic of urban high-temperature weather. So, the phenomenon of this arid atmospheric layer is a piece of indicative information for forecasting purposes as well as a materialization of the "hot island" effects of metropolitan areas. It has touched on the problem of distribution and arrangement of urban planning and construction.

What needs to be emphasized is that although the structural method has been systematically introduced in (Lin, 1998; OuYang, 1998; OuYang et al., 2002a), in terms of the analysis of abnormally high-temperature weather, it is still very sketchy on how to apply this method. So, in this section, we will explain how to use this method by looking at specific cases, and our presentation uncovers such a new significance and function of rolling currents that the human factor and human impacts on the environment also affect the appearance of abnormally high temperatures.

18.4.2 High-Temperature Weather under a Subtropical High of the West Pacific

Our focus in this section is on the structural analysis of disastrous (abnormally) high-temperature weather instead of the cause and forecast of the high temperature under subtropical high pressures. However, subtropical high-pressure systems, to a great degree, determine the disastrous weather in East Asia. So, the forecast for the advance, retreat, and maintenance of these high-pressure systems is an important problem of weather forecasting. In this section, we will illustrate the differences in the structures of high-temperature weather in either subtropical high-pressure systems or cold high-pressure systems of the temperate zone. And along with our analysis, we will resolve the problem of predicting these events.

18.4.2.1 Sustaining High Temperatures under Subtropical High Pressures

In this section, we apply the sustained high-temperature weather at Fuzhou station that occurred in the summer of 2003 and lasted for 37 days as our example of discussion. From June 28 to August 3, each day's high was above 35°C. From July 13 to 16, there appeared abnormally high-temperature weather where the temperature stayed above 40°C continuously. The corresponding V-3θ graphs show some clear structural characteristics, by which we can understand the systemic structural features for the appearance, maintenance, and round-off of high-temperature weather.

18.4.2.2 Maintenance and Reinforcement of Fuzhou's Abnormally High Temperature

On July 13, the day's high temperature in Fuzhou reached 40°C and above and stayed at that level for four days. The corresponding V-3θ graphs (Figures 18.31 to 18.34) are for July 13 to 16, when the highs reached 40°C, 40°C, 41°C, and 40°C, respectively. The basic structural characteristics of these V-3θ graphs are:

Case Studies Using V-3θ Graphs ■ 357

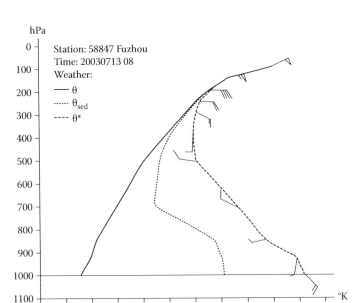

Figure 18.31 The V-3θ graph at the eighth hour on July 13.

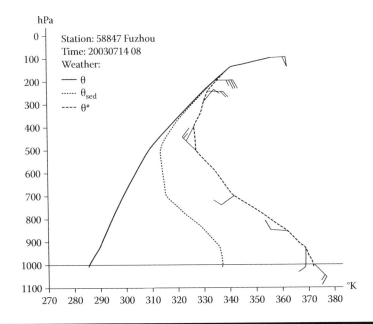

Figure 18.32 The V-3θ graph at the eighth hour on July 14.

358 ■ *Systemic Yoyos: Some Impacts of the Second Dimension*

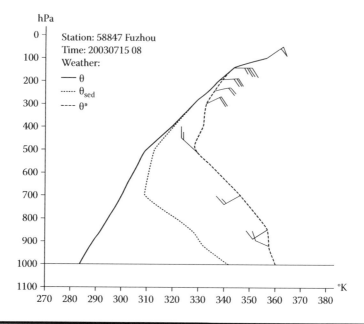

Figure 18.33 The V-3θ graph at the eighth hour on July 15.

1. The entire θ curve leaned to the right. Its angle formed with the T-axis was roughly ≥ and ≈ 45°. Especially for above 400 hPa, from the 14th day to the 16th day, the θ curve leaned further day by day to the right and approached 45°. Together with the nonexistent or weak ultra-low temperature at the top layer of the troposphere, the entire atmosphere became a stable structure.
2. At the bottom atmospheric layer, the southwest winds strengthened while the middle-level northwest winds weakened. At the high-altitude levels, there were east and northeast winds, which all turned to northeast winds on the 15th day. So, the high of that day reached 41°C. That is, the systemized northeast winds at high altitudes and southwest winds at the middle and low altitudes were the typical counterclockwise rolling current structure located to the west of the subtropical high pressure. (In this example, on the 13th and 14th days, the middle and low altitudes did not have uniform southwest winds, while starting on the 14th day, both 500 and 850 hPa turned to northward winds.) Combined with the arid air in the bottom atmospheric layers, the high temperature on the July 15 reached 41°C. During July 15 to 16, 850 to 700 hPa turned to southwest winds. Because a clockwise rolling current is formed, even though the bottom atmospheric layer was still relatively arid, the air temperature would start to drop.
3. At the middle and low altitudes, the θ* and θ_{sed} curves formed obtuse angles with the T-axis. When combined with the fact that the distance between

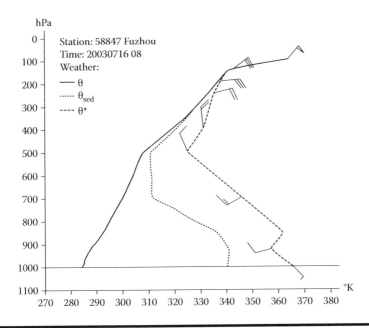

Figure 18.34 The V-3θ graph at the eighth hour on July 16.

these curves was greater than 20K, the atmosphere at the observation station contained a relatively dry and unstable structure at the middle and low altitudes. That is exactly the typical characteristic of a subtropical high pressure of suffocating hot, high temperature.

At this junction, we want to point out that in a sustained high-temperature process, predicting a sudden rise in the already high temperature is a difficult task, where the high-altitude atmosphere contains uniformly northeast winds and no ultra-low temperature. The structure is very stable. The middle- and low-level temperatures contain west or southwest winds, leaning toward dry (notice that the speed of southwest winds in a counterclockwise rolling current is smaller than the speed of southwest winds in a clockwise rolling current). The structure possesses the characteristics of quasi-stability. That is a key indicator for forecasting.

What needs to be noticed is that because Fuzhou is near the east coast of China, in its atmospheric structure of high-temperature weather, there does not exist any arid layer in the near-ground low-altitude atmosphere.

As for the forecasting of the arrival of high-temperature weather, under the condition of an existing clockwise rolling current, one should pay attention to the change of northwest or west winds in high altitudes to northeast winds. The V-3θ graphs of Fuzhou for June 25 to 26 (Figures 18.35 and 18.36), 2003, can clearly explain this end. In fact, from these graphs, it is not difficult to see that the

360 ◼ *Systemic Yoyos: Some Impacts of the Second Dimension*

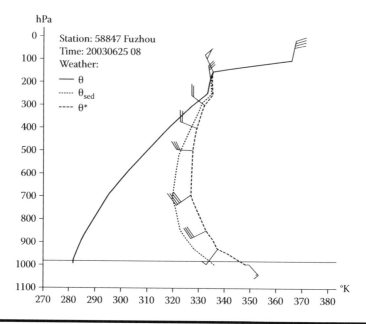

Figure 18.35 The V-3θ graph at the eighth hour on June 25, 2003.

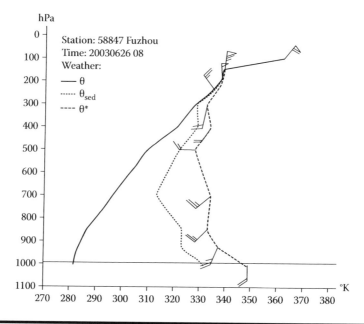

Figure 18.36 The V-3θ graph at the eighth hour on June 26, 2003.

northwest winds at the 200 hPa altitude began to turn to northeast winds. And the ultra-low temperature and the southwest winds were weakened accordingly. That indicated the start of high-temperature weather. To make sure that our forecast was correct, we should consider the atmospheric structures of the surrounding observation stations so that we would not be misguided by potentially erroneous local information. In particular, if at the surrounding stations the θ curve leaned to the right and formed a ≈ 45°C angle with the T-axis at above 300 hPa, then we could forecast that high-temperature weather was about to start.

As for forecasting the end of high-temperature weather, the key is the transformation of the counterclockwise rolling current at the high altitude into a clockwise rolling current. To limit the length of this section, all details are omitted.

In 2003 subtropical high pressures reached very deep inland in China. The ridgeline of a subtropical high appeared as far inland as the upper reaches of the Yellow River, which has rarely been seen in history. Accordingly, Nanchang, Fuzhou, Hongzhou, and other cities were located at the center or east part of the subtropical high. In particular, Hongzhou has a large lake, Xihu Lake, within the city and is very close to the ocean. Historically, it is not one of the five "ovens" (cities whose temperatures seem to be higher than those in other cities) of the Yangtze River valley. However, in recent years, high-temperature weather in Hongzhou has become very noticeable. And over Hongzhou we have seen the arid atmospheric layer near the ground at low altitude.

On August 1, 2003, the high temperatures at the stations of Nanchang, Wuhan, Hongzhou, and Nanjing (Jinagpu) reached 40°C, 40°C, 40°C, and 39°C, respectively. All these stations were located in a sustained high-temperature weather system. By carefully analyzing the **V**-3θ graphs' structures at the eighth hour of these observation stations (Figures 18.37 to 18.40), the same results as those of Fuzhou can be obtained. By comparing the **V**-3θ graphs of these four cities, it is not hard to see that Hongzhou and Nanchang were very similar. Their θ* curves reached over 290 K at 900 hPa, above those θ* curves of Wuhan and Nanjing. Generally speaking, any high-temperature region or city located in the Yangtze River valley, other than the common structural characteristics, has a **V**-3θ graph showing a θ* value of >290 K at 900h Pa and the phenomenon of relatively low humidity. These two signals can be employed to forecast abnormally high temperatures. In comparison, Nanchang is more inland than Hongzhou. And the lake surface area in Hongzhou is greater than that of Nanchang. Then, how could Hongzhou experience similar high-temperature weather as Nanchang? That is no doubt an important question. Although we talked about the problem of ultra-low temperatures in OuYang (1998) and Lin (1998), we did not develop the discussion in detail. Because θ* stands for the temperature with added saturated water vapor and condensation heat, accordingly we can add the same amount of water vapor. With the evaporation of the added water vapor, the temperature can decrease. Generally speaking, when θ* ≥ 280 K appears in the near-ground atmosphere, there always exists the problem of maximum holding capability of humidity in the environment of the arid

362 ■ *Systemic Yoyos: Some Impacts of the Second Dimension*

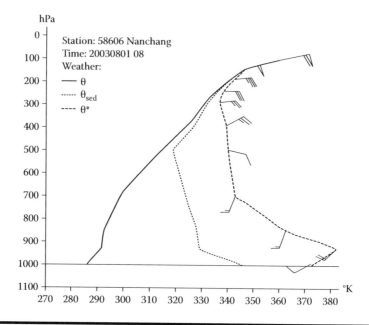

Figure 18.37 The V-3θ graph of Nanchang.

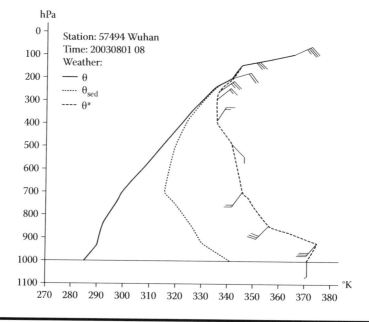

Figure 18.38 The V-3θ graph of Wuhan.

Case Studies Using V-3θ Graphs ■ 363

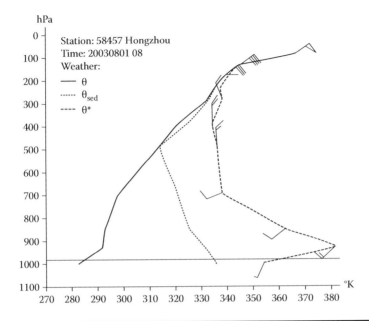

Figure 18.39 The V-3θ graph of Hongzhou.

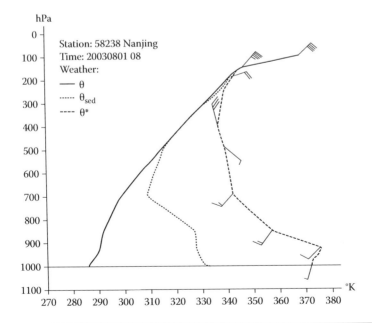

Figure 18.40 The V-3θ graph of Nanjing.

layer of the low-altitude atmosphere of the specific region. This layer will be called the maximum environmental humidity capability layer for short; see the bulging layer of the θ^* curve near 900 hPa in Figures 18.37 to 18.40. We introduce this concept based on our many years of analysis of historical data. The earliest case of this phenomenon appeared in a desert area in northwest China. Our purpose of introducing this concept is to show that abnormally high-temperature weather involves not only forecasting, but also reconsideration of urban planning and construction in inland areas as well as building structure arrangement. Traditional human interference on weather is mostly about preventing hailstones or pressing for rain, without paying much attention to how to adjust the temperature of the environment. Because the altitude of the near-ground arid layer is quite low (it is at about the top of the atmospheric boundary), humans should be able to interfere with this layer. (In comparison, reducing the air temperature by spraying water on the ground is not as efficient as spraying water over this arid layer.) Or at least we should not artificially make additional new oven cities, which should be an important problem for consideration in urban planning and construction in the future. In particular, corresponding to the current developments within cities and of satellite cities, there have appeared the chain effects of petrifaction of lights and concrete structures, which demands our understanding and new methods for handling it. What needs to be especially noticed is that the current metropolitan petrifaction of lights or concrete structures represents three-dimensional uneven effects, which are different from the quasi-level plane desertification of the northwest areas of China. These three-dimensional uneven petrifactions can more easily trigger the occurrence of suddenly appearing disastrous weather. If we compare Fuzhou City in Figures 18.31 to 18.34 with Hongzhou in Figure 18.39, we can see that both Fuzhou and Hongzhou are coastal cities; however, in Hongzhou, there even appeared an arid atmospheric layer near the ground at low altitude. This is a serious problem that deserves every bit of our attention.

18.4.3 High Temperatures under Cold High Pressures

The high-temperature weather under cold high pressures possesses structural differences from those under subtropical highs. The former weather systems are shorter, with greater variation in temperature, than the latter systems. When seen as disasters, the aftermath effects of the high temperatures under cold highs, which generally occur in the north, are much lighter than those of subtropical highs, which generally occur in the south. However, in terms of the arid atmospheric layer near the ground at low altitude, these two kinds of high temperatures are the same. In this section, we will use the high-temperature process during August 2 to 4, 2003, at Beijing station as an example to illustrate all the details. In this weather process, the respective daily highs reached 37°C, 36°C, and 36°C.

18.4.3.1 Arid Layers at Low Altitude of High Temperatures of Cold Highs

The structural characteristics of Beijing station's **V-3θ** graphs (Figures 18.41 and 18.42) can be summarized as follows (all high latitudinal observation stations have these characteristics):

1. From the middle level (500 to 400 hPa) to high levels in the atmosphere, the θ curve shows the typical stable structure, where the angle formed by the θ curve and the T-axis is roughly = or <45°.
2. This weather process contained a counterclockwise rolling current with southwest winds in the high altitudes and north or northwest winds in the low altitudes (cold highs could also have entire atmospheric counterclockwise rolling currents, made up of northwest winds). This atmospheric structure is different from that of the high temperatures under subtropical highs. What needs to be noticed is that for high-temperature weather, the wind speed at low altitudes is far slower than that of torrential rains or low-temperature weather. Even abnormally high-temperature weather at the center or east side of subtropical highs is with little clouds and light winds. However, in the near-ground atmospheric layer that is about 50 cm thick, vibrating chaotic airflows can be observed.

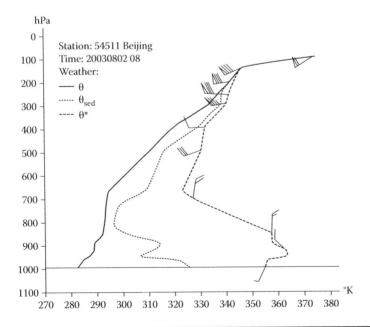

Figure 18.41 The V-3θ graph at the eighth hour on August 2.

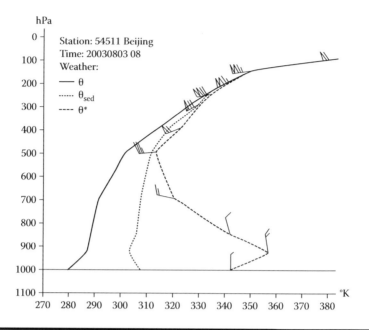

Figure 18.42 The V-3θ graph at the eighth hour on August 3.

3. The atmosphere below the middle level contains very low humidity and is dry and unstable (the θ_{sed} curve is located on the left side between the θ and θ^* curves). Near the bottom layer (about the altitude of 900 hPa), the θ^* curve shows an abnormally high temperature—the arid atmospheric layer near ground level at low altitude or the near-ground maximum environmental humidity capability layer (see the V-3θ graphs for Nanchang, Hongzhou, and other stations). No matter whether it is a relatively arid atmosphere of low humidity (cold highs) or high humidity (subtropical highs), this phenomenon exists, which we first discovered at an inland northwest region. However, in the most recent 10 years, this phenomenon appeared again and again at various areas of east China and some coastal cities. Most recently, Hongzhou, which is near the ocean, also showed an arid atmospheric layer near the ground at low altitude. That is, such a near-ocean city as Hongzhou, which has a large water surface area within the city, is already similar to inland cities such as Nanjing, Nanchang, Beijing, etc. This is no longer a problem of a single city. It has touched on the urban planning and construction of the entire Zhejiang Province. Currently, in many provinces across China, along asphalt-paved roads high rises tower. And satellite cities and towns have formed connected, concrete-based "desertification," and are no longer isolated city "points."

18.4.4 Some Final Words

High-temperature or abnormally high-temperature weather does not appear as suddenly and intensively as torrential rains or strong windstorms. So, prediction accuracies in forecasting this high-temperature weather does not seem to catch people's attention. In fact, it is not an easy task to forecast true abnormally high-temperature weather, especially transitional changes in sustained high temperatures. The desired prediction of high-temperature weather not only involves how much humans can bear, but also, and more importantly, how we can allocate the limited energy resources and arrange the overall urban planning, distribution, and construction. In other words, accurate forecasting of high-temperature weather is necessary. However, being able to forecast such weather accurately is not the same as being able to prevent and reduce the consequent disasters, and make use of and control the "resource" of high temperatures. We see the need and economic value of studying how to understand high (or abnormally high) temperatures to practically prevent and reduce the associated disastrous aftermath and to utilize the resource available with high temperatures.

The concept of an arid atmospheric layer near ground level at low altitude has specifically described the heat inland effect of cities. It is no longer only a natural cause of meteorological phenomena. Instead, it is also about human behaviors. To this end, urban planning and construction meet with new challenges. That is, the arid layer at the low altitude is not only an indicator for forecasting abnormally high-temperature weather, but also one about whether certain urban planning and construction would destroy the ecological environmental balance. Our presentation in this section indicates that the phenomenon of arid layer at low altitude should be a problem that deserves serious consideration in future urban planning. Not only should intelligent structures be placed on the order of the day, but also foresting and wet-landing cities should be considered. Although each piece of land in cities is valuable, losses from imbalanced environments and excessive consumption of energy will definitely hinder the development of commercial centers. In particular, the forests of metropolitan structures have created an uneven, rough bottom cushion for the atmosphere and can easily trigger suddenly appearing disastrous weather.

The phenomenon of an arid atmospheric layer near the ground at low altitude provides a new topic for human interference on weather and a concrete research direction, because abnormally high-temperature weather has consumed more resources of energy, medicine, drinking water, etc., than the expenses associated with the prevention of hailstones, pressing for rain. We once did an experiment for a space of 300 m^3 where the effect of lowering the temperature of the space by sprinkling water on the floor was much less efficient than spraying water mist on the ceiling from inside the space. The amount of water consumed for the mist was only about 15 percent of that used for sprinkling the floor. And under the condition of maintaining one hour of temperature control, the method of mist had produced a much better degree of comfort. As for whether this experiment and its outcomes can be used as a reference for relevant engineering implementations, only further testing can tell.

What needs to be pointed out is that using the information of an arid atmospheric layer near the ground at low altitude, combined with the existence of a counterclockwise rolling current in the troposphere, to forecast abnormally high-temperature weather has been proven effective by using over 30 years of data of real-life cases. So, this phenomenon of arid layers near ground level at low altitude deserves more detailed investigation by meteorological scientists.

References

Ackoff, R. L. 1959. Games, decisions and organizations. *Gen. Syst.* 4:145–50.
Adams, S., and D. Lambert. 2006. *Earth science: An illustrated guide to science*. New York: Chelsea House.
Akerlof, G., A. Rose, and J. Yellen. 1988. Job switching and job satisfaction in the US labor market. Brookings Papers on Economic Activity 2.
Akerlof, G., and J. Yellen. 1990. The fair wage-effort hypothesis and unemployment. *Q. J. Econ.* 105:255–84.
Akimov, A., and G. Shypov. 1997. Torsion field and experimental manifestations. *J. New Energy* 1:67.
Alchian, A. 1950. Uncertainty, evolution, and economic theory. *J. Political Econ.* 58: 211–21.
Allen, C. W. 1976. *Astrophysical quantities*. 3rd ed. New York: The Athlone Press.
Altonji, J. G., F. Hayashi, and L. J. Kotlikoff. 1997. Parental altruism and inter vivos transfers: Theory and evidence. *J. Political Econ.* 105:1121–66.
Annie E. Casey Foundation. 1999. *When teens have sex: Issues and trends*. A KIDS COUNT Special Report. Baltimore: Author.
Baland, J. M., and J. Robinson. 2000. Is child labor inefficient? *J. Political Econ.* 108: 663–79.
Bao, Y. Z., Y. Lin, and H. Z. Chen. In press. Structural analysis of irregular information and prediction of fog area. *Scientific Inquiry*.
Barro, R. J. 1974. Are government bonds new wealth? *J. Political Econ.* 82:1095–117.
Basdevant, J. L., and J. Dalibrad. 2000. Exact results for the three-body problem. In *The quantum mechanics solver: How to apply quantum theory to modern physics*, chap. 5, pp. 61–68. Berlin: Springer-Verlag.
Basu, K., and P. H. Van. 1998. The economics of child labor. *Am. Econ. Rev.* 88:412–27.
Becker, G. 1974. A theory of social interactions. *J. Political Econ.* 82:1063–93.
Becker, G. 1991. *A treatise of the family*. Cambridge, MA: Harvard University Press.
Becker, G., and K. Murphy. 1988. The family and the state. *J. Law Econ.* 31:1–18.
Bergson, H. H. 1963. L'evolution Creatrice. In *Oeuvre, Editions du Centenaire*. Paris: PUF.
Bergstrom, T. 1989. A fresh look at the rotten kid theorem and other household mysteries. *J. Political Econ.* 97:1138–59.
Berle, A., and Means, G. 1932. *The modern corporation and private property*. New York: Macmillan.

Berlinski, D. 1976. *On systems analysis*. Cambridge, MA: MIT Press.
Berlinski, S. 2000. Contracting-out and the interindustry wage structure: Do norms of internal equity matter in wage determination? Paper presented at Economic Society World Congress 2000, Contributed Papers 1053.
Bernheim, B., A. Shleifer, and L. Summers. 1986. The strategic bequest motive. *J. Labor Econ.* 4:S151–82.
Bhalotra, S. 2004. Parent altruism, cash transfers and child poverty. Discussion Paper 4/562, Department of Economics, University of Bristol, UK.
Blackburn, M., and D. Neumark. 1988. Efficiency wages, inter-industry wage differentials, and the returns to ability. Finance and Economics Discussion Series, Federal Reserve Board.
Blauberg, I. V., V. N. Sadovsky, and E. G. Yudin. 1977. *Systems theory, philosophy and methological problems*. Moscow: Progress Publishers.
Bodies, Z., and R. Merton. 2000. *Finance*. Englewood Cliffs, NJ: Prentice Hall.
Bommier, A., and P. Dubois. 2004. Rotten parents and child labor. *J. Political Econ.* 112:240–48.
Brickley, J., J. Coles, and R. Terry. 1994. Outside directors and the adoption of poison pills. *J. Financial Econ.* 35:371–90.
Brown, C., and J. Medoff. 1989. The employer size-wage effect. *J. Political Econ.* 97:1027–59.
Bruce, N., and M. Waldman. 1990. The rotten-kid theorem meets the Samaritan's dilemma. *Q. J. Econ.* 105:155–65.
Buchanan, J. 1975. The Samaritan's dilemma. In *Altruism, morality and economic theory*, ed. E. S. Phelps, 71–85. New York: Russel Sage Foundation.
Burrough, B., and J. Helyar. 1990. *Barbarians at the gate: The fall of RJR Nabisco*. New York: Harper & Row.
Canagarajah, R., and H. Coulombe. 1997. *Child labor and schooling in Ghana*. Human Development Technical Report (Africa Region). Washington: World Bank.
Chakraborty, S., and M. Das. 2005. Mortality, fertility, and child labor. *Econ. Lett.* 86:273–78.
Chen, G. Y., Y. Lin, B. L. OuYang, and L. Q. Xu. In press. Evolution engineering and technology for long-term disaster reduction. *Sci. Inquiry*.
Chen, G. Y., B. L. OuYang, and D. S. Yuan. 2005. Existence of evolution of systems. *Eng. Sci. China* 7:47–50.
Chen, G. Y., B. L. OuYang, D. S. Yuan, L. P. Hao, and L. R. Zhou. In press. Systems stability and evolution—Extended discussion on meaning and impact of stirring energy conservation. *Eng. Sci. China*.
Chen, H. Z., and Y. Lin. In press. Information structurization and applications in prediction of thunderstorms. *Sci. Inquiry*.
Chen, J., and J. Wang. 2001. Historical comparison between Yangtze and Yellow Rivers. *J. Yangtze River Sci. Res. Inst.* 19:39–41.
Chen, Z. L., L. P. Hao, and L. R. Zhou. 2003. The blown-up analysis of the regional torrential rain and heavy rain gush on July 20 in Chengdu. *Sichuan Meteorol.* 83:7–9.
Conyon, M., and L. Read. 2006. A model of the supply of executives for outside directorships. *J. Corporate Finance* 12:645–59.
Core, J., R. Holthausen, and D. Larcker. 1999. Corporate governance, chief executive officer compensation, and firm performance. *J. Financial Econ.* 51:371–406.
Dai, W. S. 1979. *The evolution of the solar system*. Vol. 1. Shanghai: Shanghai Science and Technology Press.

Dasgupta, P. 1995. The population problem. *J. Econ. Lit.* 33:1879–902.
Davin, A. 1982. Child labour, the working class family, and domestic ideology in 19th century Britain. *Dev. Change* 13:633–52.
DeLong, B. 1991. Did Morgan's men add values? An economist's perspective on financial capitalism. In *Inside the business enterprise: Historical perspectives on the use of information*, ed. Peter Temin. Chicago: Chicago University Press. pp. 205–250.
Denis, D., and D. Denis. 1995. Performance changes following top-management dismissals. *J. Finance* 50:1029–55.
Denis, D., and J. Serrano. 1996. Active investors and management turnover following unsuccessful control contests. *J. Financial Econ.* 40:239–66.
Dessy, S. 2000. A defense of compulsive measures against child labor. *J. Dev. Econ.* 62:261–75.
Dessy, S., and S. Pallage. 2001. Why banning the worst forms of child labour would hurt poor countries. Cahiers de recherche 0109, Département d'économique, Université Laval.
De Sua, F. 1956. Consistency and completeness: A resume. *Am. Math. Monthly* 63:295–305.
Dickens, W. 1986. Wages, employment and the threat of collective action by workers. Unpublished, University of California, Berkeley.
Dickens, W., and L. Katz. 1987. Inter-industry wage differences and industry characteristics. In *Unemployment and the structure of labor markets*, ed. K. Lang and J. S. Leonard. Oxford: Basil Blackwell. pp. 48–89.
Dijkstra, B. R. 2007. Samaritan vs. rotten kid: Another look. *J. Econ. Behav. Org.* 64(1):91–110.
Dirac, P. 1975. The large number hypothesis and its consequence. In *Theories and experiments in high-energy physics*. New York: Plenum.
Doepke, M., and F. Zilibotti. 2005. The macroeconomics of child labor regulation. *Am. Econ. Rev.* 95:1492–524.
Du, Y. L., Y. Lin, and Z. L. Li. In press. Structural analysis and prediction of wind and sand storms. *Sci. Inquiry*.
Easterbrook, F., and D. Fischel. 1991. *The economic structure of corporate law*. Cambridge, MA: Harvard University Press.
Einstein, A. 1976. *Complete collection of papers by Albert Einstein*. Beijing: Commercial Press.
Einstein, A. 1983. *Complete collection of Albert Einstein*. Trans. by L. Y. Xu. Beijing: Commercial Press.
Einstein, A. 1987. *The collected papers of Albert Einstein*. Princeton, NJ: Princeton University Press.
Eisenberg, M. 1976. *The structure of the corporation: A legal analysis*. Boston: Little, Brown and Co.
Engelfriet, J. 1996. A multiset semantics for the pi-calculus with replication. *Theoret. Comput. Sci.* 153:63–94.
Engelfriet, J., and T. Gelsema. 2004. A new natural structural congruence in the pi-calculus with replication. *Acta Informatica* 40:385–430.
Engels, F. 1939. *Anti-Duhring: Herr Engen Duhring's revolution in science*. New York: International Publishers.
English, J., and G. F. Feng. 1972. *Tao te ching*. New York: Vintage Books.
Ennew, J. 1982. Family structure, unemployment and child labour in Jamaica. *Dev. Change* 13:551–63.

Eswarran, M. 1996. Fertility, literacy and the institution of child labor. Manuscript, Department of Economics, University of British Columbia, Vancouver.

Eves, J. 1992. *An introduction to the history of mathematics with cultural connections.* 6th ed. Fort Worth, TX: Saunders College Publishing.

Federio, G. 2004. Samaritans, rotten kids and policy conditionality. Working Paper 0409004, Development and Comp Systems.

Fernandes, A. 2000. Altruism with endogenous labor supply. Papers 0002, Centro de Estudios Monetarios Y. Financieros.

Fjörtoft, R. 1953. On changes in the spectral distribution of kinetic energy for two dimensional non-divergent flow. *Tellus* 5:225–30.

Franks, J., and C. Mayer. 1994. The ownership and control of German corporation. Manuscript, London Business School.

Fultz, D., R. R. Long, G. V. Owens, W. Bohan, R. Kaylor, and J. Weil. 1959. Studies of thermal convection in a rotating cylinder with some implications for large-scale atmospheric motion. *Meteorological Monographs* (American Meteorological Society), vol. 21, no. 4.

Galilei, G. 2000. *Dialog Concerning the two chief world systems, copernica ptolemaic.* Trans and ed. by S. Drabe and J. A. Green. Wichita, KS: Greenwood Research.

Ghaudhuri, A., L. Gangadharan, and P. Maitra. 2005. An experimental analysis of group size and risk sharing. Research Paper 955, Department of Economics, University of Melbourne, Melbourne Victoria, Australia.

Gillan, S. 2006. Recent developments in corporate governance: An overview. *J. Corporate Finance* 12:381–402.

Gilson, S. 1990. Bankruptcy, boards, banks, and block holders. *J. Financial Econ.* 27:355–87.

Glassey, R. T. 1977. On the blowing up of solution to Cauchy problems for nonlinear Schrodinger equation. *J. Math. Phys.* 18:1794–97.

Gleick, J. 1987. *Chaos: Making a new science.* New York: Viking.

Gordon, L. 1997. Teenage pregnancy and out-of-wedlock birth. In *Morality and health*, ed. A. Brandt and P. Rozin Stanford, CA: Stanford University Press. pp. 251–270.

Gorton, G., and F. Schmid. 1996. Universal banking and the performance of German firms. Working Paper 5453, National Bureau of Economic Research, Cambridge, MA.

Grootaert, C., and R. M. Kanbur. 1995. Child labor: An economic perspective. *Int. Labor Rev.* 134:187–204.

Groshen, E. L. 1988. Sources of wages dispersion: The contribution of interemployer differentials within industry. Unpublished, Federal Reserve Bank of Cleveland.

Guo, B. L. 1995. Nonlinear evolution equations. Shanghai: Shanghai Press of Scientific and Technological Education.

Hahn, E. 1967. Aktuelle entwicklungstendenzen der soziologischen theorie. *Deutsche Z. Phil.* 15:178–91.

Hailperin, T. 1992. Herbrand semantics, the potential infinite, and ontology-free logic. *Hist. Philos. Logic* 13:69–90.

Hailperin, T. 1997. Ontologically neutral logic. *Hist. Philos. Logic* 18:185–200.

Hailperin, T. 2001. Potential infinite models and ontologically neutral logic. *J. Philos. Logic* 30:79–96.

Haken, H. 1978. *Synergetics.* 2nd ed. Berlin: Springer-Verlag.

Hamermesh, D. 1993. *Labor demand.* Cambridge, MA: MIT Press.

Hao, L., V. Hotz, and G. Jin. 2000. Games daughters and parents play: Teenage childbearing, parental reputation, and strategic transfers. JCPR Working Papers 167, Northwestern University/University of Chicago Joint Center for Poverty Research.
Hart, O. 1995. *Firms, contracts, and financial structure.* London: Oxford University Press.
Hausdorff, F. 1935. *Mengenlehve.* Watter ac. Hruyler.
Hendrix, H. 2001. *Getting the love you want: A guide for couples.* New York: Owl Books.
Hermalin, B. 2001. Economics and corporate culture. In *The international handbook of organizational culture and climate,* ed. S. Cartwright et al. New York: John Wiley & Sons. pp. 217–262.
Hermalin, B., and M. Weisbach. 2003. Boards of directors as an endogenously determined institution: A survey of the economic literature. *ERBNY Economic Policy Review,* April 7–26.
Hess, S. L. 1959. *Introduction to theoretical meteorology.* New York: Holt, Rinehart and Winston.
Hide, R. 1953. Some experiments on thermal convection in a rotating liquid. *Q. J. Roy. Meteorol. Soc.* 79:161.
Hill, M. S., and G. Duncan. 1987. Parental family income and the socioeconomic attainment of children. *Soc. Sci. Res.* 16:36–73.
Hirshleifer, J. 1977. Shakespeare vs. Becker on altruism: The importance of having the last word. *J. Econ. Lit.* 15:500–2.
Holton, J. R. 1979. *An introduction to dynamic meteorology.* 2nd ed. New York: Academic Press.
Hrbacek, K. 1978. Axiomatic foundations for nonstandard analysis. *Fund. Math.* 98:1–19.
Hrbacek, K. 1979. Nonstandard set theory. *Am. Math. Monthly* 86:659–77.
IFHP (Institute of Foreign History of Philosophy). 1962. *The philosophy of ancient Greeks* (in Chinese). Beijing: Commercial Press.
ILO. 1996. Child labor: Targeting the intolerable. ILO: Geneva.
Internal circulation. 2000. Reports on flood prevention within Chengdu City for 2001–2020.
Jarmov, G. 1981. *Development history of physics* (Chinese trans.). Beijing: Commercial Press.
Jensen, M. 1986. Agency costs of free cash flow, corporate finance, and takeovers. *Am. Econ. Rev.* 76:323–29.
Jensen, M. 1993. The modern industrial revolution, exit, and the failure of internal control systems. *J. Finance* 48:831–80.
Jensen, M., and W. Meckling. 1976. Theory of the firm: Managerial behavior, agency costs, and ownership structure. *J. Financial Econ.* 3:305–60.
Jensen, M., and R. Ruback. 1983. The market for corporate control: The scientific evidence. *J. Financial Econ.* 11:5–50.
Kahneman, D., J. Knetsh, and R. Thaler. 1986. Fairness as a constraint on profit seeking: Entitlements in the market. *Am. Econ. Rev.* 76:728–41.
Kahneman, D., and R. Thaler. 2006. Utility maximization and experienced utility. *J. Econ. Perspect.* 20:221–34.
Kanovei, V., and M. Reeken. 2000. Extending standard models of ZFC to models of nonstandard set theories. *Studia Logica* 64:37–59.
Kaplan, S., and B. Minton. 1994. Appointments of outsiders to Japanese boards: Departments and implications for managers. *J. Financial Econ.* 36:225–57.
Katz, L., and L. Summers. 1989. Industry rents and industry policy. Brookings Papers on Economic Activity.

Kawai, T. 1981. *Axiom systems of nonstandard set theory*, 57 – 64. Logic Symposia Hakone 1979, 1980 (Lecture Notes in Mathematics 891). Berlin: Springer.
Kline, M. 1972. *Mathematical thought from ancient to modern times.* Oxford: Oxford University Press.
Kline, M. 1983. *Mathematical thought from ancient to modern times* (Chinese translation). Shanghai: Shanghai Press of Science and Technology.
Klir, G. 1970. *An approach to general systems theory.* Princeton, NJ: Van Nostrand.
Klir, G. 1985. *Architecture of systems problem solving.* New York: Plenum Press.
Klir, G. 2001. *Facets of systems science.* New York: Springer.
Koyre, I. 1968. *Etudes Newtiniennes.* Paris: Gauimard.
Krueger, A. B., and L. Summers. 1987. Reflections on the inter-industry wage structure. In *Unemployment and the structure of labor markets*, ed. K. Lang and J. S. Leonard. Oxford: Basil Blackwell. pp. 17–47.
Krueger, D., and J. Donohue. 2005. On the distributional consequences of child labor legislation. Int. Econ. Rev. 46:785–815.
Kuchemann, D. F. 1965. Report on the I.U.T.A.M. symposium on concentrated vortex motions in fluids. *Fluid Mech.* 21:1–20.
Kuhn, T. 1962. *The structure of scientific revolutions.* Chicago: University of Chicago Press.
Labenne, S. 1997. The determinants of child labor in India. Manuscript, C.R.E.D., University of Namur.
Lagerlof, J. 2004. Efficiency-enhancing signalling in the Samaritan's dilemma. *Econ. J.* 114:55–69.
Lamond, O., and R. Thaler. 2003. Anomaly: The law of one price in financial markets. *J. Econ. Perspect.* 17:191–202.
Lao Tzu. 1972. *Tao te ching: A new translation*, trans. G.-f. Feng and J. English. New York: Vintage Books.
Lawrence, C., and R. Lawrence. 1985. Manufacturing wage dispersion: An end game interpretation. *Brookings Papers on Economic Activity*, pp. 45–106.
Li, C. A. 2006. Earth science. *J. Chin. Univ. Geosci.* 28:461–66.
Liang, H. M. 1996. *Lao Tzu* (in Chinese). Liaoning: Liaoning People's Press.
Lilienfeld, D. 1978. *The rise of systems theory.* New York: Wiley.
Lin, J., and S. C. OuYang. 1996. Exploration of the mystery of nonlinearity. *Research of Natural Dialectics*, pp. 12–13, 34–37.
Lin, Y. 1987. A model of general systems. *Math. Modeling Int. J.* 9:95–104.
Lin, Y. 1988. Can the world be studied in the viewpoint of systems? *Math. Comput. Modeling* 11:738–42.
Lin, Y. 1989a. Multi-level systems. *Int. J. Syst. Sci.* 20:1875–89.
Lin, Y. 1989b. A multi-relation approach to general systems and tests of applications. *Synthese Int. J. Epistemol. Methodol. Philos. Sci.* 79:473–88.
Lin, Y. 1990a. The concept of fuzzy systems. *Kybernetes Int. J. Syst. Cybernetics* 19:45–51.
Lin, Y. 1990b. A few systems-colored views of the world. In *Mathematics and science*, ed. R. E. Mickens. River Edge, NJ: World Scientific. pp. 94–114.
Lin, Y. 1995. Developing a theoretical foundation for the laws of conservation. *Kybernetes Int. J. Syst. Cybernetics* 24:52–60.
Lin, Y., guest ed. 1998. Mystery of nonlinearity and Lorenz's chaos. *Kybernetes Int. J. Syst. Cybernetics* 27:605–854.
Lin, Y. 1999. *General systems theory: A mathematical approach.* New York: Plenum and Kluwer Academic Publishers.

Lin, Y. 2000. Some impacts of the second dimension. In *Proceedings of the 15th European Meeting in Cybernetics and Systems Research*, Vienna, Austria, April 25–28, pp. 3–8.
Lin, Y. 2007. Systemic yoyo model and applications in Newton's, Kepler's laws, etc. *Kybernetes Int. J. Syst. Cybernetics* 36:484–516.
Lin, Y., and T. H. Fan. 1997. The fundamental structure of general systems and its relation to knowability of the physical world. *Kybernetes Int. J. Syst. Cybernetics* 26:275–85.
Lin, Y., and D. Forrest. 2008a. Economic yoyos and Becker's rotten kid theorem. *Kybernetes Int. J. Syst. Cybernetics* 37:297–314.
Lin, Y., and D. Forrest. 2008b. Economic yoyos and never-perfect value theorem. *Kybernetes Int. J. Syst. Cybernetics* 37:149–65.
Lin, Y., and D. Forrest. 2008c. Economic yoyos, parasites and child labor. *Kybernetes Int. Syst. Cybernetics Management.* 37:757–767.
Lin, Y., Q. P. Hu, and D. Li. 1997. Some unsolved problems in general systems theory (I). *Cybernetics Syst. Int. J.* 28:287–303.
Lin, Y., and Y. Ma. 1987. Remarks on analogy between systems. *Int. J. Gen. Syst.* 13:135–41.
Lin, Y., Y. Ma, and R. Port. 1990. Several epistemological problems related to the concept of systems. *Math. Comput. Modeling* 14:52–57.
Lin, Y., and S. C. OuYang. 1998. Invisible Tao and realistic nonlinearity propositions. *Kybernetes Int. J. Syst. Cybernetics* 27:809–22.
Lin, Y., and Y. Wu. 1998. Blown-ups and the concept of whole evolutions in systems science. *Problems Nonlinear Anal. Eng. Syst.* 4:16–31.
Lin, Y., W. J. Zhu, N. S. Gong, and G. P. Du. In press. Systemic yoyo structure in human thoughts and the fourth crisis in mathematics. *Kybernetes Int. J. Syst. Cybernetics.* 37: 387–423.
Lindbeck, A., and J. Weibull. 1988. Altruism and time consistency: The economics of fair accompli. *J. Political Econ.* 96:1165–82.
Lipton, M., and J. Lorsch. 1992. A modest proposal for improved corporate governance. *Business Layer* 48:59–77.
Luker, K. 1996. *Dubious conceptions: the politics of teenage pregnancy*. Cambridge, MA: Harvard University Press.
Lorenz, E. N. 1993. *The essence of chaos*. Seattle: Washington University Press.
Lu, D. F., trans. 2000. *Management of flood regions in the 21st century America*. Beijing: Press of Yellow River Water Powers.
MacAvoy, P., S. Cantor, J. Dana, and S. Peck. 1983. All proposals for increased control of the corporation by board of directors: An economic analysis. In *Statement of the business roundtable on the American Law Institute's proposed "Principles of Corporate Governance and Structure: Restatement and Recommendation."* New York: Business Roundtable.
Manne, H. 1965. Mergers and the market for corporate control. *J. Political Econ.* 75:110–26.
Mathematical sciences: A unifying and dynamic resource. 1985. *Notices Am. Math. Soc.* 33:463–79.
Menchik, P. L. 1980. Primogeniture, equal sharing, and the U.S. distribution of wealth. *Q. J. Econ.* 92:299–316.
Mickens, R. E. 1990. *Mathematics and science*. Singapore: World Scientific.
Milner, R. 1992. Functions as processes. *Math. Struct. Comput. Sci.* 2:119–41.

Murphy, K., and R. Topal. 1987. Unemployment, risk, and earnings: Testing for equalizing wage differences in labor market. In *Unemployment and the structure of labor markets*, ed. L. Lang and J. S. Leonard. Oxford: Basil Blackwell. pp. 103–139.

Nelson, E. 1977. Internal set theory: A new approach to nonstandard analysis. *Bull. Am. Math. Soc.* 83:1165–98.

Nerlove, M., A. Razin, and E. Sadka. 1988. A bequest constrained economy: Welfare analysis. *J. Public Econ.* 37:203–20.

Norberg-Schonfeldt, M. 2004. Children's school achievements and parental work: An analysis for Sweden. Working Paper, Umea Economic Studies 645, Department of Economics, Umea University.

Newton, I. 2008. *Newton's Principia: The mathematical principles of natural philosophy*. Trans. by A. Motte. Whitefish, MT: Kessinger Publishing.

OECD (Organization for Economic Cooperation and Development) 1995. Corporate governance environments in OECD countries. February 1.

OuYang, S. C. 1994. *Break-offs of moving fluids and several problems of weather forecasting*. Chengdu: Press of Chengdu University of Science and Technology.

OuYang, S. C. 1995. Pansystems' point of view of prediction and explosive reverses of fluids. *Appl. Math. Mechanics* 16:255–62.

OuYang, S. C. 1998a. *Weather evolutions and structural forecasting* (in Chinese). Beijing: Meteorological Press.

OuYang, S. C. 1998b. Some remarks on "explosive" evolutions of general nonlinear evolution equations and related problems. *Appl. Math. Mechanics* 19:165–73.

OuYang, S. C. In press a. Second stir and conservation of stirring energy. *Sci. Inquiry*.

OuYang, S. C. In press b. Mass-energy-gravitational waves and solenoidal field-universal gravitation. *Sci. Inquiry*.

OuYang, S. C. In press c. About rotation and evolution. *Sci. Inquiry*.

OuYang, S. C., and G. Y. Chen. 2006. End of stochastics and quantitative comparability. *Sci. Res. Monthly* 14:141–43.

OuYang, S. C., and Y. Lin. 2006. Disillusion of randomness and end of quantitative comparabilities. *Sci. Inquiry* 7:171–80.

OuYang, S. C., and Y. Lin. In press a. Probabilistic waves and torsion problem of quantum effects. *Sci. Inquiry*.

OuYang, S. C., and Y. Lin. In press b. Physical quantities, time and parametric dimensions. *Sci. Inquiry*.

OuYang, S. C., and Y. Lin. In press c. Events-quantities and digitally structurized events. *Sci. Inquiry*.

OuYang, S. C., Y. Lin, Z. Wang, and T. Y. Peng. 2001a. Evolution science and infrastructural analysis of the second stir. *Kybernetes Int. J. Syst. Cybernetics* 30:463–79.

OuYang, S. C., Y. Lin, Z. Wang, and T. Y. Peng. 2001b. Blown-up theory of evolution science and fundamental problems of the first push. *Kybernetes Int. J. Syst. Cybernetics* 30:448–62.

OuYang, S. C., Y. Lin, N. Xie, and L. P. Hao. In press. Prediction of suddenly appeared disasters and emergency measures. *Sci. Inquiry*.

OuYang, S. C., D. H. McNeil, and Y. Lin. 2002. *Entering the era of irregularity* (in Chinese). Beijing: Meteorological Press.

OuYang, S. C., J. H. Miao, Y. Wu, Y. Lin, T. Y. Peng, and T. G. Xiao. 2000. Second stir and incompleteness of quantitative analysis. *Kybernetes Int. J. Syst. Cybernetics* 29:53–70.

OuYang, S. C., H. Sabelli, Z. Wang, Y. Lu, Y. Lin, and D. McNeil. 2002. Evolutionary "aging" and "death" and quantitative instability. *Int. J. Comput. Num. Anal. Appl.* 1:413–37.
OuYang, S. C., M. Wei, and Y. Lin. In press. Abnormal high-temperature weathers, their structural characteristics, prediction and problems of urban construction. *Sci. Inquiry*.
OuYang, S. C., and Y. Wu. 1998. The transformation group of branching solutions of Lorenz's equations and "chaos." *Kybernetes Int. J. Syst. Cybernetics* 27:669–73.
OuYang, S. C., Y. Wu, Y. Lin, and C. Li. 1998. The discontinuity problem and "chaos" of Lorenz's model. *Kybernetes Int. J. Syst. Cybernetics* 27:621–35.
OuYang, S. C., K. Zhang, L. P. Hao, and L. R. Zhou. 2005. Structural transformation of irregular time-series information and refined analysis on evolutions. *Eng. Sci. China* 7:36–41.
Owen, T. 1991. *The corporation under Russian law, 1800–1917: A study in tsarist economic policy*. New York: Cambridge University Press.
Pan, N. B. 1980. Hubble constant. In *Large encyclopedia of China: astronomy*, 108. Beijing: Chinese Large Encyclopedia Press.
Patacchini, E., and Y. Zenou. 2004. Intergenerational education transmission: Neighborhood quality and/or parents' involvement? Working Paper 631 (2004), IUI, Research Institute of Industrial Economics, Stockholm, Sweden.
Popper, K. 1945. *The open society and its enemies*. London: Routledge.
Pound, J. 1988. Proxy contests and the efficiency of shareholder oversight. *J. Financial Econ.* 20:237–65.
Pratt, J., D. Wise, and R. Zechhauser. 1979. Price differences in almost competitive markets. *Q. J. Econ.* 33:189–211.
Prigogine, I. 1967. Structure, dissipation and life. Paper presented at the First International Conference of the Oretical Physics and Biology, Versailles.
Prigogine, I. 1980. *From being to becoming: Time and complexity in the physical science*. New York: W. H. Freeman and Company.
Prowse, S. 1990. Institutional investment patterns and corporate financial behavior in the United States and Japan. *J. Financial Econ.* 27:43–66.
Quastler, H. 1965. General principles of systems analysis. In *Theoretical and mathematical biology*, ed. T. H. Waterman and H. J. Morowitz. New York: Blaisdell Publishing.
Raff, M., and L. Summers. 1987. Did Henry Ford pay efficiency wages? *J. Labor Econ.* 5:S57–86.
Ranjan, P. 2001. Credit constraints and the phenomenon of child labor. *J. Dev. Econ.* 64:81–102.
Rapoport, A. 2000. Paradoxes: *The limit of science and the science of limits* (in Chinese), trans. X. Z. Li. Shanghai: Shanghai Press of Science and Technology.
Ravallion, M., and Q. Wodon. 2000. Does child labour displace schooling? Evidence on behavioural responses to an enrollment subsidy. *Econ. J.* 110:C158–75.
Ren, Z. Q. 1996. The resolution of the difficulties in the research of Earth science. In *Earth science and development*, ed. F. H. Wu and X. J. He, 298–304. Beijing: Earthquake Press.
Ren, Z. Q., and Z. W. Hu. 1989. The product relation of mass-angular momentum in celestial body systems. In *Advances of comprehensive study of mutual relations in cosmos: Earth and life*, 269–74. Beijing: Chinese Science and Technology Press.
Ren, Z. Q., Y. Lin, and S. C. OuYang. 1998. Conjecture on law of conservation of informational infrastructures. *Kybernetes Int. J. Syst. Cybernetics* 27:543–52.

Ren, Z. Q., and T. Nio. 1994. Discussions on several problems of atmospheric vertical motion equations. *Plateau Meteorol.* 13:102–5.

Robinson, A. 1964. Formalism 64, logic, methodology, and philosophy of science. In *Proceedings of the 1964 International Congress*, ed. Y. Bar-Hillel, 228–246. Amsterdam: North-Holland and Pubco.

Roe, M. 1994. *Strong managers, weak owners: The political roots of American corporate finance.* Princeton, NJ: Princeton University Press.

Rosen, S. 1986. The theory of equalizing differences. In *Handbook of labor economics*, ed. O. Ashefelter and R. Layard. Vol. 1. New York: Elsevier Science Publishers BV. pp. 641–692.

Rosenzweig, M. 1981. Household and non-household activities of youths: Issues of modeling, data and estimation strategies. In *Child work, poverty and unemployment*, ed. G. Rodgers and G. Standing. Geneva: International Labour Organisation.

Sabelli, H., and L. Hauffman. 1999. The process equation: Formulating and testing the process theory of systems. *Cybernetics Syst. Int. J.* 30:261–94.

Salop, S. 1979. A model of the natural rate of unemployment. *Am. Econ. Rev.* 69:117–25.

Saltzman, B. 1962. Finite amplitude free convection as an initial value problem. *Int. J. Atmos. Sci.* 19:329–41.

Shapiro, C., and J. Stiglitz. 1984. Equilibrium unemployment as a worker discipline device. *Am. Econ. Rev.* 27:433–44.

Shivdasani, A. 1993. Board composition, ownership structure, and hostile takeovers. *J. Accounting Econ.* 16:167–98.

Shleifer, A., and R. Vishny. 1986. Large shareholders and corporate control. *J. Political Econ.* 94:461–88.

Shleifer, A., and R. Vishny. 1989. Management entrenchment: The case of manager-specific investments. *J. Financial Econ.* 25:123–40.

Shleifer, A., and R. Vishny. 1990. Equilibrium short horizons of investors and firms. *Am. Econ. Rev. Papers Proc.* 80:148–53.

Shleifer, A., and R. Vishny. 1997. A survey of corporate governance. *J. Finance* 62:737–82.

Slichter, S. 1950. Notes on the structure of wages. *Rev. Econ. Stat.* 32:80–91.

Smith, C., and J. Warner. 1979. On financial contracting: An analysis of bond covenants. *J. Financial Econ.* 7:117–61.

Solow, R. M. 1979. Another possible source of wage stickiness. *J. Macroecon.* 1:79–82.

Soros, G. 1998. *The crisis of global capitalism: Open society endangered.* New York: Public Affairs.

Spence, M. 1973. Job market signaling. *Q. J. Econ.* 87:355–74.

Stanley, T., and W. Danko. 1996. *The millionaire next door: The surprising secrets of America's wealthy.* New York: Simon and Schuster.

Stigler, G. 1958. The economics of scale. *J. Law Econ.* 1:54–71.

Stiglitz, J. E. 1974. Alternate theories of wages determination and unemployment in L.C.D.'s: The labor turnover model. *Q. J. Econ.* 88:194–227.

Stiglitz, J. E. 1976. Prices and queues as screening devices in competitive markets. IMSSS Technical Report 212, Stanford University.

Thaler, R. 1989. Interindustry wage differentials. *J. Econ. Perspect.* 3:181–93.

Thom, R. 1975. *Structural stability and morpho-genesis.* Reading MA: Benjamin.

von Bertalanffy, L. 1924. Einfuhrung in Spengler's werk. *Literaturblatt Kolnische Zeitung*, May.

von Bertalanffy, L. 1962. *Modern theories of development*. Trans. by J. H. Woodge. Oxford: Oxford University Press.
von Bertalanffy, L. 1968. *General systems theory*. New York: George Braziller.
von Bertalanffy, L. 1972. The history and status of general systems theory. In *Trends in general systems theory*, ed. G. Klir, 21–41. New York: Wiley-Interscience.
Wang, H. Z., and X. F. Li. 2001. Computational methods for rainwater infiltration facilities in Beijing City area. *Water Supply and Drainage of China*, no. 11.
Weiner, M. 1991. *The child and the state in India*. Princeton, NJ: Princeton University Press.
Weisbach, M. 1988. Outside directors and CEO turnover. *J. Financial Econ.* 20:431–60.
Weiss, A. 1980. Job queues and layoffs in labor markets with flexible wages. *J. Political Econ.* 88:526–38.
Wigner, E. P. 1960. The unreasonable effectiveness of mathematics in the natural sciences. *Commun. Pure Appl. Math.* 13:1–14.
Wilhelm, R., and C. Baynes. 1967. *The I ching or Book of changes*. 3rd ed. Princeton, NJ: Princeton University Press.
Wood-Harper, A. T., and G. Fitzgerald. 1982. A taxonomy of current approaches to systems analysis. *Comput. J.* 25:12–16.
Wu, X. M. 1990. *The pansystems view of the world*. Beijing: Press of Chinese People's University.
Wu, Y. 1998. Blown-ups: A leap from specified local confusions to a general whole evolution. *Kybernetes Int. J. Syst. Cybernetics* 27:647–55.
Wu, Y., and Y. Lin. 2002. *Beyond nonstructural quantitative analysis: Blown-ups, spinning currents and modern science*. River Edge NJ: World Scientific.
Xian, D. C., and Z. X. Wang. 1987. Particle physics. In *Large encyclopedia of China: Physics*, 737. Chinese Large Encyclopedia Press.
Yafeh, Y., and O. Yosha. 1996. Large shareholders and banks: Who monitors and how? Manuscript, Hebrew University, Jerusalem, Israel.
Yang, J. Y. 1976. *Physical quantities and astrophysical quantities*. Shanghai: People's Press of Shanghai.
Yellen, J. 1984. Efficiency wage models of unemployment. *Am. Econ. Rev.* 74:200–5.
Yermack, D. 1996. Higher market valuation of companies with a small board of directors. *J. Financial Econ.* 40:185–212.
Yermack, D. 1997. Good timing: CEO stock option awards and company news announcements. *J. Finance* 52:449–76.
Zadeh, L. 1962. From circuit theory to systems theory. *Proc. IRE* 50:856–65.
Zelizer, V. A. 1994. *Pricing the priceless child: The changing social value of children*. Princeton, NJ: Princeton University Press.
Zeng, X. P., and Y. F. Lin. 2006. Geomagnetic irregular time-series information and prediction of disasters. *Sci. Res. Monthly* (Hong Kong) 14:126–29.
Zhu, W. J., Y. Lin, N. S. Gong, and G. P. Du. 2008a. Descriptive definitions of potential and actual infinities. *Kybernetes Int. Syst. Cybernetics Management* 37. 424–432.
Zhu, W. J., Y. Lin, G. P. Du, and N. S. Gong. 2008d. Inconsistency of uncountable infinite sets. *Kybernetes Int. Syst. Cybernetics Management* 37. 453–457.
Zhu, W. J., Y. Lin, G. P. Du, and N. S. Gong. 2008h. Mathematical system of potential infinities (I)—Preparation. *Kybernetes Int. Syst. Cybernetics Management* 37. 489–493.

Zhu, W. J., Y. Lin, G. P. Du, and N. S. Gong. 2008h. Mathematical system of potential infinities (II)—Formal system of logical basis. *Kybernetes Int. Syst. Cybernetics Management* 37. 454–504.

Zhu, W. J., Y. Lin, G. P. Du, and N. S. Gong. 2008j. Mathematical system of potential infinities (III)—Metatheory of logical basis. *Kybernetes Int. Syst. Cybernetics Management* 37. 505–515.

Zhu, W. J., Y. Lin, G. P. Du, and N. S. Gong. 2008k. Mathematical system of potential infinities (IV)—Set theoretic foundation. *Kybernetes Int. Syst. Cybernetics Management* 37. 516–525.

Zhu, W. J., Y. Lin, N. S. Gong, and G. P. Du. In 2008b. Modern system of mathematics and a pair of hidden contradictions in its foundation. *Kybernetes Int. Syst. Cybernetics Management* 37. 438–445.

Zhu, W. J., Y. Lin, G. P. Du, and N. S. Gong. 2008f. Modern system of mathematics and general Cauchy theater in its theoretical foundation. *Kybernetes Int. Syst. Cybernetics Management* 37. 465–468.

Zhu, W. J., Y. Lin, G. P. Du, and N. S. Gong. 2008e. Modern system of mathematics and special Cauchy theater in its theoretical foundation. *Kybernetes Int. Syst. Cybernetics Management* 37. 458–464.

Zhu, W. J., Y. Lin, N. S. Gong, and G. P. Du. 2008g. New Berkeley paradox in the theory of limits. *Kybernetes Int. Syst. Cybernetics Management* 37. 474–481.

Zhu, W. J., Y. Lin, N. S. Gong, and G. P. Du. 2008l. Problem of infinity between predicates and infinite sets. *Kybernetes Int. Syst. Cybernetics Management* 37. 526–533.

Zhu, W. J., Y. Lin, G. P. Du, and N. S. Gong. In 2008. The inconsistency of countable infinite sets. In this special issue. *Kybernotes Int Syst. Cybernetics Management* 37. 446–482.

Zhu, W. J., Y. Lin, N. S. Gong, and G. P. Du. 2008m. Intension and structure of actually infinite, rigid sets. *Kybernetes Int. Syst. Cybernetics Management* 37. 534–542.

Zhu, W. J., and X. A. Xiao. 1991. *An introduction to set theory.* Nanjing: Nanjing University Press.

Zhu, W. J., and X. A. Xiao. 1995. *An introduction to mathematical logic.* Nanjing: Nanjing University Press.

Zhu, W. J., and X. A. Xiao. 1996. *An introduction to foundations of mathematics.* Nanjing: Nanjing University Press.

Zhu, Y. Z. 1985. *Albert Einstein: The great explorer.* Beijing: Beijing People's Press.

Zingales, L. 1994. The value of the voting right: A study of the Milan stock exchange experience. *Rev. Financial Studies* 7:125–48.

Index

A

Absolute time and space, 135
Acceleration
   Bjerknes circulation theorem, 24
   Newton's second law of motion, 56–57
   physics of physical quantities, 125
   stirring energy and its conservation, 107
   variable
      versus average, 126
      rotations of materials, 127
Acceleration vector, Newtonian first push, 97
Ackermann, Wilhelm, 271
Action-reaction forces; *See* Newton's laws of motion, third
Actual infinities; *See* Infinities
Adaptability, system, 3
Adult child's altruism toward parents, 196–201
Advection equation, one-dimensional, 22
Advective fog, 326
Air density, meteorology, 304
Akerlof, George, 213, 216, 217, 226
Alchian, A., 238
Altitude
   and air density, 304
   three θ curves, 310, 311
   and wind speed, 305
Altruism, 177
   adult child's altruism toward parents, 196–201
   one-sided altruism model, 190–196
Altruism models, child labor, 205, 209, 210
Analysis
   economic yoyos
      dynamics of projects, 249–251
      whole evolution analysis of demand and supply, 146–149
   event, 131, 141; *See also* Structural analysis
   linear, limitations of, 22
   structural; *See* Figurative analysis
   system, 3
   systemic yoyo methodology, 13–14
Analyst, The: A Discourse Addressed to an Infidel Mathematician (Berkeley), 266–267
Analytic geometry, 265
Angular momentum
   rotation and stirring energy, 92
   solar system, conservation of informational infrastructure, 40–42
Angular speed
   versus angular speed, 95
   conservation of stirring energy and three-level energy transformation, 93–94
   rotation and stirring energy, 92
   squared, 93
   stirring energy and its conservation, 106, 107
Angular speed of rotation squared, 12
Anharmonic spin
   acting and reacting yoyo structures, 55
   colliding eddies and acting-reacting forces, 58
Archimedes, 266
Arc lengths, early calculus problems, 265
Areas
   Archimedes treatment of, 266
   early calculus problems, 265
Aristotle, 3–4, 5, 8, 27, 28, 277
Arrow of time, 135, 137
Artificial lakes; *See* Lakes, artificial
Astronomy; *See* Macro scale; Three-body problem; Universe
Atmosphere movement/dynamics
   Bjerknes circulation theorem, 24, 47
   conservation of informational infrastructure, empirical evidence, 42–43, 45
   dishpan experiments, 68–70
   law of conservation of informational infrastructure, 45

meteorology; *See* Weather prediction
stirring forces, 92
three-level circulation transformations, 104
weather prediction
Atmospheric density, conventional meteorology, 297
Atmospheric pressure; *See* Barometric pressure
Atmospheric structure
   baroclinic effects, 306–307
   unevenness as indicator of severe weather, 305
Atmospheric temperature; *See* Heat waves/abnormally high temperatures; Ultra-low temperature
Atomic scale; *See* Micro scale; Scale
Atom theory of Democritus, 8
Attraction forces
   colliding eddies and acting-reacting forces, 59
   Newton's second law of motion, 54, 56
   problems on physics of physical quantities, 130
   unevenness of materials structures and, 148
Automorphic evolution, 8–9
Automorphic structures, 8
Averages, 130
   nonquantification of events, 127–128
   physics of physical quantities and, 140
   quantitative, 131, 140, 340
Axiomatic geometry, 265
Axiomatic method, 288
Axiomatic set theory, 272, 276, 285
Axiom of choice, 282
Axis
   eddy motion model of general system, 9
   yoyo model, 148–149

B

Background radiation, 27–28
Baise, 327, 332, 333
Baland, J.M., 13, 154, 184, 186, 191, 194, 197
Bao, Youzhang, 325, 327
Baroclinic effects of atmospheric structure, 306–307
Baroclinic solenoids, 299
Baroclinic waves, 299
Barometric pressure
   conventional meteorology, 297
   high-temperature weather
      under cold high pressure, 364–366
      under subtropical high in West Pacific, 356–364
   order of information, 297–298
   prediction of severity of weather, 340
   3θ curves at different altitudes, 310

V-3θ graph design, 307
weather maps, smoothing of irregular information, 299
Barrow, Isaac, 134, 266
Base p number system, 63
Becker, Gary, 153, 154, 155, 178, 179, 258
Becker's Rotten Kid theorem, 13, 153–154, 155–165, 167, 170, 178, 181, 183, 187, 188, 233, 258
Behavior, economic; *See* Economics and finance
Beihai, 333, 334
Beijing, severe weather, 313, 314
   high-temperature weather, 364–366
   precipitation, 317–321
   sandstorm, 350–353
Bergstrom, T., 155, 167, 187, 188
Bergstrom's rotten kid theorem, 13, 154, 155, 157–165, 179–183, 188
Berkeley, George, 266–267
Berkeley paradox, 258, 268, 275, 285–288, 289, 291
Berlinski, D., 6
Berlinski, S., 217, 233
Bernays, P., 271
Big Bangs; *See* Black hole-Big Bang aspects of spinning yoyo
Big Bang theory, 27–28, 34, 45
Binary number system, 63
Binary star systems, 11, 67, 74, 75, 77, 80, 81
Biology, systems approach, 1–2
Bjerknes circulation theorem, 24–25, 33, 47, 99, 101
Black box, 3
Black hole-Big Bang aspects of spinning yoyo, 44, 75, 102
   economic yoyos, 149, 151
   eddy motion model of general system, 9, 10
Blown-ups
   concept of, 19–20
   economics and finance
      price evolution, 147–148
      whole evolution analysis of demand and supply, 148
   equal quantitative effects, 28–32
   mapping properties and quantitative infinity, 23–28
   mathematical characteristics, 20–22
   properties of, 32–34
   spinning current as physical characteristic of, 23–28
   transitional versus nontransitional, 20
   weather systems; *See* Meteorology; V-3θ graphs, case studies using
   whole evolutions of converging and diverging fluid motions, 65–66

Blown-up theory, 8
  eddy motion model of general system, 9–10
  experimental evidence, theoretical justification, mathematical foundation, and applications, 149
  introduction of, 7–8
  meteorology, 296
  Newton's first law of motion, second stir and, 52–53
  yoyo model, 10
Boards of directors, 13, 251–253, 258
  economic yoyo structure, 217, 218
  long-term versus short-term projects, 240–241
Bohr, N., 29
Bommier, A., 13
Book of Changes (I Ching), 8–9, 29, 63, 132, 302
Boolean system, 63
Boundary conditions
  quantitative unboundedness, eddy current transformations, 135
  whole evolutions of converging and diverging fluid motions, 64
Boundedness, whole evolutions of converging and diverging fluid motions, 64, 65
Brouwer, L.E.J., 269, 270
Brown, C., 231
Buchanan studies, 13

C

Calculation spills, quantitative infinities and, 22, 33
Calculus, 51, 52, 268
  creators of, 265
  origins of, 8
  physics of physical quantities, 127, 130
  seventeenth century mathematics, 266
  singularities in solutions, 26
  three-body problem open questions, 88–89
Calculus-based models
  discontinuous transitional changes, 20
  prediction of transitional changes, 33
  predictive limitations of, 51–52
Cantor, Georg, 267, 268, 269, 276
Cantor, Moritz, 291
Capital markets
  imperfect, 222–224
  perfect, 219–222
Cartesian coordinate system, 2
Cauchy, Augustin-Louis, 267, 268
Cauchy problem, 22
Cauchy theater phenomena, 283–285
Cauchy theory of limits, 268, 271, 287, 289

Causality
  multicausality, 5
  problems on physics of physical quantities, 129
Cavalieri, B., 266
Celestial mechanics, 8
  gravitational collapse, 129–130
  law of conservation of informational infrastructure, 43
  solar system angular momentum, 40–42
  three-level circulation transformations, 104
  three-ringed stability, stirring energy conservation, 105
CEOs, 258
  choices of projects, 243–253
    analysis, 249–251
    model, 247–249
    power struggle between board and CEO, 251–253
    price behavior of different investment projects, 244–246
  economic yoyo structure, 217, 218
  long-term versus short-term projects, 238–243
Change
  non-initial value automorphic evolutions, 8–9
  problems on physics of physical quantities, 129
Changeability of materials, variable acceleration and, 127
Chaos/chaos theory, 31, 181
  financial markets, 62
  material versus parametric dimensions, 139–140
  mathematics, 265, 284
  three-body problem, 78, 79, 87, 89
  time, 139, 140–141
  uniform flow transition into chaotic currents, 258, 275
Characteristic analysis of V-3θ graphic structure for wind- and sandstorms, 344–345
Characteristics of systems, 3
Chen, G.Y., 106
Chen, J., 110
Chengdu City, 114, 118
Child labor, 13, 154, 187, 188, 189–213, 258
  conclusion, 212–213
  different efficiencies and potentially different outcomes, 201–203
  disutility of child's work, 190–201
    adult child's altruism toward parents, 196–201
    one-sided altruism model, 190–196
  marginal bans on child labor, 204–212
China; *See also* V-3θ graphs, case studies using big river cultures and water-based environment, 8
  flood management; *See* Flood disasters, principles of flood evolution and development

historical approaches to water management, 108–109
three-ringed energy transformation, Bei Wei period, 92
time, concept of, 132–133
Chinese science, systems as wholes, 3–4
Chongqing, 355
Circular definitions, 269
Circular time trajectory, 66
Circulation levels, fourth, fifth, and higher, 104
Circulation theorem of Bjerknes, 24–25, 33, 47, 99, 101
City planning, modification of atmospheric convection, 325
Classical mechanics, 8
  attributes of materials, 139
  Bjerknes circulation theorem and, 25
  conservation laws, 44
  quantitative physics, 134–135
  and time, 123–124
Classical theory
  as first dimension of two-dimensional science, 6
  need for modification, 6–7
Classifications of systems, 3
Clockwise eddies, coexistence with counterclockwise eddies, 26
Clockwise rolling currents, 301, 303
  high-temperature weather, 359
  wind- and sandstorm weather, 353
Clockwise rotation, 29
Closed line integral, squared speed, 95
Closed systems
  Bjerknes circulation theorem, 24
  yoyo model, 10
Cloud clusters, V-3θ graphs, 311
Coexistence, clockwise and counterclockwise eddies, 26
Cognitive concepts, irregular information, 339
Coincidentia oppositorum (Nicholas of Cusa), 4
Colliding eddies and acting-reacting forces, 57–61
Communications, and scope of science, 6
Competition
  competing tendencies in nature, 8
  and corporate governance, 238–243
  free market model, 12–13
Complexities, unevenness of space and time of evolutionary materials structures and, 9
Computational instabilities/spills
  equal quantitative effects and, 30, 34
  quantitative infinities and, 22, 33
Computer technology, Boolean system, 63
Concave curvature spaces, 106
Concrete multiset semantics, 259

Condensation, atmospheric water vapor, 299–300
Condensed materials and energies, twist-up contents, 31–32
Conflicts of interest, and corporate governance, 238–243
Confrontational movement, meteorology, 296, 297, 298
Congruence, middle, 259
Conservation between Two New Sciences (Galileo), 134
Conservation laws
  informational infrastructure
    empirical evidence, 43–44
    impacts of, 44–46
  kinetic energy, 12
  law of conservation of informational infrastructure, 43–44
  mass-angular momentum products, 42
  problems on physics of physical quantities, 128
  solar system angular momentum, 40
  stirring energy; *See also* Stirring energy and its conservation
    flood disasters, principles of flood evolution and development, 110
    meteorology, 296
    physical significance of energy transformations, 103–106
  yoyo model, 10
Conservation of informational infrastructure, 10, 35–47
  angular momentum of solar system, 40–42
  atmosphere movement, measurement analysis, 42–43
  Dirac's large number hypothesis, physical essence of, 36–40
  laws of
    empirical evidence, 43–44
    impacts of, 44–46
  other empirical evidence for yoyo structures, 46–47
Conservation of speed kinetic energy, 95
Conservation of stirring energy, 93–95, 104
Consistency problems, formalist school and, 271
Constant coefficient equation, 20
Content of information, uniformity of all imaginable things, 36
Continuity model of material structure, 8, 26
  blown-ups and, 19
  and weather, 296–297
Continuous solutions, well-posedness of nonlinear evolution equations, 20
Contractors, 233
Contradiction, proof by, 280
Control, system, 3

Controllability, system, 3
Convection model of Salzmann, 139
Convective weather systems, sudden severe, 301, 313–325
 background, 314
 blown-ups, 314–315
 disaster prevention measures, 325
 discussion, 324–325
 over major metropolitan areas, 317–322
 regional heavy rain, 322–324
 structural analysis, 315–316
 structural characteristics, 316–317
Convergence/divergence
 creation of small yoyo structures, 150
 eddy motions, 9
 evolution of systems, blown-ups, 33
 fluid motions, and concept of time, 11
 interactions of systems, 86
 Newton's laws of motion
  acting and reacting yoyo structures, 55
  colliding eddies and acting-reacting forces, 57, 59, 60, 61
  evolutions of converging and diverging fluid motions, 64–66
  second law, 56
 one-dimensional advection equation, 22
 three-body problem; See Three-body problem
 universal gravitation, 75–76
Convex curvatures, 106
Coordinate system (Cartesian), 2
Coriolis force, 112, 113
Corporate governance, 238–243, 258
Cosmic scale; See Celestial mechanics; Macro scale; Universe
Cosmos, Chinese versus English terms, 133
Cost, minimum, 6
Counterclockwise eddies, coexistence with clockwise eddies, 26
Counterclockwise rolling currents
 fog prediction, 327–329, 333
 high-temperature weather, 368
 meteorology, 301, 302
 no-fog weather prediction, 337
 wind- and sandstorm weather, 353
Counterclockwise rotation, 29
Cross-disciplinary studies, systems science, 1–2
Currents
 chaotic
  in rotating fluids, 258
  uniform flow transition into, 275
 eddy, 33
  and Newton's second law, 53–57
  quantitative unboundedness, 135
 rolling; See Rolling currents
 spinning, blown-up characteristics, 23–28
Curvature space
 blown-ups in, 33
 problems on physics of physical quantities, 129
 Riemann ball mapping relations, 22–23
 rotation and stirring energy, 92
 stirring energy and its conservation, 106–107
 time, Western concept of, 135
Curvature tensors, 102
Cutoff, 104
Cybernetics, 3

D

d'Alembert, Jean le Rond, 267, 289
Dalton, John, 8
Damping, 104
Dams, water control, 12, 103, 107, 109, 111
Decision making
 corporate governance
  by CEO, 248
  economic yoyos, 217, 218
  investor control, 242
  by management, 244
 households, 160
 investment, 13
Decision theory, 3
Decomposition, wind vectors, 304
Dedekind, Richard, 267, 268, 269
Deduction, 280
Definitions, circular, 269
Demand-supply, whole evolution analysis, 146–149, 150
Democritus, atom theory of, 8
Density
 infinite, gravitational collapse, 129
 law of conservation of informational infrastructure, 45
 law of conservation of informational infrastructure and, 44
 Newton's definition of mass, 125
 uneven, twisting force production, 26, 53
Density pressure, pressure gradient force as stirring gradient force of, 25
Derivation force, computational uncertainty, 30
Descartes, Rene, 265
Descartes second principle, 4, 5, 6
Descriptive-metaphysical science (Aristotelian), 5
Desertification, concrete-based, 366
Detection, singularities, 129

Determinism, irregularities as, 34
Dew point, V-3θ graph design, 306
Dialectic structure, 4
Dickens, W., 232
Differentiability, well-posedness of nonlinear evolution equations, 20
Differential equations, 26, 235
Dijkstra, B.R., 156, 157
Dimensions
 material; *See* Material dimensions, time and parametric; *See* Parametric dimensions, time and yoyo structures, measurement of, 86
Dirac's large number hypothesis, 36–40
Direction, time, 135, 136, 137, 138
Directionality of events, quantities and, 315
Direction of motion, meteorology, 297, 300
 nonuniformity, 303
 rolling currents, predictive value of, 306
 troposphere, vertical differences, 303
 windstorms and sandstorms, 340
Directorates, interlocking, 243
Disaster prediction, 14, 52; *See also* V-3θ graphs; Weather prediction
Disaster reduction (floods), 12, 14, 108, 109, 120
 analysis of facilities, 116–119
 evolution engineering and technology for, 107
 feasible technology, 113–116
Discontinuity, 19
 in nonlinear differential equations, 26
 nonlinear evolutions and, 8
 prediction of transitional changes, 33
 singularities in calculus solutions, 26
 three-body problem, open questions, 88
Discrete objects and forces, twist-ups versus, 32
Dishpan experiments, 68–70, 88, 160–161, 243, 258, 290
 and mathematic foundations, 275
Dispersion, energy transformations, 104
Dissipation, 104
Disutility of child's work, 190–201
Divergence; *See* Convergence/divergence
Divisiblity of magnitudes, 265–266
Donohue, J., 213
Drainage, flood water system structure, 111
Drainage systems, urban, 108, 109, 325
Dredging, 119–120
Du, G.P., 277, 281, 285, 288, 292
Du, Y.L., 339, 342
Dualism, Western, 125
Duality of eddies, 33
Dubois, P., 13
Duration of events; *See* Time frame/sequence/duration of phenomena

Duststorms; *See* Windstorms and sandstorms
Dynamic equilibrium, 92
Dynamic models, economic yoyos, 151
Dynamic systems, and eddy motions, 26–27

E

Earth
 atmospheric dynamics; *See* Atmosphere movement/dynamics
 Coriolis force, 103, 112, 113
 planetary scale; *See also* Meso scale; Scale
 structure of, law of conservation of informational infrastructure, 44–45
Earth-moon system, angular momentum, 41–42
Earthquake prediction, 46
Eastern thought, environmental basis, 8
Economics and finance, 2, 52, 62, 258
 child labor and its efficiency, 189–213
  adult child's altruism toward parents, 196–201
  conclusion, 212–213
  different efficiencies and potentially different outcomes, 201–203
  disutility of child's work, 190–201
  marginal bans on child labor, 204–212
  one-sided altruism model, 190–196
 complexity of social, political, economic, technological systems, 6
 economic eddies and different industry sizes, 215–224
  economic yoyos and their flows, 217–219
  simple model for imperfect capital markets, 222–224
  simple model for perfect capital markets, 219–222
 economic yoyos, 145–151
  whole evolution analysis of demand and supply, 146–149
  yoyo evolution of single cycle, 149, 151
 flood control methods, 112, 114
 households, 153–188
  Becker's Rotten Kid theorem, 155–165
  Bergstrom's rotten kid theorem and Samaritan's Dilemma, 179–183
  discussion, 186–188
  maximization of family income and child labor, 183–186
  never perfect value systems and parasites, 170–179
  two other mysteries of the family, 165–170
 long-term versus short-term projects, 237–253
  CEO choices of projects, 243–253
  conclusion, 253

price behavior of different investment projects, 244–246
yoyo model foundation for empirical discoveries, 238–243
stirring energy applications, 107–108
wage differentials, interindustry, 225–235
  companies with limited resources, 229–231
  financially resourceful companies, 227–229
  law of one price, 234–235
  literature review, 231–233
yoyo model applicability, 12–14
Economic yoyos, 13, 145–151
  economic eddies and different industry sizes, 217–219
  whole evolution analysis of demand and supply, 146–149
  yoyo evolution of single cycle, 149, 151
Eddies
  economic, 215–224
  nested and multiplicity of, 29; *See also* Sub-eddies/sub-sub eddies
  sub-eddy/sub-sub eddy creation, 31–32
  tendon of moving fluids is eddies, 47
Eddy current fields, prediction of, 26
Eddy currents, 33
  meteorology, fog, 338
  and Newton's second law, 53–57
  quantitative unboundedness, 135
Eddy effects
  atmospheric baroclinicity as, 305–306
  Bjerknes circulation theorem and, 25
  and Newton's laws
    colliding eddies and acting-reacting forces, 57–61
    eddy currents and second law, 53–57
    whole evolution analysis of demand and supply, 148
Eddy motions
  acting and reacting yoyo structures, 55
  Bjerknes circulation theorem, 47
  energy transformations, 92
  general system model, 9
  irregularities, formalization of, 30
  meteorology, baroclinic waves, 299
  Newton's second law of motion, 54
  Newton's third law generalization, 59
  nonlinearity as, 23
  stirring forces, 92
  time, 123
Eddy sources, 33
  Bjerknes circulation theorem, 25
  material structure, unevenness, 8
Effectors, systems components, 3

Efficiencies, child labor, 201–203
Efficiency, system, 3
Efficiency wage models, 233
Einstein, Albert, 106
  conservation of stirring energy and three-level energy transformation, 95
  general relativity theory, 27
  mass-energy formula, 12, 98–99
    law of conservation of informational infrastructure and, 44
    problems on physics of physical quantities, 128–129
    stirring energy and its conservation, 98–99, 106
  problems on physics of physical quantities, 128–129
  relativity theory, 10
  space-time unevenness
    and assumptions about materials structure unevenness, 31
    eddy motions, 26–27
    materials evolution, 148
  stirring energy, 110
    mass-energy formula, 98–99, 106
    rotationality in nonlinearity, 102
    rotation and, 92
  time, Western concept of, 134, 135, 136, 137
    limitations of quantitative physics, 123–124
    whole evolutions of converging and diverging fluid motions, 66
Elastic pressure effects, 25
Electrical force, Dirac's large number hypothesis, 36–37, 38
Electromagnetic interaction
  law of conservation of informational infrastructure, 43
  unification of basic forces, 39, 40
Ellipses, law of (Kepler's first law), 67
Elliptical orbits, 68, 71, 74
Energy
  law of conservation of informational infrastructure, 43, 44, 45
  problems on physics of physical quantities, 128, 129
  twist-up contents, 31–32
  yoyo model constants, 10
Energy conservation law, 43, 44, 128
Energy density, law of conservation of informational infrastructure and, 44
Energy technology, and scope of science, 6
Energy transformation
  conservation of stirring energy and three-level energy transformation, 94–95

eddies and, 26, 29, 92
flood water system structure, 112
quantitative methods and, 124
stirring energy and its conservation, 95–97
Engels, F., 29
Engineering, evolution, 12, 107–108; *See also* Flood disasters
Entering the Era of Irregularity (OuYang), 130
Entropy, 124
Environmental conditions, West versus East, 8
Environments, white boxes, 3
Epimenides, 269
Epistemology, 27, 29, 31, 63
　meteorology, 298, 300, 339
　quantities as postevent measurements, 339
Equal areas, law of (Kepler's second law), 67
Equal quantitative effects, 11, 62
　blown-ups, 28–32
　Newton's laws of motion, 61–64
　three-body problem open questions, 87–88
　whole evolution analysis of demand and supply, 148
Equations of motion, three-body problem, 78
Equilibria
　dynamic, 92
　eddy heat-kinetic forces, 26, 92
　redistribution of separable energy transformations, 104
　subeddy motion description, 87
Eubulides, 269
Euclid, 264
Euclidean space, 51, 107
　blown-ups in, 33
　momentum representation in, 92
　problems on physics of physical quantities, 129–130
　stirring energy and its conservation, 106
　time and, 139
　unboundedness problem with quantities, 128
Eudoxus, 262–264, 265, 288
Eudoxus construction, 263–264
Event analysis, 141
Events
　defining, 121–122
　evolution of, 124
　generality of, 127, 131, 140
　nonquantification of, 130–131
　physics of, 127
　quantities/quantitation and, 125–126, 127, 130, 131, 141, 339, 340
　structural characteristics, weather prediction, 354
　time origins, 136, 138

Evolution engineering and technology, 12, 107–108; *See also* Flood disasters
Evolution of systems
　automorphic, 8–9
　blown-up properties, 32, 33
　blown-up theory, 7–8, 19–20
　economics
　　corporate governance, 238–243
　　demand-supply analysis, 146–149, 150
　inherence of forces in structures of evolving objects, 27
　mathematics applicability, 30
　meteorology, 300
　　fog, 338
　　suddenly appearing severe convective systems, 314
　　V-3θ graphs, 341
　origins of systemic yoyo, 7–11
　space-time distribution, 86
　stirring energy and its conservation, 95–96, 105, 106
　time, 122–123, 124
　transitionality of, 34
　whole evolutions of converging and diverging fluid motions, 64–66
Excluded middle principle, 270, 280
Executive investment decision making, 13
Exhaustion, method of, 266
Extended congruence, 259
Extensionality, theory of, 272
External forces
　elimination of requirements for, 11
　first push; *See* First push, Newtonian

F

Fair-wage models, 226–227
Family yoyos; *See* Household economies
Fechner, Gustav, 4
Fermat, Pierre de, 266
Field function, and gravitational pull and acceleration equations (Newton's second law generalization), 87
Fields
　eddy, kinetic energy transformations, 25–26
　eddy current, prediction of, 26
　spin; *See* Spin fields
　spinning current, 26
Fields of flows, stirring energy and its conservation, 106
Figurative (structural) analysis, 34, 52, 131
　applications of systemic yoyo model, 11
　Newton's laws of motion, 61–64

three-body problem, 79–86
  all visible, 79–83
  with at least one invisible, 83–86
  n-nary star system possibility, 82–83
  open questions, 87–88, 89
  universal gravitation, 75
  V-3θ graphs, 298; See also V-3θ graphs, case studies using
Figurative structures
  equal quantitative effects, 31
  quantity abstraction into, 30
Finance; See Economics and finance
Finite constructability, 270
Finite sets, 270
First-level circulation, 104
First-order equations, simplification of higher-order nonlinear evolution systems, 21
First push, Newtonian, 53
  evolution engineering and, 119
  flood prevention, 109, 110, 119
  governance law of slaving energy, 97–98
  meteorology, 296, 298, 300, 302
  and second stir, 31
  stirring energy and its conservation, 110
  weather prediction, 296–297, 298, 300, 302
Fjörtoft, R., 95
Flood disasters
  facility analysis, 116–120
    computation on artificial lake at Fuhe Bridge along northern outer-ring road, 116–118
    estimate for artificial lake along Qingshu River on upper reaches of Nan River, 118–119
    forecasting severe precipitation, 325
  prevention and disaster reduction in urban areas, feasible technology, 12, 113–115
    artificial urban lake design and river course drainage problem, 115
    computational method for Q_1, 114–115
  principles of flood evolution and development, 109–113
    law of conservation of stirring energy, 110
    stirring energy and its conservation, 103
    three-ring quasi-stability problem of urban flood movements, 110–113
  stirring energy applications, 12
Floodwater storage in lakes; See Lakes
Flow fields
  rolling current effects, 306
  stirring energy and its conservation, 106
Flow function, angular speed representation, 93–94
Flow rates, stirring energy and its conservation, 103
Flows
  rotational, 95
  transitions into chaotic currents, 258, 275
Fluid models, 63
Fluid motions
  and Asian thought, 8
  Bjerknes circulation theorem, 24
  eddy motion model of general system, 9
  equations of, 97
  time, 123
Fluid resistance, evolution engineering, 119
Fluids, Western epistemology, 29, 31
Fluxions, 266–267
Fog, 325–339
  background, 326
  case studies, 327–338
    after-rain fog prediction, 330–335
    counterclockwise rolling current analysis of graphs, 327–328
    description of fog situation, 327
    heavy fog prediction problem, 328–329
    light fog prediction problem, 329–330, 331
    no-fog weather and fog dissipation prediction, 335–338
  discussion, 338–339
  rolling currents and, 296
Force(s)
  and acceleration, Newton's second law, 99
  inherence in structures of evolving objects, 27, 28
  meteorology, 297
  moments of, 31
  Newton's second law of motion, 53–54
  physics of physical quantities, 125, 128
  structural changes as sources of, 316
  structural unevenness and, 86
  twist-ups, 32
  unification of basic forces, 39, 40
  Western dualistic model, 125
Forcing systems, 27
Forecasting; See Prediction/forecasting
Formalist school, 269, 271
Formalization of eddy irregularities, 30
Formal logic, 131, 265
Fourth dimension, 134, 135, 139
Fourth law on state of motion, 61
Fouzhou, 355, 356–361, 364
Free market model, 12–13
Frege, F.L.G., 269
Fronts, meteorological, 31–32
Fultz dishpan experiment, 68–69, 160–161
Functionalities
  physical quantities and, 126
  time as attribute of materials, 136, 137
Fuzhou, 361

G

Galilei, Galileo, 4, 5, 6, 134
Game theory, 3
Gas constant, 307
Gauss, C.F., 267
Geiger method, hydrology, 114
General dynamic systems, and eddy motions, 26–27
Generality
  of events
    average acceleration description of, 127
    quantitative averages and, 131, 140
  from general to particular and back, 5–6
Generalization of laws of motion, 11, 12
General observ-control, blown-ups, 23
General relativity theory, 27, 106
General system, 36
Geometry, 34, 261–265
  analytic, 265
  equal quantitative effects, 34
  pure, 265
  yoyo model, 10
Godel, Kurt, 271
Gong, N.S., 277, 281, 285, 288, 292
Governance, corporate, 238–243, 258
Governance law of slaving energy of Newtonian first push, 97–98
Governing boards; *See* Boards of directors
Government systems
  marginal bans on child labor, 204–212
  regulation of corporations, 238–243
Gradients
  atmospheric pressure, 45, 297
  stirring, 25
  unevenness of materials structures, 31, 86
Grand unification theory, 39, 40
Graphic representations of numerical relationships, 63–64; *See also* Figurative analysis
Graphic structure, weather; *See* V-3θ graphs
Gravitation
  black hole formation, 44
  field function for describing, 87
  Newton's second law of motion, 54–56
  space-time unevenness, 27
  three-body problem; *See* Three-body problem
  unevenness of materials structures and, 148
  unification of basic forces, 39
  universal, 11
    Dirac's large number hypothesis, 36–37, 38, 39
    stirring energy and its conservation, 99–102
Gravitational anomalies, 102
Gravitational collapse, 129–130
Gravitational constant, 37, 72, 101, 286
Gravitational waves, 130
Greek mathematics, 261–265
Groshen, E.L., 231
Grundlagen der Geometrie (Hilbert), 271
Guilin, 333, 334
Guiyang, 327, 330, 331

H

Hailperin, T., 259
Halley, Edmund, 266
Hamiltonian, three-body problem, 78
Happy family; *See* Family yoyos
Harmonics
  Kepler's third law
    defined, 67
    Newton's generalization, 72–74
  and three-level energy transformation, 94
Harmonic spinning patterns
  acting and reacting yoyo structures, 55, 56, 57, 58, 59
  defined, 54
  and financial markets, 240
  three-body problem, 79, 82, 83
  universal gravitation, 75
Harmony, family, 187; *See also* Household economies
Harriot, Thomas, 265
Hausdorff, Felix, 273
Heat/heat energy
  eddy motion energy transformations, 25–26, 29
  entropy, 124
  internal, of eddy motions, 92
  weather phenomena
    photochemical reactions, atmospheric, 309–310
    V-3θ graphs, specific heat capacity, 308
Heat waves/abnormally high temperatures, 354–368
  background, 355–356
  under cold high pressure, 364–367
  under subtropical high in West Pacific, 356–364
Hegel, G.W.F., 4
Hess, S.L., 149
Hide dishpan experiment, 69–70, 160, 161
Hierarchy of monads (Leibniz), 4
Higher dimensions, yoyo model, 10–11
Higher-order nonlinear evolution systems, simplification of, 21
High-pressure conditions, high-temperature weather
  under cold high pressure, 364–366
  under subtropical high in West Pacific, 356–364
Hilbert, David, 269, 271

Hippasus paradox, 261–262
Historical review, systems, 2–7
Holding capacity of artificial lake, 116
Holding capacity of second circulation, 113
Hongzhou, 355, 361, 363, 364, 366
Horizontal field of masses, rolling currents, 302
Horizontal weather scale, meteorology, 301
Hot island effect, urban areas, 356
Household economies, 153–188, 258
　Becker's Rotten Kid theorem, 155–165
　Bergstrom's rotten kid theorem and Samaritan's Dilemma, 179–183
　discussion, 186–188
　maximization of family income and child labor, 183–186
　never perfect value systems and parasites, 170–179
　two other mysteries of the family, 165–170
Hu, Q.P., 257
Hu, Z.W., 40, 41
Hubble constant, 37
Huhhot sandstorm weather, 347–348
Human behavior; *See* Household economies
Human capital, 13, 154; *See also* Child labor; Wage differentials, interindustry
Humanities, yoyo model applicability, 12–13
Human-machine relationships, 6
Human relationships; *See also* Household economies
　spinning yoyo structure, 46–47
　three-dimensional spinning yoyo structure, 10–11
Human thought; *See also* Mathematics
　systemic yoyo structure in, 14, 258–259
　time concept, 125, 126, 135, 136
　　in China, 132–133
　　in West, 134–136
　understanding actual versus potential infinities, 14
Humidity, atmospheric water vapor; *See* Water vapor/humidity, atmospheric
Hydraulics phase, rolling currents, 300
Hydrogen atom, Dirac's large number hypothesis, 36–37
Hydrology, stirring energy and its conservation, 103, 104, 107–120; *See also* Flood disasters

I

I Ching (*Book of Changes*), 8–9, 29, 63, 132, 302
Identification of systems, 3
Images, Tao of, 31, 34
Imperfect capital markets, 222–224
Indeterminacy problems, 124
　evolution science and, 127
　narrow observ-control transformation into determinant situations of more general observ-control systems, 23
Indirect (secondary) circulations
　flood water system structure, 12, 111, 112, 113, 120
　laws of physics and, 104
Individual behavior; *See* Household economies; Human relationships
Industry size and economic eddies, 215–224
　economic yoyos and their flows, 217–219
　simple model for imperfect capital markets, 222–224
　simple model for perfect capital markets, 219–222
Inertial system
　classical mechanics, 134
　foundations of modern science, 126–127
　variable acceleration and, 126
Inevitability of orderlessness, 9
Infinite, theory of, 268
Infinite divisibility of magnitudes, 265–266
Infinite sets, existence of, 270, 273, 281–283
Infinitesimal change, 266
Infinitesimals, 264, 266, 291
Infinities
　actual and/or potential, 14, 258–259, 275, 276–281, 292
　Berkeley paradox, 286, 287, 288
　concepts, 276–279
　equivalence of, 280–281, 289
　problems on physics of physical quantities, 130
　quantitative, 129
　　blown-ups, 23–28
　　resolution of problems, 33
　　V-3θ graphs, 315, 341
　time, 123
Information
　irregular; *See* Irregular information
　small-probability, 31–32, 34
　as symbols of events, 131
Informational discontinuity, V-3θ graphs, design of, 306
Informational infrastructure, conservation of; *See* Conservation of informational infrastructure
Information content, uniformity of all imaginable things, 36
Information order, meteorology, 297–298, 340, 341
Information theory, 3
Infrastructure, informational; *See* Conservation of informational infrastructure
Initial field, whole evolutions of converging and diverging fluid motions, 64

Initial force; *See* First push, Newtonian
Initial movement, Newton's cannonball, 68
Initial state
 convergence versus divergence, 33
 whole evolutions of converging and diverging fluid motions, 66
Initial value problem
 evolutions of converging and diverging fluid motions, 64
 mathematical characteristics of blown-ups, 21
Initial values
 blown-ups, 355
 nonlinear system evolution, 8–9
 time, 123
 transitional changes and, 326
Initiation, competing tendencies in nature, 8
Instabilities, computational, 22, 33
Integers, 261
Integral of motion (Jacobi integral), 78
Integrals, 131
Integration
 Archimedes and, 266
 seventeenth century mathematics, 266
Interactions of materials, problems on physics of physical quantities, 128
Interactions of systems, 3
 calculations, 86
 economic yoyos, 151
 family yoyo, 162
 three-body problem, 79, 80, 81
Interconnectedness, 1–2
Interdisciplinary studies, systems science, 1–2
Interference
 Kepler's laws of planetary motion, 71
 observation/measurement effects, 28
Interindustry wage differentials; *See* Wage differentials, interindustry
Interlocking directorates, 243
Internal heat, eddy motions, 92
Internal set theory, 275
Intervals; *See* Periodicity; Time span/intervals
Intuition, 13
Intuitionist school, 269, 270–271, 292
Investment decision making, 13
Invisible organization of blown-up systems, 29
Inwardly spinning pools, 31
Irrational numbers, 261–262, 265
Irregular events
 nonquantification of events, 130–131
 physics of physical quantities and, 140
Irregular information, 31–32, 124
 changes, 127
 meteorology, 298, 302, 324, 339
 V-3θ graph design, 306
 V-3θ graph predictive value, 315, 316–317
 nonquantification of events, 128
 physics of physical quantities and, 141
 structural characteristics, 316
Irregularities
 eddy/eddy motion, 29
 eddy motion model of general system, 9
 formalization of, 30
 evolution, mathematics applicability, 30
 regularization, concept of, 123
 transformations of rotations, 32
 unevenness of space and time of evolutionary materials structures and, 9
Irregular structural analysis, meteorology, 298
Irrotational kinetic energy, 12
 conservation of, 97–98
 interactions, 98–99
 nonconservation of, 102
 stirring energy and its conservation, 97

J

Jacobi integral, 78
Jacobi operator, 64
Jie, 133
Jinzhou flying dust weather, 345–347

K

Kahneman, Daniel, 216
Katz, L., 231, 232
Kawai set theory, 275
Kepler, J., 266
Kepler's laws of planetary motion, 2, 11, 67–76, 258
 law of harmonics and Newton's generalization, 72–74
 Newton's cannonball, 67–72
 universal gravitation, 74–76
Kinematic equation, eddy motions, 26–27
Kinetic energy
 eddy motion energy transformations, 25–26, 29, 92
 irrotational, 12
 law of conservation of, 12
 nondivergent flows, 95
 rotational flows, 95
 squared speeds, angular and linear, 93
 stirring; *See* Stirring energy and its conservation
 stirring energy and, 106, 110
Klir, George, 1, 2, 6, 7, 36, 105

Kronecker, L., 270
Krueger, A.B., 216
Krueger, D., 213
Kuchemann, D.F., 25, 47
Kuhn, T., 4
Kunming, 329, 330, 335, 337

L

Lagrange, J.L., 267
Laissez faire triple of child labor, 13; *See also* Child labor
Lakes
  artificial
    along Qingshu River on upper reaches of Nan River, 118–119
    design of, 115
    at Fuhe Bridge along northern outer-ring road, 116–118
  and flooding, 103
  floodwater storage, 92
  structure of floodwater system, 111
  Yangtze River, 110, 111
Land-ocean breezes, 25, 33, 47
Lao Tzu, 8–9, 27, 29, 31, 103, 132
Large number hypothesis of Dirac, 36–40
Law of harmonics, 72–74
Law of one price, interindustry wage differentials, 234–235
Lawrence, C., 232
Lawrence, R., 232
Laws of conservation; *See* Conservation laws
Laws of motion
  generalization of, 11
  Newton's; *See* Newton's laws of motion
Laws of myriad things, 135
Laws of physics, and second-level circulation, 104
Lazy rotten kids, 13, 157–165
Learning and adaptation, system, 3
Leeward wave theory, 31
Leibniz, G., 4, 8, 135, 136, 138, 265, 266, 269, 286
Li, D., 257
Li, X.F., 115
Li, Z.L., 339
Liability, system, 3
Li Bai, 133
Li Bing, 109
Life span of movements
  Earth atmosphere, 42–43
  economic yoyos, 151
Light, speed of, 128
Lijiang, 335, 337, 338
Limits, 291

Limits theory, Cauchy, 268, 287
Lin, Y., 10, 11, 20, 22, 24, 31, 34, 35, 51, 52, 61, 62, 64, 86, 88, 131, 148, 149, 154, 180, 181, 189, 258, 277, 281, 285, 288, 292, 295, 300, 326, 339, 341, 342, 354, 361
Linear analysis, limitations of, 22
Linearity
  predictive ability, 19
  as special case of nonlinearities, 29
Linear manifolds, and nonlinear tensors, 129
Linear models, eddy motions, 29
Linear speed
  angular momentum representation, 92
  versus angular speed, 95
  stirring energy and its conservation, 107
Line segments, 261, 262
Linhe sandstorm weather, 347–348
Literature review, wage differentials, 231–233
Li Yizhi, 108
Local characteristics of eddy motions, 29
Local conditions, blown-up absence, 33
Location
  physics of physical quantities, 125
  time as attribute of materials, 136
Logarithms, 265
Logic
  formal, 131, 265
  symbolic, 269
Logical calculus, two-value, 280
Logistic school, 269, 270
Long-term and long-effect technology, stirring energy and its conservation, 119–120
Long-term versus short-term business projects, 237–253
  CEO choices of projects, 243–253
    dynamics of projects, 247–253
    price behavior of different investment projects, 244–246
  yoyo model foundation for empirical discoveries, 238–243
Long waves, Rossby's, 31
Lorenz chaos model, 139
Lu, Shi-jia, 26

M

Macro (cosmic) scale; *See also* Scale; Universe
  Big Bang theory, 27–28, 34, 45
  black holes, 44, 71
  eddy motions, 25, 29
  three-body problem; *See* Three-body problem
  three-ringed stability, stirring energy conservation, 105

Magnitudes, infinite divisibility of, 265–266
Makeup coefficient, 115
Management, long-term versus short-term projects, 238–243
Manufacturing, stirring energy applications, 107–108
Mapping
 blown-ups, 23–28
 Riemann ball, 22–23
 weather, smoothing of irregular information, 299
Marginal bans on child labor, 204–212
Market competition, and corporate governance, 238–243
Markets; *See* Capital markets
Marx, K., 4
Mass
 Dirac's large number hypothesis, 37
 as function of time, 126
 Kepler's law of harmonics, 73–74
 law of conservation of informational infrastructure, 43, 44, 45, 46
 Newton's definition, 125
 three-body problem; *See* Three-body problem
 variable acceleration and, 126
 yoyo model constants, 10
Mass conservation laws, 44
Mass density, law of conservation of informational infrastructure and, 44
Mass-energy conservation, problems on physics of physical quantities, 128
Mass-energy formula, Einstein
 law of conservation of informational infrastructure and, 44
 problems on physics of physical quantities, 128–129
 stirring energy and its conservation, 98–99, 106
Material dimensions, time and, 121, 138–140, 295
 Chinese concept of time, 132, 133
 quantities as parametric dimensions, 125
 Western concepts of time, 134–135
Material flows, time, 123
Material invariance, average acceleration and, 127
Materials
 evolution science, 126–127
 time as attribute of, 136, 137
 twist-up contents, 31–32
 Western dualistic model, 125
Materials evolution
 evolution engineering, 120
 narrow observ-control, 29
 stirring energy and its conservation, 105
 time, 122–123, 124, 136, 137, 138
 time as attribute of materials, 139

Materials science, and scope of science, 6
Materials structures
 evolutionary, unevenness of space and time of, 9
 evolution of Western thought, 8
 evolution science, 127
 force origination in, 125, 316
 human thought structure and, 257–258
 meteorology, 297, 300
 and nonlinearity, 34
 time, 122
 unevenness, as sources of nonlinearities/eddys, 8
Materials transformations
 physics of physical quantities and, 141
 time as attribute of materials, 139, 140
Material world, human thought structure and, 257–258
Mathematical-positivistic science (Galilean), 5
Mathematical Principle of Natural Philosophy (Newton), 134
Mathematical structures of systems, 3
Mathematics
 blown-ups, 20–22, 33
 chaos, 89
 definition of system, 36
 eddy motions, 29
 evolution of Western scientific principles, 8
 hidden contradictions in system, 275–292
  actual versus potential infinities, 276–279
  actual versus potential infinities, equivalence of, 280–281
  Berkeley paradox, 285–288
  Cauchy theater phenomena, 283–285
  fourth crisis, 288–292
  infinite sets, existence of, 281–283
 history of, 257–273
  beginning, 260–261
  first crisis (classical philosophers), 261–265
  second crisis (17th-19th centuries), 265–268
  third crisis (19th-20th centuries), 268–273
 irregularities of materials evolutions, 62
 nonquantification of events, 130–131
 time, 123–124
 time quantification, 134
 twist-up zone description, 32
 yoyo model; *See* Circulation theorem of Bjerknes
 yoyo structure, 14
Mathesis universalis (Leibniz), 4
Matter, creation of, 37
Maximization of family income and child labor, 183–186
Maximum economic effect and minimum cost, 6
Measurement, quantities as post-event measurements, 130, 339, 340

Measurement of yoyo structure, 86
Measurement uncertainty, 28–29
Mechanics
  laws of
    generalization of, 11, 12
    Newtonian; *See* Newton's laws of motion
    three-body problem; *See* Three-body problem
Medicine, Chinese, 3–4
Medoff, J., 231
Mengzi, 331
Merit goods, 13, 154–155
Meso (planetary) scale; *See also* Atmosphere movement/dynamics; Planetary scale; Scale
  Earth atmosphere, analysis of movements, 42; *See also* Atmosphere movement/dynamics
  Earth rotation and river flows, 103
  eddy motions, 25
  flood control, 112, 113
  stirring forces, 92
  weather prediction; *See* Weather prediction
Meteorology, 52; *See also* V-3θ graphs; Weather prediction
  Bjerknes circulation theorem, 24–25, 47
  Earth atmosphere, analysis of movements, 42–43
  equal quantitative effects, 30
  law of conservation of informational infrastructure, 46
  prediction of zero probability disasters, 14
  Rossby's long waves, 31
  three-ringed stability, stirring energy conservation, 105
  twist-ups, 31–32
Method of exhaustion, 266
Methodology, systems science, 3
Micro (atomic) scale
  eddy motions, 25, 29
  equal quantitative effects, 28
Middle congruence, 259
Moment of momentum conservation, 43, 44
Moments of forces, 31, 32
Momentum
  law of conservation of, 43, 44, 128
  rotation and stirring energy, 92
Monad hierarchy (Leibniz), 4
Motion/movement
  clockwise and counterclockwise rotation, 29
  equal quantitative movement, 28
  law of conservation of informational infrastructure, 46
  Newton's laws of; *See* Newton's laws of motion
  orbital
    angular momentum, 41

Newton's first law rewriting, 53
  uniformity of all imaginable things, 36
Motivators, systems components, 3
Mo Tzu, 125, 133
Mo Zi, 125, 133
Multicausality, 5
Multidimensionality, 9, 10
  yoyo model, 148–149
Multilayered irregular structure, 3θ curves, 311
Multiplex eddies, 29; *See also* Sub-eddies/sub-sub eddies
Multiplicities
  evolution of universe, 28
  unevenness of space and time of evolutionary materials structures and, 9
Multirelationship, 5
Multiset semantics, concrete, 259
Murphy, K., 226
Mutual interference, observation/measurement effects, 28

N

n-nary star systems, 11, 14, 77, 79–83, 84, 86, 89
Naive set theory, 273, 276, 289
Nanchang, 355, 361, 362, 366
Nanjing, 355, 361, 363, 366
Nanning, fog prediction, 330–331, 332
Nan River, 118–119
Napier, John, 265
Narrow observ-control
  blown-ups, 23
  discontinuous singularities in calculus solutions, 26
  eddy motions, discontinuous singular evolutionary characteristics of, 23–24
  multiplicity of materials evolution, 29
Natural numbers, 268, 283–285
Natural sciences, unification with social sciences, 31, 34, 63
Negative vorticity, 66
Nested eddies, 29; *See also* Sub-eddies/sub-sub eddies
Never perfect value systems, 170–179
Newton, Isaac, 8, 124, 266, 286
  conservation of stirring energy and three-level energy transformation, 95
  and Kepler's laws of planetary motion
    cannonball, 67–72
    generalization of third law, 72–74
    universal gravitation, 74–76
  prime mover; *See* First push, Newtonian
  rotation and stirring energy, 92
  and rotations, 92
  time, 136, 137

Western concept of, 134
whole evolutions of converging and diverging fluid motions, 66
universal gravitation, 11, 74–76, 102, 258;
See also Universal gravitation
Newton's laws of motion, 2, 11, 12, 28, 51–66
basis of, 8
calculus creation, 265
equal quantitative effects and figurative analysis, 61–64
first
rewriting based on yoyo model, 53
second stir and, 52–53
generalization of, 11
meteorology, 296–297
particle assumption, eddy evolutions and, 27
problems on physics of physical quantities, 129
second, 148, 186
Bjerknes circulation theorem, 24
eddy effects and, 53–57
and eddy motions, 26
force-acceleration equivalence, 99
generalization of, 56, 87
problems on physics of physical quantities, 129
with uneven structures, 33
variable acceleration and mass, 126
stirring energy concept and, 110
third, 87
colliding eddies and, 57–61
and Kepler's law of harmonics, 73
problems on physics of physical quantities, 128
stirring energy and its conservation, 98–99, 110
whole evolutions of converging and diverging fluid motions, 64–66
Nicholas of Cusa, 4
Nightlight controversy, 13, 167–169, 181–183
Nio, T., 45
Nondifferentiability, three-body problem open questions, 88
Non-Euclidean space; See Curvature space
Nonexpansion method, whole evolutions of converging and diverging fluid motions, 64–65
Noninertial systems, evolution science, 126–127
Non-initial value automorphic evolutions, 8–9
Nonlinear equations
discontinuities in, 26
whole evolutions of, 19
Nonlinear evolution

blown-up theory of, 7–8
constant coefficient equation, 20
initial values and, 8–9
mathematical characteristics of blown-ups, 21
Nonlinearity
Bjerknes circulation theorem and, 25
blown-ups, 33
eddy motions, 9
and eddy motions, 33
eddy motions, discontinuous singular evolutionary characteristics of, 23–24
origins in figurative structure, 28
problems on physics of physical quantities, 129
rotationality in nonlinearity, 102
rotations of materials, 127
as singularities, 23
solenoidal fields and problems on universal gravitation, 101, 102
supply-demand relationship, 148
Nonparticle materials evolution, 34
Nonparticle spinning yoyos, stirring energy, 12
Nonquantification of events, time and its dimensionality, 127–131
nonquantification, 130–131
physics of physical quantities, problems on, 128–130
Nonstandard set theory, 275–276
Nontransitional blow-up, defined, 20
Not-well posed quantities, 128
Number systems, 63
Numerical instability, resolution of problems, 33
Numerical values; See also Quantification
abstraction into figurative structures, 30
calculus, 51

O

Objects
calculus-based analysis, 51
Western dualistic model, 125
Object set, 36
Objects of systems, 3
Observation, particle assumption, 28–29
Observ-control; See Narrow observ-control
Ocean movements, Bjerknes circulation theorem, 24
One-dimensional advection equation, 22
One-sided altruism model, 190–196
Opposites, coincidence of (Nicholas of Cusa), 4
Optimization, systems, 3
Orbital angular momentum, 41
Orbits, three-body problem, 68, 71, 74, 77, 78, 89
Order, systems theory, 5

Orderlessness, inevitability of, 9
Order structure
  meteorology, 297
  prediction of severity of weather, 340
  windstorms and sandstorms, 341
Organismic biology, 2
Organization
  of blown-up system, 29
  order structure, 297, 340
  systems theory, 5
Organization, theories of, 3
Organization levels, systems science, 1–2
Orthogonal plane divergences, 94
Oughtred, William, 265
Outside contractors, 233
Outwardly spinning pools, 31
OuYang, Shoucheng, 2, 7, 10, 25, 29, 30, 31, 35, 47, 52, 62, 63, 92, 94, 99, 104, 106, 107, 113, 126, 129, 130, 131, 138, 139, 141, 148, 149, 295, 296, 314, 315, 326, 339, 340, 341, 342, 354, 361
Ownership of companies, comparison of different countries, 243–244
Oxyhydrogen atoms, 309, 310

P

Pairwise interactions, three-body problem, 80
Panrelativity theory (Xuemou Wu), 29
Paradoxes
  set theory, 268–273
  ZFC system and, 289
Parametric dimensions, time and, 121, 122, 124, 127, 138–140
  chaos theory, 141
  classical mechanics, 134–135
  physical quantities and, 125, 141
  quantities as parametric dimensions, 125
  rotations and, 124
  Western concept of, 134
Parasites, 170–179
Pareto improvement/optima, 157, 170, 180, 181, 190, 203, 206, 207, 209, 210, 211, 212, 213
Particle assumption
  calculus-based analysis, 51
  eddy evolutions and, 27
  observation effects, 28–29
Particle mechanics, 30, 126
  Bjerknes circulation theorem, 47
  quantitative space and time and, 134
Part-whole relationships, 3–4
Pascal, Blaise, 265, 266

Peak flows, flood evolution and development, 112–113
Peano, Giuseppe, 267, 269
Pearson type III curves, 112
Peng, T.Y., 31
Perfect capital markets, 219–222
Periodicity
  crises in foundation of mathematics, 291
  period between conjunctions, three-body problem, 78
  transitional blow-ups, constant coefficient equation, 20
Phase space, P-T, 315
Philosophiæ Naturalis Principia Mathematica (Newton), 134
Photochemistry, atmospheric, 299–300, 309, 310
Physical quantities
  figurative analysis, 62
  physics of, 125–127
    parametric dimensions in, 139–140
    time and its dimensionality, 125–127
    time quantification and, 134
  time as, 124
Pi, graphic representation of, 63, 64
Pingliang sandstorm weather, 349–350
Planar infinity, Riemann ball mapping relations, 22–23
Plane divergences, orthogonal, 94
Planetary motions
  eddy motions, 25
  Kepler's laws; *See* Kepler's laws of planetary motion
  three-level circulation transformations, 104
Planetary scale; *See* Meso scale; Scale
Plato, 134, 276, 277
Platonic school, 265
Poincare, Henri, 270, 272–273
Policy
  complexity of social, political, economic, technological systems, 6
  sensor and effector performance, 3
Political systems
  complexity of social, political, economic, technological systems, 6
  marginal bans on child labor, 204–212
Pools, inwardly and outwardly spinning, 31
Popper, Karl, 5
Potential infinities; *See* Infinities, actual and/or potential
Power, Newtonian first push, 98
Power set axiom, 281–282
Power struggle, between boards of directors and CEOs, 218–219, 251–253

p-plane, Bjerknes circulation theorem, 24–25, 47
Practice, from practice to theory and back, 5–6
Prediction/forecasting
  calculus-based model limitations, 51–52
  discontinuous transitional changes and eddy field appearances, 26
  economic yoyos, 13
  evolution of systems, 33, 106
  figurative analysis, 34
  flood prevention/reduction, 113
  law of conservation of informational infrastructure, 46
  measurement of yoyo systems, 86
  physics of event processes, 127
  reversal and transitional changes, 19
  structural analysis and, 131
  time, 124
  weather; *See* V-3θ graphs; Weather prediction
  zero probability disasters, 14
Pressure; *See also* Barometric pressure
  conventional meteorology, 297
  order of information, 297–298
  3θ curves at different altitudes, 310
  V-3θ graph design, 307
Pressure gradient force, as stirring gradient force of density pressure, 25
Price behavior, 13, 234–235
  investment projects; *See* Long-term versus short-term business projects
  whole evolution analysis of demand and supply, 146–149, 150
Prigogine, I., 107, 136, 137
Primary circulation, flood water system structure, 111
Primary resonance, 78
Prime mover, 11, 27; *See also* First push, Newtonian
Principia Mathematica (Whitehead and Russell), 270
Principle of excluded middle, 270, 280
Processes, problems on physics of physical quantities, 129
Prodigal son, 13, 169–170
Profit maximization model, 13
Programming, systems, 3
Proof by contradiction, 280
Properties
  of blown-ups, 32–34
  of materials
    physical quantities and, 126
    time, attributes of, 136
Psychophysical law (Fechner), 4
P-T coordinate, V-3θ graph elements, 306

P-T phase space, 315
Public policy, 6
Pulling forces
  colliding eddies and acting-reacting forces, 58, 59
  Newton's second law of motion, 54–56
  universal gravitation, 75–76
Pushing forces
  colliding eddies and acting-reacting forces, 58, 59, 60
  Newton's second law of motion, 54–56
  universal gravitation, 75–76
Pythagoreans/Pythagorean school, 261–262, 263, 264, 288
Pythagorean theorem, 265, 288

Q

Qingshu River, 118–119
Qualitative means (intuition), 13
Qualities, quantification by Newton, 125
Quantification
  applicability of mathematics to irregular evolution, 30
  figurative analysis, 62, 63–64
  nonquantification of events, 130–131
  physics of physical quantities, 125–127
  time, 122
Quantitative analysis
  linear versus angular speed concepts, 95
  and materials evolution, 107–108
  processes and paths, 124
Quantitative averages, 140
Quantitative effects, equal, 11, 28–32
Quantitative forces
  conversion back to uneven structures of materials, 296
  meteorology, 300
Quantitative infinity; *See* Infinities
Quantitative irregularities
  multiplicity of materials evolution, 29
  unevenness of space and time of evolutionary materials structures and, 9
Quantitative parametric dimensions, 138–140
Quantitative singularities, 130
Quantitative time, 134, 137
Quantities, physical
  events and, 127, 131
  forms of complications, 303
  meteorology, 298
  physics of, 125–127
  as post-event measurements, 130, 339, 340
  time, 122, 131

time quantification, 134
Western dualistic model, 125
Quantum events, 131
Quantum mechanics, 8, 123–124
Quasi-closed systems, three-level circulation transformation as, 104
Quasi-equal computational uncertainty, 30
Quasi-stability
 stirring energy and its conservation, 106
 three-level circulation transformations, 104
Quasi-three-ringed stirring energy, 103
Quasi-vertical structures, wind- and sandstorm weather, 353
Quastler, H., 3

R

ρ
 eddy motions, 27
 and mutual reactions of systems, 87
ρ plane, Bjerknes circulation theorem, 24–25, 47
Rains/rainstorms, 2, 46, 314
 after-rain fog prediction, 330–335
 flood evolution and development, 112
  floods versus disastrous floods, 111, 112
  maximum intensity Fuhe River, 116–118
  runoff computation, 114
 order of information, 298
 regional thunderstorms, 325–339
 severe
  Beijing, 317–321
  forecasting, 325
  prediction of, 2, 309
  regional heavy rain, 322–324
Randomness, 124, 127
Rational numbers, 261
Real number system, 256–268
Real number theory, 268, 271
Regional heavy rain, suddenly appearing severe convective systems, 322–324
Regional short-lived fog and thunderstorms, 325–339
 background, 326
 case studies, 327–338
  after-rain fog prediction, 330–335
  counterclockwise rolling current analysis of graphs, 327–328
  description of fog situation, 327
  heavy fog prediction problem, 328–329
  light fog prediction problem, 329–330, 331
  no-fog weather and fog dissipation prediction, 335–338
 discussion, 338–339

Regularization, concept of, 123
Regulations, sensor and effector performance, 3
Relativity theory, 8, 10, 44, 123–124
Ren, Z.Q., 10, 35, 40, 41, 46, 149
Repellence, universal gravitation, 75–76
Research problems, as black boxes, 3
Resistance, fluid, 119
Resonance, primary, 78
Resourceful father example, family yoyos, 174–178
Reversals, blown-up theory, 8
Ricatti equation, 20
Riemann, Georg Bernhard, 267
Riemann ball mapping relations, 22–23
Riemann geometry, 106, 129, 133
Right-hand rules, 303
Rivers
 direction of course (left versus right bending), 113
 flood control; *See* Flood disasters
 stirring energy and its conservation, 103
Robinson, A., 273
Robinson, J., 13, 154, 184, 186, 191, 194, 197
Rolling currents, 338–339
 changes in direction, 315
 effects of, 306
 fog prediction, 327–329, 333
 high-temperature weather, 356, 359, 368
 no-fog weather prediction, 337
 3θ curves, 311
 vertical, 296
 weather evolution, 299–305
 wind- and sandstorm weather, 353
Rosen, S., 226
Rossby's long waves, 31
Rotating fluids
 chaotic currents in, 258
 uniform flow transition into chaotic currents, 258, 275
Rotation
 angular speed of rotation squared, 12
 Bjerknes circulation theorem, 25, 47
 as common form of movement in universe, 34
 conservation of stirring energy and three-level energy transformation, 93
 curvature space, 33
 duality of eddies, 33
 evolution engineering, 120
 kinetic energy of rotational flows, 95
 meteorology, 296, 297
  rolling current direction, 303–305
  rolling current evolutions, 300–301, 302
 moments of forces, 32
 nonlinearity as rotationality of solenoidal fields, 129

physics of physical quantities and, 141
problems on physics of physical quantities, 129
rotationality in nonlinearity, 102
solenoidal stirs, 101–102
and stirring energy, 91–92
stirring energy and its conservation, 105
time, 123, 124, 138
  attributes of, 136
  origins of, 138
  transformations of, 34
Rotational angular speed squared, closed line integral, 95
Rotational flows, kinetic energy of, 95
Rotational materials, prediction of zero probability disasters, 14
Rotten kid theorem, 13, 258
  of Becker, 153–154, 155–165
  of Bergstrom, 13, 154, 155, 157–165, 179–183, 188
Rouqiang sandstorm weather, 345–347
Rules, sensor and effector performance, 3
Runoff, flooding and floodwater system structure, 111, 113, 114–115, 117
Runoff coefficient, 117
Russell, Bertrand, 268–269, 270, 272
Russell paradox, 268, 271

S

Saint-Vincent, G., 266
Salzmann's convection model, 139
Samaritan's dilemma, 13, 155, 179–183
Sandstorms; *See* Windstorms and sandstorms
Scalar product operation, 97
Scale
  Bjerknes circulation theorem, 47
  colliding eddies and acting-reacting forces, 57, 58
  Earth atmosphere, analysis of movements, 42
  eddy motions, 25, 29
  law of conservation of informational infrastructure, 43, 45
  Newton's second law of motion, 56
  stirring forces, 92
  three-ringed stability, stirring energy conservation, 105
  time; *See* Time and its dimensionality
  uniformity of structural information from microcosm to macrocosm, 38–39
Scale systems, weather, 301
Science, synthetic development in, 5
Scientific Revolution (16th-17th centuries), 5

Secondary (indirect) circulations
  flood water system structure, 12, 111, 112, 113, 120
  laws of physics and, 104
Second-level circulation
  energy transformations, 95–96
  instability generation, 103
  laws of physics and, 104
  nonconservative evolution of stirring energy, 96–97
Second-order nonlinear evolution equations, 21
Second stir, 27, 31, 33
  evolution of universe, 28
  first push and, 31
  and Newton's first law, 52–53
Semantics, concrete multiset, 259
Sensors, systems components, 3
Sequence of events, meteorology, 297–298
Sets
  infinite, 281–283
  systems components, 36
  ZFC system and, 289
Set theory, 268, 269
  internal, 275
  naive, 289
  nonstandard, 275–276
  third crisis in mathematics, 268–273
  vase puzzle, 279
Severe weather; *See also* V-3θ graphs, case studies using; Weather prediction
  convective; *See* Convective weather systems, sudden severe
  determination of severity, 304
Shanghai, severe weather, 313, 314, 321–322, 355
Shapes of spin fields, three-body problem, 80, 81
Shareholders/investors
  comparison of different countries, 243–244
  long-term versus short-term projects, 242
  and wage differentials, 227
  yoyo model, 238–243
Shear effects, and rolling currents, 304
Shear forces, evolution engineering, 119
Shi-jia Lu, 26
Shi jie, 133
Shi Zheng, 133
Shi Zi (Shi Zheng), 133
Short-term business projects; *See* Long-term versus short-term business projects
Shoucheng OuYang, 2, 7, 25, 29, 30, 31, 47
Sichuan, 313, 323
Signal representation, 3
Singularities
  Big Bang theory, 27–28

blown-ups, 7–8, 23
   discontinuous, in calculus solutions, 26
   prediction of transitional changes, 33
   problems on physics of physical quantities, 129, 130
Sjoberg-Martenssson method, 114, 115
Skolem, T.A., 272
Slaving energy of Newtonian first push, 97–98
Slichter, S., 216, 232
Small, regional, short-lived fog and thunderstorms; *See* Regional short-lived fog and thunderstorms
Small-probability events, 34, 127–128
   twist-ups, 31–32
Small-probability information, 124
Small-scale weather phenomena, forecasting, 316–317, 324–325
Social changes, and technology development, 6
Social phenomena
   equal quantitative effects, 31
   spinning yoyo structure, 46–47
Social sciences
   unification with natural sciences, 34, 63
   yoyo model applicability, 12–13
Social system
   complexity of social, political, economic, technological systems, 6
   equal quantitative effects, 63
   West versus East, 8
Solar system
   angular momentum, conservation of informational infrastructure, empirical evidence, 40–42
   eddy motions, 25
   Kepler's laws of planetary motion, 70–71
Solenoidal fields
   nonlinearity as rotationality of, 129
   stirring energy and its conservation, 99–102
Solenoid circulation, Bjerknes circulation theorem, 24–25
Solenoids, baroclinic, 299
Solids, 31
   as fluids, 44
   Western epistemology, 29
Solow, Robert, 216, 226
Solution, well-posedness of nonlinear evolution equations, 20
Space
   absolute time and space, 135
   blown-ups, 20
   Chinese versus English terms for cosmos and universe, 133
   law of conservation of informational infrastructure, 43, 45, 46
   time and, 139
   unevenness, 9
   Western dualistic model, 125
   yoyo model constants, 10
Space-time
   blown-up characteristics, 32, 33
   law of conservation of informational infrastructure and, 44
   Newton's second law of motion, 54
   problems on physics of physical quantities, 129
   whole evolutions of converging and diverging fluid motions, 66
Space-time curvature, stirring energy and its conservation, 106
Space-time unevenness
   and assumptions about materials structure unevenness, 31
   eddy motions, 27
   evolution of materials, 86
   Newton's second law of motion, 54
Spatial changes, evolution equations containing, 22
Spatial decomposition of field occupied by materials, material dimension as, 138
Spatial distribution of speed, unevenness, 93–94
Spatial dynamics, blown-ups as representations of transformations, 23
Spatiality, time, 121–122
Spatial level measures, Earth atmosphere, analysis of movements, 42–43
Spatial location
   blown-up characteristics, 32, 33
   one-dimensional advection equation, 22
Spatial properties, of dynamic implicit transformations, 23
Spatial scale, eddy motions, 25
Spatial structures, uniformity of all imaginable things, 36
Specific heat capacity, 307
Specificity of events, variable acceleration and, 127
Speed
   angular, rotation and stirring energy, 92
   flow, one-dimensional advection equation, 22
   kinetic energy, 12
   linear versus angular speed concepts, 95
   Newton's second law of motion, 56
   relativity theory, 44
   squared, 93
   stirring energy and its conservation, 107
Speed energy, stirring energy special case, 106
Speed kinetic energy, stirring energy and its conservation, 95, 106
Speed of light, 37, 106, 128

Spin
  economic yoyos, 151
  yoyo model, 148–149
Spin angle, 56
Spin directions
  colliding eddies and acting-reacting forces, 57, 60
  measurement of, 86
  Newton's second law of motion, 56
  universal gravitation, 76
Spin fields
  colliding eddies and acting-reacting forces, 57, 58, 59, 60
  dishpan experiments; See Dishpan experiments
  economic yoyos, 13
  family yoyo, 162
  figurative analysis, 11, 52; See also Figurative analysis
  interactions of, 11
  law of conservation of informational infrastructure and, 44
  Newton's second law of motion, 56
  relationship with other known fields, 86
  three-body problem, 80, 81, 87
  universal gravitation, 75
  yoyo model, 10, 149
  yoyo structure and Newton's laws; See Newton's laws of motion
Spinning current, blown-ups, 23–28, 33
Spinning current fields, 26
Spinning fluids, inwardly and outwardly spinning pools and their discontinuity, 31
Spinning vortex, transformations of internally spinning vortex to externally spinning vortex, 101
Spins, applications of systemic yoyo model, 11
Spraying currents, 29
Squared speeds
  angular, 93
  kinetic energy representation, 93
  rotational angular speed, 95
Square root of two, graphic representation of, 63, 64
Stability
  stirring energy and its conservation, 106
  three-ringed, 105
Stable solutions, well-posedness of nonlinear evolution equations, 20
Standard congruence, 259
Star systems
  binary, 11, 67, 74, 75, 77, 80, 81
  multiplex, 11
  n-nary, 11, 14, 77, 79–83, 84, 86, 89

States
  physical quantities and, 126
  time as attribute of materials, 136, 137
Static electrical force, Dirac's large number hypothesis, 36–37, 38
Statistical analysis, limitations of, 22
Statistics, 124
Stigler, George, 216, 238
Stiglitz, J., 217, 226
Stirring energy and its conservation, 91–120
  conservability of stirring energy and physical significance of energy transformations, 103–106
  conservation of stirring energy and three-level energy transformation, 93–95
  energy transformation process and nonconservative evolution of stirring energy, 95–97
  evolution engineering and technology for long-term disaster reduction, 107
  flood disasters, principles of flood evolution and development, 109–113
    law of conservation of stirring energy, 110
    three-ring quasi-stability problem of urban flood movements, 110–113
  flood prevention and disaster reduction facility analysis, 116–120
    computation on artificial lake at Fuhe Bridge along northern outer-ring road, 116–118
    estimate for artificial lake along Qingshu River on upper reaches of Nan River, 118–119
  flood prevention and disaster reduction in urban areas, feasible technology for, 113–115
    artificial urban lake design and river course drainage problem, 115
    computational method for Q_1, 114–115
  governance law of slaving energy of Newtonian first push, 97–98
  interactions and Einstein's mass-energy formula, 98–99
  long-term and long-effect technology, 119–120
  meteorology, 296, 297
  rotation and stirring energy, 91–92
  solenoidal fields and problems on universal gravitation, 99–102
Stirring energy/stirring forces, 2, 12
  discovery of, 95
  equal quantitative effects, computational uncertainty, 30
  gradients, 25
  law of conservation of, 104
  second stir; See Second stir
Stirring motions, time, 123

Stochastics, 123, 127
Stockholders; *See* Shareholders/investors
Strong interaction
  law of conservation of informational infrastructure, 43
  unification of basic forces, 39, 40
Structural analysis; *See* Figurative analysis
Structural changes, twist-ups, 32
Structural evolutions
  eddy motions, 23–24, 27
  force inherence in, 27, 28
Structural method, 131
Structural representation of equal quantitative effects, 32
Structure(s)
  atmosphere, baroclinic effects, 306–307
  characteristics of events, 131
  figurative, 62
  figurative analysis, 63–64
  of materials, origins of forces, 316
  of materials, origins of nonlinearity, 9, 34
  measurement of, 86
  meteorology, 296
  suddenly appearing severe convective systems, 316–317
  time as attribute of materials, 137
  unevenness, second stir, 31
  yoyo model constants, 10
Sub-eddies/sub-sub eddies, 29
  colliding eddies and acting-reacting forces, 59, 60, 61
  creation of, 31–32
  description of motion in, 87
Subtropical high pressures, high temperature weather under, 356–364
Sudden severe convective systems; *See* Convective weather systems, sudden severe
Sudden torrential storms, flood evolution and development, 112
Summations of numbers, 291
Summers, L., 216, 231, 232
Supply-demand concepts, 12–13
  whole evolution analysis, 146–149, 150
Su-Shi principle, 29
Symbolic logic, 269
Symbols, information as symbols of events, 131
Synthesis, system, 3
Synthetic development, in science, 5
System analysis, 3, 13–14
Systemic yoyo
  economics and finance, supply-demand relationship, 148–149
  history of mathematics as, 290
  methodology, 13–14
  origins of, 7–11
Systems, historical review, 2–7
Systems science, 1–3

T

Tao of images, 31, 34
Tao of yin and yang, 29
Tao Te Ching (Lao Tzu), 27, 132–133
T-axis-θ curve angles, 308, 310
  high-temperature weather, 358, 361, 365
  sudden convective systems, 317, 319, 320, 321, 322
  wind- and sandstorm weather, 345, 346, 349, 351, 352, 353
Technology, 6
Teleology, 4
Temperature
  dishpan experiments, 68–70, 88
  meteorology, 304
    conventional, 297
    ground level, rising, 302
    heat waves; *See* Heat waves/abnormally high temperatures
    prediction of severity of weather, 340
    V-3θ graph design, 306, 307
  ultra-low
    fog prediction, 329, 330, 333
    reduction of, 325
    severe convective weather, 317
    wind- and sandstorm weather, 353
  ultra-low, tropospheric
    discovery of, 296
    role in weather evolution, 299–305
    and rolling current structure, 302
    V-3θ graph design, 307–308
    wind- and sandstorm weather, 353
Tensors, 102, 129
Thaler, R., 216, 217, 226, 227, 231, 233, 234, 253
Thales, 264
Theaetetus, 264
Theodorus, 264
Theorem of orthogonal plane divergences, 94
Theoretical foundation of yoyo model; *See* Blown-up theory
Theories of organization, 3
Theory, from practice to theory and back, 5–6
Theory of extensionality, 272
Theory of limits, Cauchy, 268, 271, 287, 289
Theory of Social Interactions, A (Becker), 153
Theory of types, 272

θ curves; *See* T-axis-θ curve angles; V-3θ graphs, case studies using
Third-level circulation, 96, 104
Thought; *See* Human thought
Three-body problem, 2, 5, 11, 77–89, 149, 258
   current results, 77–78
   open questions, 86–89
   three visible bodies with at least one invisible, 83–86
Three-ringed circulation/energy transformation, 104
   Chinese Bei Wei period, 92
   flood evolution and development
      law of conservation of, 110
      quasi-stability problem of urban flood movements, 110–113
   nonconservative evolution of stirring energy, 96–97
   second level circulation, 95–96
Three-ringed stabilities, 105
Threes, Lao Tzu on, 103
Three-stage lifespan, economic yoyos, 151
Thunderstorms, 325–339
   background, 326
   discussion, 338–339
   Shanghai, 321–322
Time
   absolute time and space, 135
   blown-up characteristics, 32, 33
   blown-ups, 20
   Chinese versus English terms for cosmos and universe, 133
   concept of, 11
   Dirac's large number hypothesis, 37
   Earth atmosphere, lifespan of movements, 42–43
   law of conservation of informational infrastructure, 43, 45, 46
   mass as function of, 126
   one-dimensional advection equation, 22
   quantitative, 134, 137
   unevenness, 9
   Western dualistic model, 125–126
   whole evolutions of converging and diverging fluid motions, 66
   yoyo model constants, 10
Time and its dimensionality, 121–144
   concepts of time, 131–138
      in China, 132–133
      definition, 136–138
      problem of, 131–132
      in West, 134–136
   material and quantitative parametric dimensions, 138–140

nonquantification of events, 127–131
   nonquantification, 130–131
   physics of physical quantities, problems on, 128–130
physics of physical quantities, 125–127
problems, 121–125
Time frame/sequence/duration of phenomena
   business projects; *See* Long-term versus short-term business projects
   Earth atmosphere, 42–43
   economic yoyos, 151
   long-term and long-effect technology, stirring energy and its conservation, 119–120
   meteorology, 297–298
      rainfall, runoff computation, 114
      regional short-lived fog and thunderstorms, 325–339
      suddenly appearing severe convective systems, 313–325
Time moment, whole evolutions of converging and diverging fluid motions, 66
Time scales, laws of physics, 135
Time-space distribution, eddy motions, 26–27
Time span/intervals; *See also* Periodicity
   between crises in foundation of mathematics, 291
   period between conjunctions, three-body problem, 78
   between rains, runoff computation, 114, 115
Topal, R., 226
Topographic leeward wave theory, 31
Torrential rains; *See* Rains/rainstorms
Torricelli, E., 266
Transfinite numbers, 268
Transformation of energies
   eddies and, 26, 29
   rotation and stirring energy, 92
   stirring energy and its conservation, 110
   stirring energy conservation and physical significance of, 103–106
Transformations of changes of materials, evolution science, 126–127
Transitional blown-up, defined, 20
Transitional changes of materials and events
   blown-ups, 8, 19–20, 33, 34; *See also* Blown-up theory; Blown-ups
   nonlinear differential equations and, 26
   prediction of, 33, 124
   structural analysis, 315
   weather prediction; *See* Meteorology; V-3θ graphs, case studies using
Treatise on the Family, A (Becker), 155
Triangles, 262–264

Trinary star systems, 11; *See also* Three-body problem
Troposphere
   airstream directions, 303
   local measures to control convection, 325
   rolling currents, 301, 302
   ultra-low temperatures; *See* Temperature, ultra-low, tropospheric
Turbulence, time, 122–123
Turnover models, wages, 226
Twisting forces
   blown-ups, 33
   and eddies, 26
   evolution engineering, 119
   time, 123
   uneven densities of materials and, 53
Twist-ups, contents of, 31–32
Two-body problem
   Kepler's law of harmonics, 73–74
   solar system angular momentum, 41–42
Two-dimensional nonlinearity, solenoidal fields and problems on universal gravitation, 101
Two-dimensional science, 6
Two-star systems, 11, 67, 74, 75, 77, 80, 81
Two-value logical calculus, 280
Types, theory of, 272

U

Ultra-low temperature; *See* Temperature, ultra-low
Unboundedness, 130, 135
Uncertainty, measurement, 28–29
Uncertainty model of equal quantitative movements, 29
Underground water levels, 120
Unequal quantitative effects, 31, 34
Unequal quantitative movements, 28
Uneven eddy motions, Bjerknes circulation theorem, 25
Uneven evolutions, eddy motions, 23–24
Unevenness of materials structure
   eddy irregularities from, 29
   force inherence in, 86, 148
   meteorology, 296, 297
   and nonlinearity, 34
   space-time unevenness and, 31
   troposphere, 310
   V-3θ graph design, 307
Unevenness of space and time
   and eddy motions, 26–27
   and evolutionary materials structures, 9
   Newton's second law of motion, 54
Unevenness of spatial distribution of speed, 93–94

Unification of basic forces, 39
Unified spin fields, 75
Uniformity of spatial structures, 35–36
Unionization, 232
Unique solutions, well-posedness of nonlinear evolution equations, 20
Universal gravitation, 11, 258
   Dirac's large number hypothesis, 36–37, 38, 39
   Kepler's law of harmonics, generalization by Newton, 72–74
   Kepler's laws of planetary motion, 74–76
   problems on physics of physical quantities, 129, 130
   stirring energy and its conservation, 99–102
Universal gravitation constant, 72
Universe; *See also* Celestial mechanics
   age of, 39
   Chinese versus English terms, 133
   eddy current composition of, 31
   evolution of, 27–28, 34; *See also* Big Bang theory
   law of conservation of informational infrastructure, 43
   scale; *See* Macro scale; Scale
Universes, multiple, 45
Urban areas
   concrete-based desertification, 366
   high-temperature weather, 355, 356, 364
   severe weather, suddenly appearing severe convective systems, 317–322
   three-ring quasi-stability problem of urban flood movements, 110–113
Utility functions, 17
   CEO, 248
   child labor, 189, 192, 196, 197, 202
   family yoyos
      child labor and its efficiency, 184, 185
      never perfect value systems and parasites, 171, 172, 173, 174, 175, 176, 177
      rotten kid theorems, 157, 158, 162, 163, 165, 168, 169–170, 178, 180, 181

V

Value systems, never perfect, 155, 170–179
Variable acceleration, rotations of materials, 127
Vase puzzle, 278–279
Vector operations
   Newtonian first push, 97–98
   rotationality in nonlinearity, 102
   solenoidal fields and problems on universal gravitation, 99–101, 102
Vectors/vectoricities
   angular speed representation, 93–94

eddies and eddy evolutions, 9, 29
time, 137
wind, 304
V-3θ graphs, 306, 316, 340
Velocity
   Newton's second law of motion, 56
   vertical
      Earth atmosphere, 42–43
      whole evolutions of converging and diverging fluid motions, 64
Vertical rolling currents, 296
Vertical structure
   Earth atmosphere, 42–43
   meteorology
      rolling currents, 302
      troposphere, 303
      vorticity, 299, 300
      V-3θ graph design, 307
   wind- and sandstorm weather, 353
   whole evolutions of converging and diverging fluid motions, 64
Vicious circle principle, 269
Volume
   Archimedes treatment of, 266
   early calculus problems, 265
   Newton's definition of mass, 125
von Bertalanffy, L., 1–2, 3, 4, 5
von Neumann, J., 271
von Neumann's principle of program storage, 29
Vortical kinetic energy, 95
Vortical vectoricity, materials structures, 29
Vorticity
   meteorology, 300
      rolling currents and, 301
   transformations of internally spinning vortex to externally spinning vortex, 101
   whole evolutions of converging and diverging fluid motions, 64, 66
Vorticity tensors, 102
V-3θ graphs, 295–311
   design of, 306–311
   fundamentals, 295–298
   rolling currents and ultra-low temperature in weather evolution, 299–305
V-3θ graphs, case studies using
   abnormally high temperatures, 354–368
      background, 355–356
      under cold high pressure, 364–367
      under subtropical high in West Pacific, 356–364
   small, regional, short-lived fog and thunderstorms, 325–339
      background, 326
      discussion, 338–339

small, regional, short-lived fog and thunderstorms, case studies, 327–338
   after-rain fog prediction, 330–335
   counterclockwise rolling current analysis of graphs, 327–328
   description of fog situation, 327
   heavy fog prediction problem, 328–329
   light fog prediction problem, 329–330, 331
   no-fog weather and fog dissipation prediction, 335–338
suddenly appearing severe convective systems, 313–325
   background, 314
   blown-ups, 314–315
   discussion, 324–325
   over major metropolitan areas (Beijing and Shanghai), 317–322
   regional heavy rain, 322–324
   structural analysis, 315–316
   structural characteristics, 316–317
windstorms and sandstorms, 339–354
   background, 340–342
   structural characteristics of graphs, 353–354
windstorms and sandstorms, practical applications, 342–353
   Beijing sandstorm and Xi'an flying dust weather, 350–353
   characteristic analysis of graphic structure for wind- and sandstorms, 344–345
   Huhhot and Linhe sandstorm weather, 347–348
   path and affected areas of sandstorms, 343–344
   Rouqiang sandstorm and Jinzhou flying dust weather, 345–347
   Yinchuan and Pingliang sandstorm weather, 349–350

W

Wage differentials, interindustry, 225–235
   companies with limited resources, 229–231
   financially resourceful companies, 227–229
   hypotheses, 216–217
   law of one price, 234–235
   literature review, 231–233
Wage structure, 258
Wang, H.Z., 115
Wang, J., 110
Wang, S., xiv, xix
Wang, Z., 31
Wang, Z. X., 39, 42
Water circulation; *See* Hydrology

Water levels, underground, 120
Water resources, artificial flood control storage lakes, 113, 120
Water vapor/humidity, atmospheric, 304
  condensation, temperature and, 299–300
  conventional, 297
  conventional meteorology, 297
  fog prediction, 329, 330
  high-temperature weather, 355, 361, 364, 366, 367, 368
  no-fog weather prediction, 337, 338
  order of information, 298
  prediction of severe weather, 300–301
  prediction of severity of weather, 340
  V-3θ graph design, 306
    design of, 306
    3θ curves at different altitudes, 310
  wind- and sandstorm weather, 353, 354
Wave motions, 29, 123
  meteorology
    baroclinic effects, 299
    rolling currents and, 301
Weak interaction, 39
Weather prediction, 46, 52, 295–311
  equal quantitative effects, 30
  torrential rains, 2
  V-3θ graphs
    case studies using; *See* V-3θ graphs, case studies using
    design of, 306–311
    fundamentals, 295–298
    rolling currents and ultra-low temperature in weather evolution, 299–305
  zero probability disasters, 14
Weather systems, analysis of atmospheric movements, 42–43
Weierstrass, Karl, 267
Well-posedness, evolutions of nonlinear evolution equations, 20
Western concepts of time, 134–136
Western philosophy, dualism of, 125–126
West wind systems, 301
Weyl, Hermann, 269
Whirlpool, 87
White box, 3
Whitehead, Alfred North, 269, 270, 272
White source, 102
Whole evolutions
  converging and diverging fluid motions, Newton's laws of motion, 64–66
  demand-supply analysis, 146–149, 150
  economics, demand and supply, 146–149
  mathematics; *See* Mathematics

  meteorology, 296
  three-body problem open questions, 88
Wholes, systems as, 3–4, 19
  stirring energy and its conservation, 105–106
Wigner, Eugene, 257
Wind
  conventional meteorology, 297
  order of information, 298
  west wind systems, 301
Wind direction, 338
  fog, case studies, 330
  high-temperature weather, 358, 359, 361
  rolling currents, 303–305
  structural analysis, 316
  V-3θ graphs
    design of, 306
    3θ curves, 311
    windstorms and sandstorms, 340, 341, 342
Wind speed
  meteorology, 304
  prediction of severity of weather, 340
  V-3θ graph design, 306
Windstorms and sandstorms, 339–354
  background, 340–342
  forecasting, 325
  Shanghai, 321–322
  V-3θ graph applications, 342–353
    Beijing sandstorm and Xi'an flying dust weather, 350–353
    characteristic analysis of graphic structure for wind- and sandstorms, 344–345
    Huhhot and Linhe sandstorm weather, 347–348
    path and affected areas of sandstorms, 343–344
    Rouqiang sandstorm and Jinzhou flying dust weather, 345–347
    Yinchuan and Pingliang sandstorm weather, 349–350
  V-3θ graph structural characteristics, 317, 353–354
Wind vectors, V-3θ graph design, 306
Wu, Xuemou, 29, 34
Wu, Y., 11, 20, 22, 24, 34, 51, 52, 61, 62, 64, 86, 88, 139, 141, 148, 149, 154, 180, 181, 189, 258, 295, 300
Wuhan City, severe weather, 314, 320, 361, 362
Wuzhou, 333, 335

X

Xian, D.C., 39, 42
Xi'an, severe weather

flying dust, 350–353
fog, case studies, 328, 329
Xichiang, 335, 336
Xihu Lake, 361
Xuemou Wu, 29

Y

Yangtze River, 103, 110, 111, 113
Yangtze River valley, 361
Yellen, J., 216, 217, 226
Yellow River, 103, 111, 112–113
Yibing, 335, 336
Yinchuan sandstorm weather, 349–350
Yin-Yang concept, 63, 302
Yoyo model
   economic yoyos
     evolution of single cycle, 149, 151
     long-term versus short-term business projects, 238–243
   eddy motion model of general system, 10
   history of mathematics as, 290
   meteorology, 297

Yu shi ju hua, 133
Yu zhou (space time; universe), 133

Z

Zadeh, L., 3
Zeng, X.P., 138, 141
Zeno, 261, 264
Zeno paradoxes, 264, 265, 289
Zermelo, Ernst, 272, 276
Zermelo-Fraenkel axiom of choice (ZFC), 272, 275–276, 281–283, 285, 289
Zero, physics of physical quantities, 130
Zero-curvature (Euclidean) space, 92, 107
Zero probability disaster prediction, 14, 52
Zero-th order tensors, 102
ZFC; *See* Zermelo-Fraenkel axiom of choice (ZFC)
Zhejiang Province, 366
Zhou (time), 133
Zhu, W.J., 277, 281, 285, 288, 292
Zhuang Zi, 133